F. Kraft

Thomas Heinzel
**Mesoscopic Electronics
in Solid State Nanostructures**

Thomas Heinzel

Mesoscopic Electronics in Solid State Nanostructures

WILEY-VCH

WILEY-VCH GmbH & Co. KGaA

Author

Prof. Dr. Thomas Heinzel
Universität Freiburg, Institut für Physik
Thomas.Heinzel@physik.uni-freiburg.de

This book was carefully produced. Nevertheless, authors, editors and publisher do not warrant the information contained therein to be free of errors. Readers are advised to keep in mind thar statements, data, illustrations, procedural details or other items may inadvertently be inaccurate.

Library of Congress Card No.: applied for
British Library Cataloging-in-Publication Data:
A catalogue record for this book is available from the British Library

Bibliographic information published by
Die Deutsche Bibliothek
Die Deutsche Bibliothek lists this publication in the Deutsche Nationalbibliografie; detailed bibliographic data is available in the Internet at <http://dnb.ddb.de>.

© 2003 WILEY-VCH GmbH & Co. KGaA, Weinheim

All rights reserved (including those of translation into other languages). No part of this book may be reproduced in any form – nor transmitted or translated into machine language without written permission from the publishers. Registered names, trademarks, etc. used in this book, even when not specifically marked as such, are not to be considered unprotected by law.

Printed in the Federal Republic of Germany
Printed on acid-free paper

Printing betz-druck GmbH, Darmstadt

Bookbinding Litges & Dopf Buchbinderei GmbH, Heppenheim

ISBN 3-527-40375-2

To my daughter Carola Sophia

Foreword

Any student who will read and understand this book all the way through will be an excellent candidate for an outstanding thesis in mesoscopic physics. Starting with a crash course through elementary solid state physics, Heinzel's book covers all the important subjects an experimentalist encounters in this field, such as technology and fabrication of devices, experimental techniques for low temperature measurements, a discussion of the electronic properties of two-, one- and zero-dimensional systems followed by a thorough presentation of single electron devices and quantum dots. The text is clearly written and explained by many explanatory figures. The physics problems as well as the specially selected publications will lead the students directly to the discussion of important and modern physics questions in the field. This book is a useful tool also for the experienced researcher to get a summary about recent developments in solid state nanostructures. I applaud the author for a marvelous contribution to the scientific community of mesoscopic electronics.

K. Ensslin
Solid State Physics Laboratory, ETH Zurich

Preface

The desire to explore new orders of magnitude seems to be a part of human nature. This tendency is reflected even in the language. If a particular megadeal is really stunning, we feel more and more obliged to call it a gigadeal. It won't take long before the first Terastar pops up. Each new generation of particle colliders, telescopes, or lasers, in fact of almost any scientific device or technique you can think of, extends the accessible interval of a physical quantity. It is rather the rule than the exception that novel phenomena are discovered in such a process, some of which have been anticipated, others have not. Such an evolution seems to gain speed as soon as it gets classified as useful, besides pure scientific interest. It is of course debatable what exactly should be considered as "useful".

In any case, the miniaturization of electronic circuits during the past 50 years has been both scientifically rewarding and useful. It is certainly not necessary to support this statement by examples, since microprocessors are part of daily life. Scientifically, however, the expression "microstructure" no longer fuels our imagination. Meanwhile, the really exciting electronic circuits are *nanostructures*. As this name already suggests, nanostructures are objects with structures in the nanometer regime, which can mean just 1 nanometer, but also just a little less (in fact, in some cases even somewhat more) than 1000 nanometers. The point of nanostructure science is that within the last two decades, tremendous progress has been made in fabricating, controlling and understanding structures in this size regime. This is true for a wide variety of fields, including, for example, gene technology, crystal growth or microchip - excuse me, nanochip - fabrication. The resulting novel possibilities at hand are really breathtaking and get heavily explored by a significant fraction of the scientific community. In many cases, having control over the size and shape of an object in the nanometer regime means being able to control its chemical and/or physical properties. For example, the size of a semiconductor nanocluster determines its optical emission spectrum via size quantization, while from introductory solid state physics lecture, we have learned that this property is related to the band gap, an intrinsic feature of the material. By now, the size reduction has actually reached the dimensions which are of interest for chemistry and biology, and there is a rapidly growing overlap. For example, you can think of an ionic channel (they reside in cell membranes and control the electrochemical potential of cells by selectively transferring certain types of ions across the membrane) as a molecular transistor. On the other hand, nanostructure physicists have started to use DNA strands as wires and templates for nanocircuit fabrication.

One branch of nanoscience deals with the electronic transport properties of solid state nanostructures. This field is often referred to as "mesoscopic transport", an expression which indicates that the explanations for the observed transport phenomena must be sought somewhere in between microscopic and macroscopic models. The purpose of the present book is to introduce the reader to this topic from an experimental point of view. "The reader" is hereby assumed to be a student of physics or a related field, who has just finished the introductory courses, in particular those on solid state physics and on quantum mechanics, and plans to study nanosciences more closely. The reader is picked up at the knowledge he/she is likely to have, and a ride is given to ongoing research activities in the field of mesoscopic transport. Along the way, the elementary concepts and nanostructures get introduced.

Selecting illustrative experiments for such a purpose is of course a highly subjective matter. The author has tried to pick particular instructive examples well-known to him, which can furthermore be explained within the scope of this book. These examples have thus not necessarily been of high relevance for the evolution of the field, and I apologize for this shortcoming. It should be remarked that in some of the figures, the original data have been redrawn for better reproduction quality and for a consistent presentation.

The text contains a somewhat unusual feature, namely "papers" in the exercises sections. Their purpose is to encourage the student to go through selected, usually quite recent, original publications. They are referred to by [P chapter.number] in the text. The student should be able to summarize the beef of such a paper in a 15 minutes talk. The reader is strongly encouraged to actually do this. Besides collecting complementary information and getting exposed to different styles of presentation, the experience of being able to understand the stuff written down not in a textbook but in an original paper can be highly motivating.

I hope that after going through the text, the reader is not only able to join with some confidence an experimental research group working in this field, but also feels well prepared for more advanced theoretical lectures on mesoscopic physics.

This book, and with it his author, has enjoyed a lot of encouragement and support from many sides. I would like to thank particularly Hermann Grabert and Wolfgang Häusler, who read through parts of the manuscript and made many valuable comments. I am grateful to my colleagues Klaus Ensslin, Andreas Fuhrer, Miha Furlan, Ryan Held, Thomas Ihn, Silvia Lüscher, Jörg Rychen, and Volkmar Senz for countless fruitful discussions and stimulating ideas. Special thanks go to those who supplied figures for this book, namely Günther Bauer, Mildred Dresselhaus, Andreas Fuhrer, Adam Hansen, Roland Ketzmerick, Anupam Madhukar, Andy Sachrajda, and Elke Scheer. Furthermore, I thank my students, whose critical but always constructive comments have shaped and improved the presentation of the material. Last, but not least, I thank my wife Ulrike. With her tremendous energy and selfless support, she managed to supply the refuge I needed to transform my disorganized lecture notes into a book.

Thomas Heinzel

Freiburg, January 2003

Contents

1	**Introduction**		**11**
	1.1	Preliminary remarks	11
	1.2	Mesoscopic transport	12
		1.2.1 Ballistic transport	13
		1.2.2 The quantum Hall effect and Shubnikov - de Haas oscillations	15
		1.2.3 Size quantization	16
		1.2.4 Phase coherence	17
		1.2.5 Single electron tunnelling and quantum dots	17
		1.2.6 Superlattices	19
		1.2.7 Samples and experimental techniques	19
2	**An Update of Solid State Physics**		**23**
	2.1	Crystal structures	24
	2.2	Electronic energy bands	26
	2.3	Occupation of energy bands	33
		2.3.1 The electronic density of states	33
		2.3.2 Occupation probability and chemical potential	34
		2.3.3 Intrinsic carrier concentration	35
	2.4	Envelope wave functions	36
	2.5	Doping	40
	2.6	Diffusive transport and the Boltzmann equation	43
		2.6.1 The Boltzmann equation	45
		2.6.2 The conductance predicted by the simplified Boltzmann equation	47
		2.6.3 The magneto-resistivity tensor	49
	2.7	Scattering mechanisms	50
	2.8	Screening	53
3	**Surfaces, Interfaces, and Layered Devices**		**59**
	3.1	Electronic surface states	60
		3.1.1 Surface states in one dimension	60
		3.1.2 Surfaces of 3-dimensional crystals	66
		3.1.3 Band bending and Fermi level pinning	68
	3.2	Semiconductor-metal interfaces	69
		3.2.1 Band alignment and Schottky barriers	70
		3.2.2 Ohmic contacts	74

		3.3	Semiconductor heterointerfaces	75

    3.3   Semiconductor heterointerfaces 75
    3.4   Field effect transistors and quantum wells 78
            3.4.1   The silicon metal-oxide-semiconductor FET (Si-MOSFET) 78
            3.4.2   The Ga[Al]As high electron mobility transistor (GaAs-HEMT) 81
            3.4.3   Other types of layered devices 83
            3.4.4   Quantum confined carriers in comparison to bulk carriers 87

4   Experimental Techniques   **93**
    4.1   Sample fabrication 93
            4.1.1   Single crystal growth 95
            4.1.2   Growth of layered structures 96
            4.1.3   Lateral patterning 101
            4.1.4   Metallization 108
            4.1.5   Bonding 110
    4.2   Elements of cryogenics 110
            4.2.1   Properties of liquid helium 111
            4.2.2   Helium cryostats 117
    4.3   Electronic measurements on nanostructures 121
            4.3.1   Sample holders 122
            4.3.2   Application and detection of electronic signals 122

5   Important Quantities in Mesoscopic Transport   **131**

6   Magnetotransport Properties of Quantum Films   **137**
    6.1   Landau quantization 137
            6.1.1   2DEGs in perpendicular magnetic fields 137
            6.1.2   The chemical potential in strong magnetic fields 140
    6.2   The quantum Hall effect 143
            6.2.1   Phenomenology 143
            6.2.2   Origin of the integer quantum Hall effect 145
            6.2.3   The quantum Hall effect and three dimensions 149
    6.3   Elementary analysis of Shubnikov-de Haas oscillations 150
    6.4   Some examples of magnetotransport experiments 153
            6.4.1   Quasi-two-dimensional electron gases 153
            6.4.2   Mapping of the probability density 155
            6.4.3   Displacement of the quantum Hall plateaux 155
    6.5   Parallel magnetic fields 157

7   Quantum Wires and Quantum Point Contacts   **165**
    7.1   Diffusive quantum wires 167
            7.1.1   Basic properties 167
            7.1.2   Boundary scattering 169
    7.2   Ballistic quantum wires 171
            7.2.1   Phenomenology 171
            7.2.2   Conductance quantization in QPCs 172

	7.2.3	Magnetic field effects	177
	7.2.4	The "0.7 structure"	181
	7.2.5	Four-probe measurements on ballistic quantum wires	182
7.3	The Landauer-Büttiker formalism		184
	7.3.1	Edge states	185
	7.3.2	Edge channels	189
7.4	Further examples of quantum wires		190
	7.4.1	Conductance quantization in conventional metals	190
	7.4.2	Carbon nanotubes	192
7.5	Quantum point contact circuits		195
	7.5.1	Non-ohmic behavior of collinear QPCs	195
	7.5.2	QPCs in parallel	197
7.6	Concluding remarks		198

8 Electronic Phase Coherence — 203
8.1 The Aharonov-Bohm effect in mesoscopic conductors — 203
8.2 Weak localization — 206
8.3 Universal conductance fluctuations — 209
8.4 Phase coherence in ballistic 2DEGs — 213
8.5 Resonant tunnelling and S - matrices — 216

9 Singe Electron Tunnelling — 225
9.1 The principle of Coulomb blockade — 225
9.2 Basic single electron tunnelling circuits — 227
    9.2.1 Coulomb blockade at the double barrier — 229
    9.2.2 Current-voltage characteristics: the Coulomb staircase — 232
    9.2.3 The SET transistor — 236
9.3 SET circuits with many islands; the single electron pump — 241

10 Quantum Dots — 249
10.1 Phenomenology of quantum dots — 250
10.2 The constant interaction model — 253
10.3 Beyond the constant interaction model — 261
10.4 Shape of conductance resonances and current-voltage characteristics — 269
10.5 Other types of quantum dots — 270

11 Mesoscopic Superlattices — 277
11.1 One-dimensional superlattices — 277
11.2 Two-dimensional superlattices — 279

A SI and cgs Units — 289

Appendices — 289

B Correlation and Convolution — 291
B.1 Fourier transformation — 291

B.2	Convolutions	291
B.3	Correlation functions	292

C Capacitance Matrix and Electrostatic Energy — 295

D The Transfer Hamiltonian — 299

E Solutions to Selected Exercises — 301

References — 323

Index — 335

1 Introduction

1.1 Preliminary remarks

Over the past 30 years, the miniaturization of electronic devices has strongly influenced the technological evolution. Just think of the progress made in communication technology, or of the improvements of personal computers. For the money spent on a pocket calculator (which barely managed to carry out the 4 basic arithmetic operations) 30 years ago, you can buy today a desktop computer able to solve quite sophisticated numerical tasks, which in the 1970's could be tackled only by supercomputers. *Moore's law* states that roughly every 3 years, the number of transistors per microchip doubles. This law has been valid remarkably well in the past 3 decades, and it is expected to hold for some more years to come, although probably with a slightly reduced rate. This process, however, requires an ongoing reduction of the feature sizes, which up to now is essentially achieved by using smaller wavelengths for the optical lithography (the wavelengths determines the resolution limit via diffraction). This is much more challenging as it may sound, for several reasons.

First of all, a quick glance in an optics textbook reveals that the index of refraction of all common glasses diverges rapidly as the wavelength gets reduced to about 200 nm. In addition, metals get transparent at their plasma frequencies, which typically falls in the same range of wavelengths. Hence, constructing both lenses and mirrors for the 100 nm regime is not that easy. Currently, the wavelengths used for lithography are of the order of 250 nm. Alternative lithographic techniques are able to pattern significantly smaller feature sizes. Although electron beam lithography is used in industry for some fabrication steps, it is too expensive for mass production of microchips. Novel patterning schemes, such as self-assembly or lithography with scanning probe microscopes, are presently subject of extensive studies in research labs all around the world. It is, however, very unlikely that these techniques will replace optical lithography within the foreseeable time. Second, the patterns illuminated in an optical photoresist have to be transferred into a structured device. Processes like developing the photoresist, semiconductor etching, metal evaporation, alloying or selective doping must be carried out without losing the resolution. Furthermore, the devices must be connected to wires, and the inevitable heating generated during operation must be kept under control.

Suppose the nano-scientists will find adequate solutions to all these technological problems; there is in fact little doubt that they will. Then, however, another issue will become more and more important: all the above considerations implicitly assume that the components of a microchip can be scaled down arbitrarily without changes in their performance. This is not the case! Conventional transport theory makes presumptions about certain length and energy scales. For example, it is assumed that the electron mean free path is small compared to the feature size of the device, like the gate length of a transistor. The concept of resistivity is based on this assumption. Within the Boltzmann theory of electronic transport, it is assumed that the acceleration of the Fermi sphere by external electric fields gets compensated by many kinds of relaxation processes, which form a generalized friction. In a stationary state, these friction forces balance the effects of the external field, and the resistivity of the sample can be defined.

What happens for device sizes comparable to the mean free path, or to other relevant length scales? Well, we then enter the regime of *mesoscopic transport*. Novel effects occur, which may profoundly change the device performance. Introducing these effects is the major goal of this book. In the following section, we will look at the specific length and energy scales somewhat more closely and give examples for typical transport properties of samples in the mesoscopic regime.

1.2 Mesoscopic transport

What characterizes the mesoscopic regime? The answer depends on the particular quantity under study. For the above example, the criterion would be that the device size must be larger than the electronic mean free path ℓ_e. Other length scales are the de Broglie wavelength of the electrons that carry the currrent, which in almost all cases studied in this book are those electrons at the Fermi edge. Their de Broglie wavelength is the Fermi wavelength $\lambda_F = h/\sqrt{2m^* E_F}$, where m^* denotes the effective electron mass (see chapter 2 for details), and E_F the Fermi energy. If the feature sizes of the sample are comparable to λ_F, the wave character of the electrons will become essential, and their kinetic energies will quantize. This fact is often referred to as *size quantization*, which is nothing but elementary quantum mechanics. If size quantization takes place in one spatial direction only, the electron system is confined to two dimensions, and we speak of *quantum films*, which are the topic of chapters 3 and 6. Suppose we confine our electrons in a second spatial direction. Their motion then becomes one-dimensional, and we have a *quantum wire*. The basic properties of quantum wires are discussed in chapter 7. Finally, we can confine the electron in all directions, like in an atom. The resulting objects are known as *quantum dots* or *artificial atoms*, see chapter 10. Another important length scale is the *phase coherence length*. Most of us are aware of the diffraction pattern electrons produce as they traverse a double slit setup in a vacuum tube. However, we usually do not think of electronic interference effects in solid state devices. Nevertheless, these effects do occur and become particularly important in devices with dimensions of the order of the phase coherence length. Phase coherent electrons are the topic of chapter 8. Furthermore, it has turned out that the granular character of the electrons, which even in macroscopic samples plays an important role since it is responsible for the shot noise, becomes increasingly important in nanostructures. The point here is that the energy needed to charge a small island with a single electron may become significant. The resulting effects are summarized by *single electron tunnelling*, which is the title of chapter 9.

Fig. 1.1 gives an overview of the most important mesoscopic regimes, and we continue with a brief survey of the phenomena to be discussed.

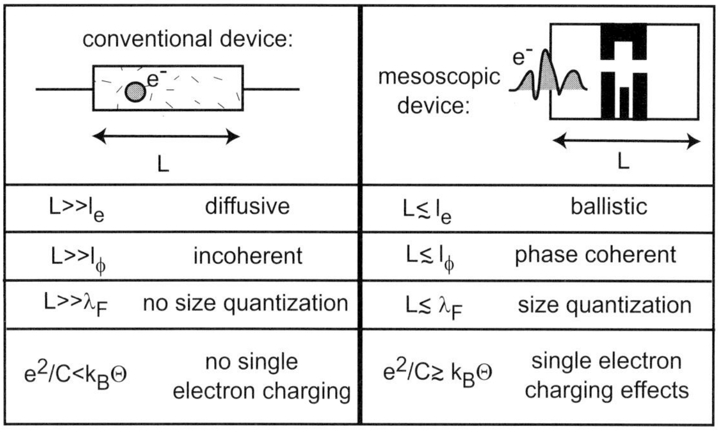

1.2 Mesoscopic transport

conventional device:		mesoscopic device:	
$L \gg l_e$	diffusive	$L \lesssim l_e$	ballistic
$L \gg l_\phi$	incoherent	$L \lesssim l_\phi$	phase coherent
$L \gg \lambda_F$	no size quantization	$L \lesssim \lambda_F$	size quantization
$e^2/C < k_B \Theta$	no single electron charging	$e^2/C \gtrsim k_B \Theta$	single electron charging effects

Figure 1.1: The left column summarizes what we mean by a "conventional device", like the resistor sketched to the top. Electrons can be thought of strongly localized wave packets, which move through a disordered device with the drift velocity. Their mean free path is much smaller than the device size L. Transport is *diffusive*. Since the phase coherence length ℓ_ϕ is also small compared to L, the transport is *incoherent*. Furthermore, the Fermi wavelength is much smaller than L, and consequently, size quantization is absent. Finally, the capacitance of the device is so large that the energy needed to charge it with a single electron is negligibly small. In the right column, the conditions necessary to enter the mesoscopic regime are shown. The cartoon to the top indicates a sample free of scatterers, except a man-made non-conducting structure (black). Transport through the sample is therefore ballistic. If $\ell_\phi \geq L$, the electrons pass through the sample coherently, and we can expect interference effects. Furthermore, the feature sizes may by comparable to λ_F, such that size quantization occurs. And finally, the capacitances may be sufficiently small, such that single electron charging effects may become observable.

1.2.1 Ballistic transport

In order to enter the ballistic regime, the mean free path ℓ_e, which roughly speaking is the average distance an electron travels before getting scattered,[1] must be small compared to the relevant sample length L. At room temperature, a major source of scattering is electron-phonon interaction, with a mean free path of the order of the order of 20 nanometers. How are we supposed to describe, for example, the electron transport through a wire with $L < \ell_e$? Elementary solid state physics tells us that Bloch electrons in a perfect crystal lattice experience no resistance at all. We are therefore tempted to expect an infinite conductance. This would mean that in such small circuits, there is no dissipation, no heat generation and no energy loss as the electronic signal is transferred, like in a superconductor! Surprisingly, we cannot avoid resistances as we transfer electrons across ballistic wires, although strictly

---
[1] A more accurate definition will be given in chapter 2.

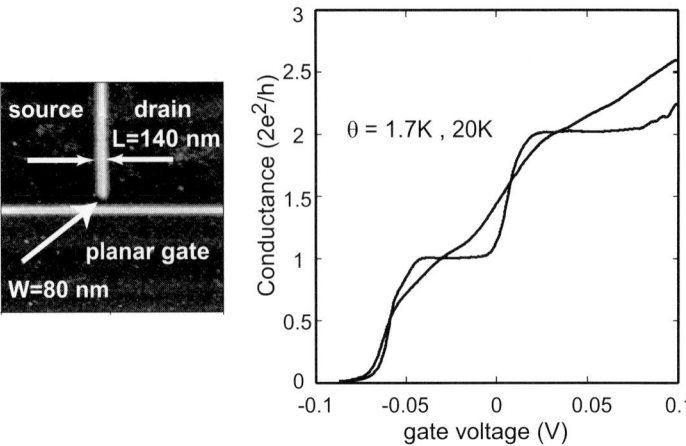

Figure 1.2: Ballistic transport through a quantum point contact. To the left, the surface topography of a GaAs microchip is shown. The picture has been taken with an atomic force microscope. The chip hosts a quantum film about 30 nm below its surface, which is removed underneath the bright lines. A small and short wire of length 140 nm and width 80 nm connects source and drain. By applying voltages to the planar gate electrode, the width of the wire can be tuned. The measurement to the right shows the conductance of the wire as a function of the gate voltage. At low temperatures, a conductance quantization in units of $2e^2/h$ is visible, which vanishes around 20 K.

speaking, the wire itself does have an infinite conductance. It should surprise you even more that the conductance G we measure is in fact quantized in multiple integers of $2e^2/h$, see Fig. 1.2. We do not worry too much about the sample details for now. After going through chapters 3 and 4, we will know that below the sample surface shown in the picture to the left, a quantum film of electrons resides. It has been removed underneath the bright lines, which are oxide lines on a semiconductor (GaAs) surface. You will then, hopefully, accept the fact that the whole area shown here is free of scatterers, at least at low temperatures. The structure can thus be thought of as a *three-terminal device*. If a voltage is applied between the *source* and the *drain* terminal, a current will pass the narrow constriction defined by the two oxide lines. Such ballistic constrictions with size quantization in two directions are called *quantum point contacts*. The width of this constriction is of the order of the Fermi wavelength and can be tuned by applying an additional voltage to the third terminal, labelled as *planar gate*. This works because of the *field effect*, which should again have become clear after reading chapters 3 and 4. The conductance of this device as a function of the planar gate voltage is shown to the right. At temperatures Θ of a few Kelvins, the conductance shows steps in units of $2e^2/h$, which vanish at more elevated temperatures. This effect is one of the most fundamental observations [Wees1988], [Wharam1988] in mesoscopic transport.

Where is the resistance and where does the voltage drop? After all, there are no scatterers.

What determines the energy scale of thermal smearing? How do we model transport through ballistic samples in the first place? These issues are discussed in chapter 7.

1.2.2 The quantum Hall effect and Shubnikov - de Haas oscillations

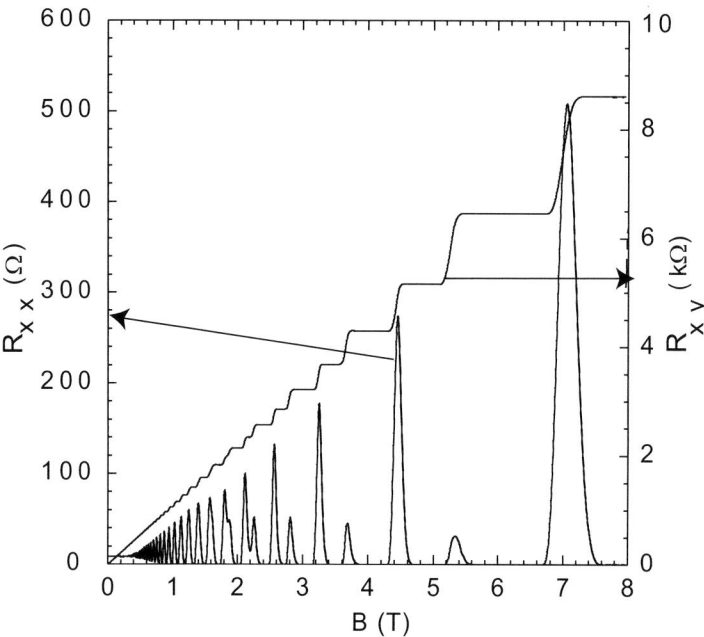

Figure 1.3: Shubnikov-de Haas oscillations and the quantum Hall effect. We look at a measurement of the longitudinal and the Hall resistance (R_{xx} and R_{xy}, respectively), of a two-dimensional electron gas, as a function of a magnetic field applied perpendicular to the plane of the quantum film. The experiment has been performed at a temperature of 100 mK.

Fig. 1.3 shows the resistance of a homogeneous electronic quantum film along the direction of the current flow (the *longitudinal resistance* R_{xx}), as well as perpendicular to it (the *Hall resistance* R_{xy}). Apparently, the Hall resistance quantizes in units of $h/(je^2)$ at strong magnetic fields, and in units of $h/(2je^2)$ at smaller magnetic fields (j is an integer). This is the *quantum Hall effect*, to be discussed in chapter 6. It has been discovered by [Klitzing1980]. Soon afterwards, it became clear that this quantization of the Hall resistance is independent of the material system, as long as the electron gas is two-dimensional. In 1982, these observation were supplemented by the discovery of the fractional quantum Hall effect by Tsui and coworkers [Tsui1982]. This variation is observed only for very high electron mobilities, and has its origin in strong electron-electron interactions. We will not discuss the fractional quantum Hall effect in this book, though. It is tempting to suspect that the quantum Hall effect is somehow related to the conductance steps in quantum point contacts, which quantizes in the same units. But how can this be? The sample size here is hundreds of micrometers, which is certainly larger than the mean free path. Second, the sample is two-dimensional. Also, we are now looking at the Hall resistance, while in the previous example, we looked at the two-terminal conductance ($G_{xx} + G_{xy}$, strictly speaking), and the magnetic field was zero. As we shall see in chapter 7, there is in fact a close, although by no means obvious, relation between these two effects.

Note that the behavior R_{xx} is strongly correlated to the quantum Hall effect. We observe longitudinal resistance peaks at the steps in R_{xy}, while R_{xx} becomes *zero* in the regions of quantized Hall resistances. These oscillations are known as *Shubnikov - de Haas oscillations*. Any explanation for the quantum Hall effect should therefore also explain these oscillations, in particular the remarkable fact of vanishing resistance! It should be remarked that the quantum film does *not* become superconducting. You may further wonder why the resistance of a diffusive two-dimensional electron gas can vanish, while that one of a ballistic one-dimensional electron gas remains non-zero. It is an essential part of this book to answer these questions and reveal their interconnections. For now, we leave it at the statement that in quantum films placed in strong magnetic fields, the scattering of electrons is strongly suppressed, and the transport develops a one-dimensional character.

1.2.3 Size quantization

In modern semiconductor heterostructures (see chapter 3 for more on this), the Fermi wavelength can become as large as 100 nm, and may thus be well comparable to the size of the device. This strongly modifies the electronic density of states and changes the dimensionality of the electron system. Size quantization plays an essential role for the phenomena presented above. We will see that in these semiconductor structures, many of the model potentials treated in elementary quantum mechanics can be tailored, such that we have some sort of a quantum mechanics construction kit at hand. We will meet, for example, parabolic quantum wells, square wells, and triangular potentials.

If you think about this, a non-trivial question probably springs to your mind: the electrons are in a crystal, after all. The wave functions must obey Bloch's theorem. Why can we speak of simple potentials and wave functions as encountered in elementary quantum mechanics? The answer is actually well-established in solid state physics and is most frequently used in relation to the potential and energy levels of doping atoms in semiconductors. It consists of the *envelope function approximation* and the concept of *effective masses*. The effects of the crystal potential are thereby taken into account by a dielectric constant, and by assigning an effective mass to the electron, which then moves in the superimposed potential. This approximation is used throughout this book, after its introduction in chapter 2.

Size quantization and the corresponding change of the dimensionality (see Table 1.1) are already sufficient to change the properties of an electron gas profoundly. For example, the quantum Hall effect is absent in three-dimensional electron gases.

1.2 Mesoscopic transport

Table 1.1: Effect of size quantization on the electronic properties.

Dimension	Energy dependence of the density of states	Unit of the resistivity
3	$\propto \sqrt{E}$	Ωm
2	constant	Ω
1	$\propto 1/\sqrt{E}$	Ω/m
0	δ-functions	n.a.

1.2.4 Phase coherence

When we speak of an electron, we refer to a wave packet which certainly has some phase coherence length ℓ_ϕ. We expect interference effects of the electronic waves to play a role on length scales smaller than the phase coherence length. The phase coherence is destroyed by inelastic scattering events, such as electron-phonon-scattering and electron-electron scattering, both of which depend strongly on temperature. At low temperatures, ℓ_ϕ may actually become as large as 100 μm. A prominent example of electronic interference is the *Aharonov-Bohm effect* in small quantum rings (Fig. 1.4). It will be explained in more detail in chapter 8. An important consequence of phase coherence is that the resistance becomes a non-local quantity. Suppose we apply a current to a sample and measure the longitudinal voltage drop. This setup yields the longitudinal resistance R_{xx}. In conventional devices, it would just be the resistivity of the sample, multiplied by a geometrical factor. In a phase-coherent sample, however, scattering events outside the probed region may influence the local electron density between the voltage probes. Just think of the increase of complexity in the operation of a circuit of transistors within the phase coherence length, which then mutually influence each other. It should be remarked that on the other hand, the option of building electronic circuits in a phase coherent electron gas offers fascinating possibilities, which are outside the scope of this book, though.

A phase coherent electron gas is not necessarily ballistic, since elastic scatterers do not cause dephasing. Situations can be established where the electronic dephasing is governed by electron-electron scattering, which does not, or only marginally, show up in the resistance. This is the case because electron-electron scattering events do not modify the total momentum of the electron gas. We can therefore ask how phase coherent, diffusive systems behave. Such systems show in fact some interesting phenomena, which are presented in chapter 8 as well.

1.2.5 Single electron tunnelling and quantum dots

Size reduction goes along with a reduction of capacitances. Consider a parallel plate capacitor of area L^2 at a separation L. Its capacitance C scales with $1/L$. At small sizes, the energy required to store an additional electron on it, $E = e^2/2C$, may become larger than the thermal energy. As a consequence, the quantization of charge can dominate the behavior of suitabe circuits, in which tunnelling of single electrons across leaky capacitors carry the current. This so-called single electron tunnelling can be used to design new types of devices,

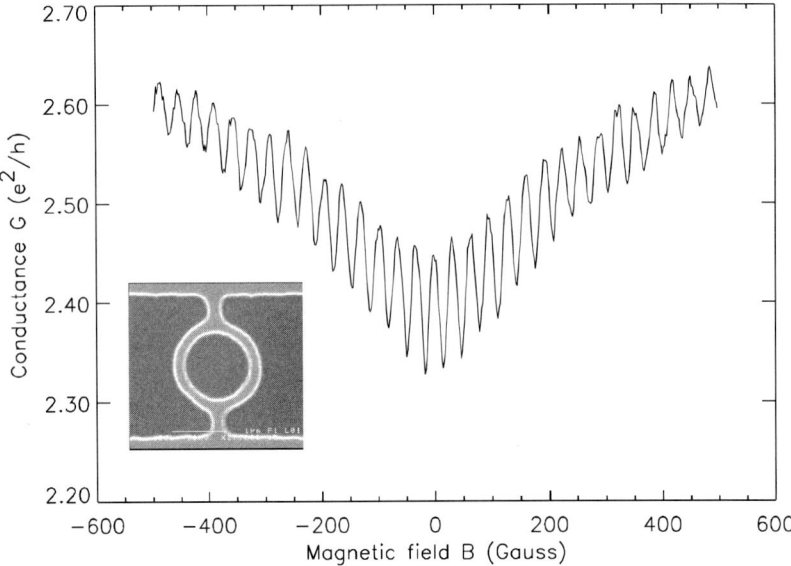

Figure 1.4: The resistance of a small ring with a diameter of about 1 μm (the light gray areas in the inset) as a function of a magnetic field applied perpendicular to the ring plane shows periodic oscillations, known as Aharonov-Bohm oscillations. They indicate that a significant fraction of the electrons traverse the ring phase coherently. Taken from [Pedersen2000].

in particular the single electron tunnelling transistor. Probably, it is Gorter who deserves the credit for giving birth to the field mesoscopic physics [Gorter1951]. In 1951, he suggested to explain transport experiments by van Itterbeek and coworkers [Itterbeek1947] on metal grains embedded in an isolated matrix by single electron charging. The first transistor that exploits this effect was built by Fulton and Dolan in 1987, [Fulton1987]. Fig. 1.5 shows an experimental realization of such a transistor in a semiconductor structure. We can call it a transistor since the gate voltage controls the current flowing between two further contacts. Single electron tunnelling is a very important member of the family of mesoscopic effects and will be presented in chapter 9. The structure shown in Fig. 1.5 actually represents also an example of a quantum dot. The electrons in the island are confined in all spatial directions, while their Fermi wavelength is comparable to the dot size. Quantum dots have been discovered by chance during the investigations of disordered quantum wires, which segregated into small islands [Scott-Thomas1989]. Soon afterwards, they have been fabricated on purpose [Meirav1989, Meirav1990]. The particular properties of such quantum dots are discussed in chapter 10, where we will also explain the data shown here in somewhat more detail.

1.2 Mesoscopic transport

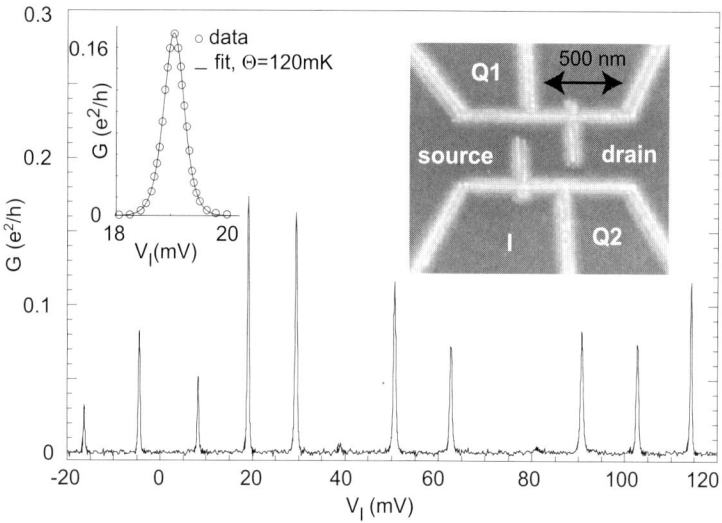

Figure 1.5: The right inset shows again the surface topography of a semiconductor with a two-dimensional electron gas underneath. Here, the bright lines enclose a small island. It is coupled to source and drain via two quantum point contacts, which in this case are closed, i.e., they form tunnel barriers for the electrons. This can be achieved by adjusting the voltages applied to gates Q1 and Q2 accordingly. The main figure shows the conductance through the island as a function of the gate voltage V_I applied to region I. V_I tunes the potential of the island. The conductance peaks indicate that only for a particular island potential, electrons can be transferred between the island and the leads. The left inset shows a fit to a function one would expect for peaks that are governed by thermal smearing of the Fermi function.

1.2.6 Superlattices

An interesting type of superpotentials are artificial crystals. They can be manufactured by patterning periodic structures on top of a semiconductor, followed by a transfer of the pattern into the electron gas. The resulting artificial lattices are either one-dimensional or two-dimensional, and have been patterned for many experiments, after [Warren1985], and [Bernstein1987], respectively. An alternative route is to grow layers of different semiconductor materials on top of each other [Esaki1974]. In contrast to laterally patterned samples, these lattices are almost exclusively one-dimensional.

We can not only build model potentials this way, but also study effects that occur in principle in periodic potentials, but remain unaccessible in natural crystals. Some prominent examples of such effects are treated in chapter 11.

1.2.7 Samples and experimental techniques

You have certainly noticed that the temperature has been quite low in all the examples given. The highest temperature encountered so far was 20 K, at which the remarkable conductance quantization in Fig. 1.2 was no longer visible. Also, all the samples were patterned semiconductors, so-called Ga[Al]As *heterostructures*, to be more precise. This seems to be a very narrow range of materials and temperatures. We live at room temperature, and the semiconductor industry makes its living on silicon. There is no doubt that these material systems and effects are fascinating from a purely scientific point of view. But are they really relevant for applications?

As far as the material is concerned, it is true that the Ga[Al]As system is sort of a workhorse for research in mesoscopic transport. Many groups work exclusively with Ga[Al]As heterostructures. This material is very versatile, and the electron gases can reach an almost incredible quality. For example, the electronic mean free path can exceed 100 μm at low temperatures. The foundation for achieving the corresponding ultra-high electron mobilities was laid by Dingle et al. in 1978, who invented a technique called *modulation doping*. As we shall see, this technique allows to spatially separate the doping ions from the mobile carriers; scattering is therefore greatly reduced. The details will be presented in chapter 3. Silicon, however, is the material of choice for fabricating microprocessors. First of all, silicon is readily available in large quantities. A major advantage of Si is that it has a natural oxide with excellent mechanical and electronic properties. It is therefore easy to fabricate high-quality insulators on-chip. Also, the advantages of Ga[Al]As are particularly striking at low temperatures; they are smaller at room temperature, although still quite relevant. In fact, Ga[Al]As systems fill certain niches at the market. They are used in optoelectronics (which is outside the scope of this book), since they have a direct band gap, in contrast to silicon. Also, Ga[Al]As is used in certain applications where high speed and low noise are essential. You sometimes hear that Ga[Al]As is, and will always remain, the material of the future. On the other hand, Si has played, and still does, an important role in mesoscopic research as well. It is probably fair to state that transistor structures have entered the field of mesoscopic transport in 1966, when it was observed that the electron gas in a Si -*MOSFET* (metal-oxide-semiconductor field effect transistor) has in fact a two-dimensional character, due to size quantization at the interface between the silicon and its oxide [Fang1966]. The quantum Hall effect, for example, has been discovered in a Si-MOSFET [Klitzing1980]. In chapters 2 and 3, we will therefore predominantly discuss these two material systems. However, other systems are by no means irrelevant in mesoscopic research! Each material has its particular strengths and weaknesses, and sometimes, more exotic systems are best. For example, InAs offers an extremely high effective g-factor and a very small effective mass. Hole gases in SiGe, on the other hand, have large effective masses. The choice of the material thus often depends on the particular experiment one has in mind. It should further be stressed that metallic nanostructures play a very important role in the field as well. Several seminal mesoscopic experiments have been actually performed in small metallic structures. For example, Aharonov-Bohm oscillations have been observed in metal loops, several years before they were seen in semiconductor rings. Also, the first single electron tunneling transistor has been made out of aluminum. We will meet metallic samples throughout the book.

Within the past few years, novel materials have moved in the focus of attention. One example are *carbon nanotubes*, see Fig. 1.6. They were discovered in 1991 by Iijima [Iijima1991]. These rolled-up graphite sheets can be thought of extremely small quantum wires. We will study some of their properties in chapters 7 and 10. Furthermore, electronic quantum films can also be generated in organic polymers. These systems, which are among the potential

1.2 Mesoscopic transport

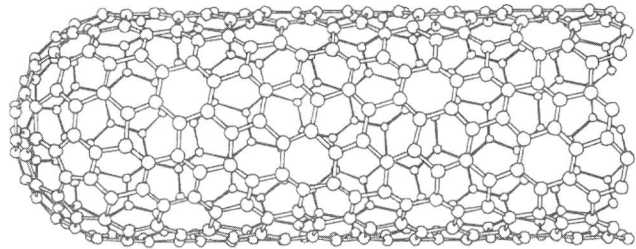

Figure 1.6: Structure of a carbon nanotube. The circles denote carbon atoms in a graphite sheet, which is rolled up and forms a tube with a diameter of a few nanometers. The ends are supposedly capped by a carbon hemisphere. After [Saito1992].

"materials of the future", are briefly presented in chapter 3. Finally, it should be said that along with the advance of nanotechnology, transport experiments on single molecules have become possible. Conductance quantization at a single hydrogen molecule has been reported by [Smit2002], for example. We will occasionally mention such examples of *molecular electronics*. It should be pointed out that the concepts exemplified on semiconductor nanostructures can be transferred to molecules in a straightforward way.

Now, how about the relevance of the mesoscopic transport effects for applications? Well, there are already applications. For example, the resistance quantization by the quantum Hall effect is so accurate that it is used as the resistance standard in many countries. Second, there is no fundamental reason why the mesoscopic effects should not occur at room temperature. This is in contrast to superconductivity, for example, since the critical temperature is well below room temperature for all superconductors known up to now. The temperature at which a mesoscopic effect vanishes, on the other hand, is essentially determined by the feature size. Just scale down your structure to, say, the phase coherence length or the mean free path, and you will see the mesoscopic behavior at room temperature. There are in fact several examples for mesoscopic behavior at room temperature, some of which we will meet later on. In many (not all) cases, the samples are cooled down just because we are not yet able to pattern them at sufficiently small length scales. Phonons are major obstacles for the electronic motion, and often limit the mean free path at room temperature. Optical phonons have typical energies in the range of a few ten meV, and are thus frozen out below about 30 K. Also, the density of acoustic phonons is greatly reduced by cooling the samples.

Smaller feature sizes mean stronger size quantization and larger separations between adjacent discrete energy levels. We can resolve the quantized structure as soon as the thermal smearing of the Fermi function is small compared to this energy level spacing. This is certainly the case for atoms at room temperature, but not for the artificial atom of Fig. 1.5, for example. Cooling the samples can therefore be regarded as a convenient way to look in the future, i.e., how the devices to come will behave at room temperature once their size has been sufficiently reduced. Table 2 gives some typical length scales.

Table 1.2: Typical length scales at which the mesoscopic regime is reached depends for different temperatures. The numbers just give an order of magnitude.

Temperature (K)	L (nm)
4.2 (liquid helium)	<5000
77 (liquid nitrogen)	<100
300 (room temperature)	<10

Both the technology of patterning nanostructures as well as performing transport experiments at very low temperatures are therefore very important issues. It is furthermore of great help to have an idea of experimental and technological boundary conditions to appreciate the measurements and the conditions under which they have been performed. Chapter 4, which deals with such issues, is therefore one of the central chapters of this book.

Finally, it should also be said what is *not* contained in this book. First of all, a very limited selection had to be made, just because of the plethora of mesoscopic phenomena out there. Three missing topics should probably be singled out. The large and extremely active field of interacting electron gases (besides elementary screening and single electron tunnelling) has been omitted. This concerns issues such as the fractional quantum Hall effect, the metal-insulator transition in two dimensions, Luttinger liquids or Kondo correlations. Also, *spintronics*, the rapidly growing field of spin transport as well as the fascinating topic of mesoscopic noise are not included. The reader is referred to the more advanced literature available.

2 An Update of Solid State Physics

Mesoscopic systems are fabricated from various bulk materials, which are often, but not always, semiconductors. Some basic knowledge of their bulk properties is important and represents the major part of this chapter. Although this is in many respects just a polishing up of solid state physics at an introductory level, we introduce many specifics of the materials of interest along the way, in particular of Si and GaAs. Occasionally, conventional metals and carbon crystals are mentioned as well, although the reader is supposed to know their basics.

We begin by a brief recapitulation of the most relevant crystal structures in section 2.1, and proceed by looking at the corresponding electronic band structures of the materials in section 2.2. Here, it is of particular importance to model the valence and the conduction bands around their maximum and minimum, respectively. As always, we can approximate the energy dispersions near the band extremal points by parabolas, which leads to the concept of *effective masses*. We shall see that within this approximation, the crystal properties can be "put aside" in many cases. Instead, the charge carriers behave like free electrons with a modified mass. The properties of electrons and holes within the effective mass approximation are looked at in section 2.3. Also, the effective mass approximation allows us to work with *envelope wave functions*. With this approach, superpotentials like those frequently met in nanostructures, can be treated with a Schrödinger equation for just this superpotential. The crystal potential enters only via the effective masses as well as via its dielectric constant. This is a very elegant concept, which simplifies our life substantially in subsequent chapters. Developing this approximation is the topic of section 2.4.

Doping is the standard way to fill the bands of a semiconductor with a significant and temperature-independent carrier density. The important issues concerning doping are reviewed in section 2.5. In the subsequent section, we look at the transport properties of electron gases within the simplest version of the Boltzmann model. We will occasionally use these results when looking at diffusive samples later on. Furthermore, it is of help to know the approximations that enter this model, in order to appreciate the deviations we will look at in subsequent chapters. A non-vanishing resistance indicates that some sort of scattering mechanism must be present, which are the topic of section 2.7. Finally, we spend a few words on screening in section 2.8.

Readers who discover that parts of this chapter are white spots on their map of solid state physics knowledge are encouraged to consult one of several excellent introductory textbooks for further information, e.g., [Ashcroft1985, Ziman1995]. If everything sounds familiar, please consider this chapter as a warm-up exercise!

2.1 Crystal structures

Many elements and compounds crystallize in a face centered cubic (fcc) lattice. This is not surprising, since this crystal structure represents one of the two closed packings possible, which one might naively expect to occur when identical or very similar spheres are piled up. Both Si and GaAs have this lattice structure. The lattice constant a is the length of one edge of a unit cell. Si is composed of two fcc lattices shifted relative to each other by $(a/4, a/4, a/4)$. This crystal structure is also known as the diamond structure. GaAs also has a two-atom base, except that here, the one base atom is Ga, the other one As. This is the zincblend lattice. The lattice constants are 0.565 nm for Si and 0.543 nm for GaAs (both numbers hold for room temperature). Fig. 2.1 shows the Si and the GaAs structure.

The reciprocal lattice of an fcc lattice is a body-centered cubic (bcc) lattice. Since the crystal

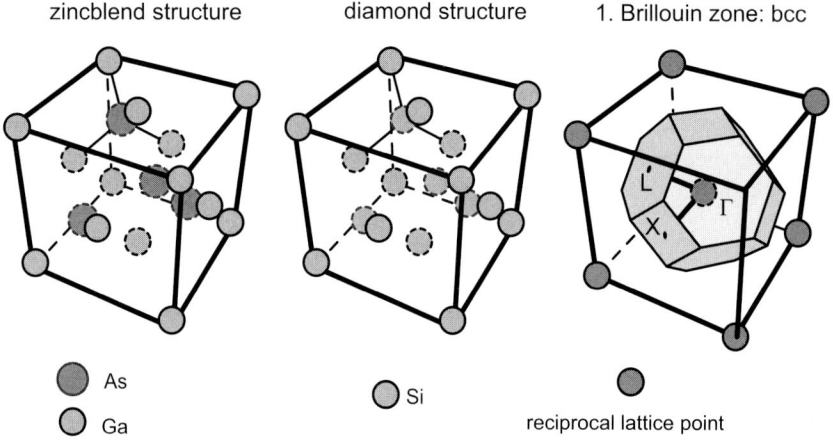

Figure 2.1: Crystal structures of GaAs (left) and Si (center), as well as their first Brillouin zone (right), a truncated octahedron. Points of high symmetry are labelled as K, Γ and L, see text.

momentum is invariant under translations by reciprocal lattice vectors, we can represent the behavior of electrons and phonons within one elementary cell of the reciprocal lattice, which is always chosen as the first Brillouin zone. For an fcc lattice, this is a truncated octahedron, composed of 6 squares and 8 hexagons, see Fig. 2.1. The center of the first Brillouin zone is labelled the Γ-point, while the centers of the hexagons and squares are referred to as L- and X-points, respectively. Occasionally, one hits upon more exotic directions of lower symmetry, such as K,U, and W, which are located at the center of the edges and at the corners of the first Brillouin zone.

Germanium crystallize in a diamond structure like silicon. This is also the case for many compound semiconductors. Any binary combination of Al, Ga, or In with As, Sb or P (the so-

2.1 Crystal structures

called III-V compounds) will result in a zincblend lattice. Combining these group III elements with nitrogen can lead to both a fcc lattice or a hexagonal lattice, depending on the crystallization process and the subsequent treatment. This is also the case for most II-VI compounds, such as CdSe or ZnS (which gave the zincblend structure its name, after all). Thus, when working with semiconductors, you will barely ever meet any further crystal structures. To finish this section, let us have a look at a particular simple lattice, namely a sheet of graphite, the second crystal structure carbon forms besides diamond. It consists of a hexagonal, lattice of sp^2 - hybridized carbon atoms (Fig. 2.2).

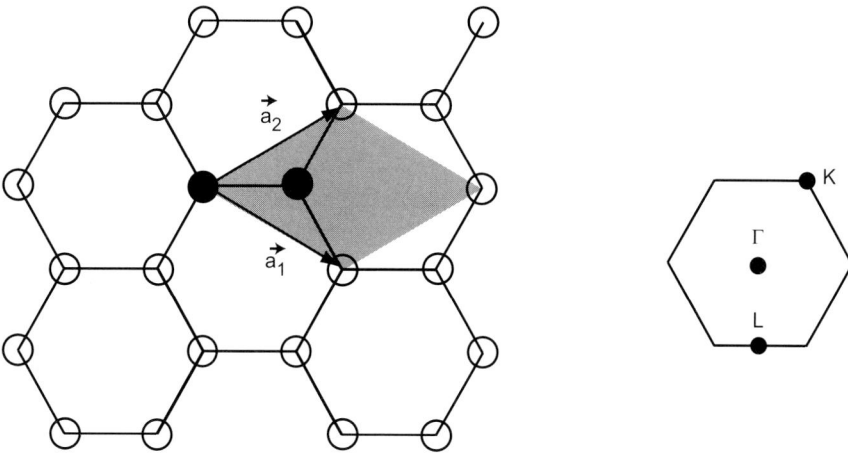

Figure 2.2: Structure of a graphite sheet. Left: The unit cell (gray) of this hexagonal lattice is spanned by the lattice vectors $|\vec{a}_1|$ and $|\vec{a}_2|$, with a lattice constant of $|\vec{a}_1| = |\vec{a}_2| = 0.246$ nm. It contains a basis of two carbon atoms (full circles) occupying non-equivalent sites. The distance of two neighboring atoms is 0.142 nm. Right: the first Brillouin zone of the graphite sheet with points of high symmetry.

Question 2.1: Calculate the reciprocal lattice of the graphite sheet and construct its first Brillouin zone.

The reciprocal lattice is again hexagonal. The center of the first Brillouin zone is denoted by Γ, the corners by K, and the centers of the edges are labelled L, respectively (Fig. 2.2).

2.2 Electronic energy bands

An electronic energy band is an energy interval in which electronic states are allowed in the crystal. The bands are separated by *band gaps*. This energy structure is obtained by solving the Schrödinger equation for electrons in the crystal

$$[-\frac{\hbar^2}{2m}\Delta + V_{crystal}(\vec{r})]\phi(\vec{k},\vec{r}) = \epsilon(\vec{k})\phi(\vec{k},\vec{r}) \qquad (2.1)$$

Here, the electronic wave functions depend on both the wave vector $\vec{k}$ and the spatial coordinates $\vec{r}$. They are denoted by $\phi(\vec{k},\vec{r})$, while $V_{crystal}(\vec{r})$ is the crystal potential. Elementary solid state physics tells us that the wave functions have to obey Bloch's theorem, which states that they are of the form

$$\phi(\vec{k},\vec{r}) = u_{\vec{k}}(\vec{r})e^{i\vec{k}\cdot\vec{r}} \qquad (2.2)$$

where $u_{\vec{k}}(\vec{r})$ has the periodicity of the crystal lattice. Such wave functions are *Bloch functions*. The task is to determine the eigenvalues $\epsilon(\vec{k})$ and eigenvectors, which is usually done by transforming the differential equation into an algebraic equation. An exact solution, though, is only possible for some special cases. Some reasonable approximation is therefore called for. How eq. (2.2) is then solved in detail depends on the model. The *nearly free electron model* starts from a free electron gas and treats a weak periodic crystal potential within perturbation theory. Here, the band gaps emerge from interferences of the electronic waves that get scattered at the crystal potential, which results in standing waves at the edges of the Brillouin zones. The reader is referred to the extensive literature on solid state physics for details. Here, we look at a different approach, which constructs the electronic eigenstates from those of the individual atoms that form the crystal. This approach is known as the *tight binding model*. Within this picture, the energy bands and the band gaps are remainders of the discrete energy spectrum of the atoms.

The tight-binding model is based on the assumption that the atomic orbitals $\xi_j(\vec{r})$ are a good starting point for constructing Bloch waves $\xi_j(\vec{k},\vec{r})$. Let us assume there is only one atom per unit cell. We can obtain Bloch functions via

$$\xi_j(\vec{k},\vec{r}) = \frac{1}{\sqrt{N}}\sum_{\vec{R}_n} e^{i\vec{k}\vec{R}_n}\xi_j(\vec{r}-\vec{R}_n) \qquad (2.3)$$

Here, the lattice vectors are denoted by $\vec{R}_n$. The crystal wave functions can be expanded in these Bloch functions, such that

$$\phi(\vec{k},\vec{r}) = \sum_j d_j(\vec{k})\xi_j(\vec{k},\vec{r}) \qquad (2.4)$$

The Schrödinger equation for the Bloch functions is now multiplied by $\xi_i^*(\vec{k},\vec{r})$ and integrated over space. The emerging algebraic equation has a nontrivial solution only for

$$det[T_{ij}(\vec{k}) - \epsilon O_{ij}(\vec{k})] = 0 \qquad (2.5)$$

2.2 Electronic energy bands

Here, $T_{ij}(\vec{k})$ and $O_{ij}(\vec{k})$ denote the "transfer matrix elements" and the "overlap matrix elements", respectively. They are defined as

$$T_{ij}(\vec{k}) = \langle \xi_i(\vec{k},\vec{r})|H|\xi_j(\vec{k},\vec{r})\rangle$$

and

$$O_{ij}(\vec{k}) = \langle \xi_i(\vec{k},\vec{r})|\xi_j(\vec{k},\vec{r})\rangle$$

These matrix elements are often approximated by inserting the known atomic orbitals, and choosing a suitable crystal potential, which can be used as a parameter to fit the experimentally determined properties of the crystal.

Question 2.2: Determine the energy dispersion for the simplest case, namely for a single band in one dimension, with a constant (and negative) transfer integral γ, and a vanishing overlap integral. Show that the energy dispersion in that case reads $E(k) = E_0 + 2\gamma \cos(ka)$!

As an example, we consider the graphite sheet, in which atomic s - and p - orbitals generate the bands of relevance, shown in Fig. 2.3. For the p_z orbitals of the carbon atoms arranged in a honeycomb configuration, a bonding and an antibonding π band results [Wallace1947]. To a first approximation, its tight-binding energy dispersion is

$$E_\pi(\vec{k}) = \pm T\sqrt{1 + 4\cos\left[\frac{1}{2}\sqrt{3}k_x a\right] \cos\left[\frac{1}{2}k_y a\right] + 4\cos^2\left[\frac{1}{2}k_y a\right]} \qquad (2.6)$$

Solids are usually classified as metals, semiconductors, and insulators. In a metal, at least one of the bands is partly occupied with electrons. These bands are called conduction bands in metals. In semiconductors and insulators, all bands are either full or empty at zero temperature. Here, the full band with the highest energy is the valence band, while the conduction band is the empty band with the lowest energy. In a semiconductor, a significant density of electrons can be transferred from the valence band into the conduction band by thermal excitation, which requires a band gap of less than 4 eV. Consequently, insulators have larger band gaps.

It turns out that the graphite sheet is a very special case in this classification scheme. The bonding π band is in fact the valence band, while its antibonding counterpart is the conduction band. As can be seen from eq. (2.6), the valence band can be mapped onto the conduction band by a reflection at the planes defined by the K-points. The conduction band and the valence band of a graphite sheet, represented by bold lines in Fig. 2.3, touch each other at the K-points. It can thus be regarded as a semiconductor with zero band gap.[1]

By adopting the tight-binding method appropriately, the band structure of other materials, like Si and GaAs, can be calculated. Naively, one might assume that due to the similar crystal structures, the band structures of the two semiconductors should be very similar as well. However, this is not the case, mainly because the Ga-As base is polar, while the Si base is

---
[1] In bulk graphite, the interaction between adjacent graphite sheets causes small energy shifts of both π - bands, such that they overlap somewhat around the K - points. It is therefore a metal with an extremely small carrier density.

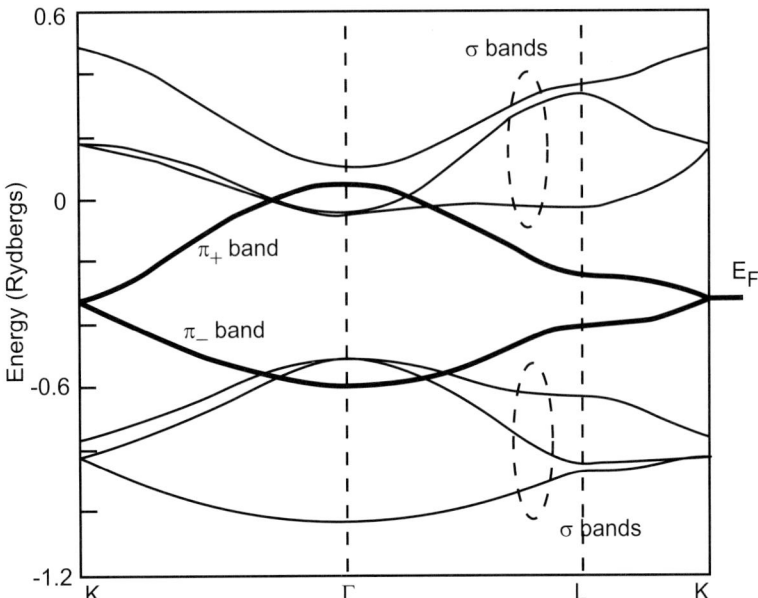

Figure 2.3: Band structure of a graphite sheet. Valence band and conduction band touch each other at the K-points. After [Painter1970].

covalent. Fig. 2.4 shows the structures of the valence and conduction bands of both crystals. The extremal points of the bands shown here dominate both the electronic and optical properties. The number of electrons in the conduction band, as well as that one of the holes in the valence bands, is small compared to the number of available electronic states in all cases of relevance, and the few carriers will find themselves in close proximity to the band extremal points. Around these extremal points, we can expand the energy dispersion in a Taylor series up to second order:

$$E(\vec{k}) = E_0 + \frac{1}{2}\vec{k} \cdot \left(\frac{\partial^2 E}{\partial k_i \partial k_j}\right) \cdot \vec{k} \tag{2.7}$$

By comparing this expression with the energy dispersion of the free electron gas $E(\vec{k}) = \hbar^2 \vec{k}^2 / 2m$, we see that the tensor of second derivatives of the energy can be identified with effective masses,

$$\frac{1}{\hbar^2}\left(\frac{\partial^2 E}{\partial k_i \partial k_j}\right) = \left(\frac{1}{m^*}\right)_{ij} \tag{2.8}$$

which is therefore also known as the *effective mass tensor*. It can be diagonalized, such that the extremal points of energy bands can be characterized by its three effective masses along the principal axes. Carriers in semiconductors therefore usually behave free-electron like, except that their masses have been changed by the crystal structure. Throughout the rest of the book, we will use effective masses to describe the behavior of carriers.

2.2 Electronic energy bands

Question 2.3: What is the effective mass around the minimum of the the energy band obtained in Question 2.2?

Let us have a somewhat closer look at these band structures. Si has a conduction band minimum at 0.85 $\vec{\Gamma X}$. Around this minimum, two different effective masses exist, a transverse mass in all directions perpendicular to the $\vec{\Gamma X}$- direction, $m_{e,t} = 0.19\,m$, and a longitudinal mass along the $\vec{\Gamma X}$- direction, $m_{e,l}= 0.92\,m$. Since there are 6 X-points, the conduction band minimum in Si shows a 6- fold degeneracy known as "valley degeneracy". In GaAs, the conduction band minimum is located at the Γ-point. Here, the 3 effective electron masses are identical: $m^*_{e,1}(GaAs) = m^*_{e,2}(GaAs) = m^*_{e,3}(GaAs) = 0.067m$. In both materials,

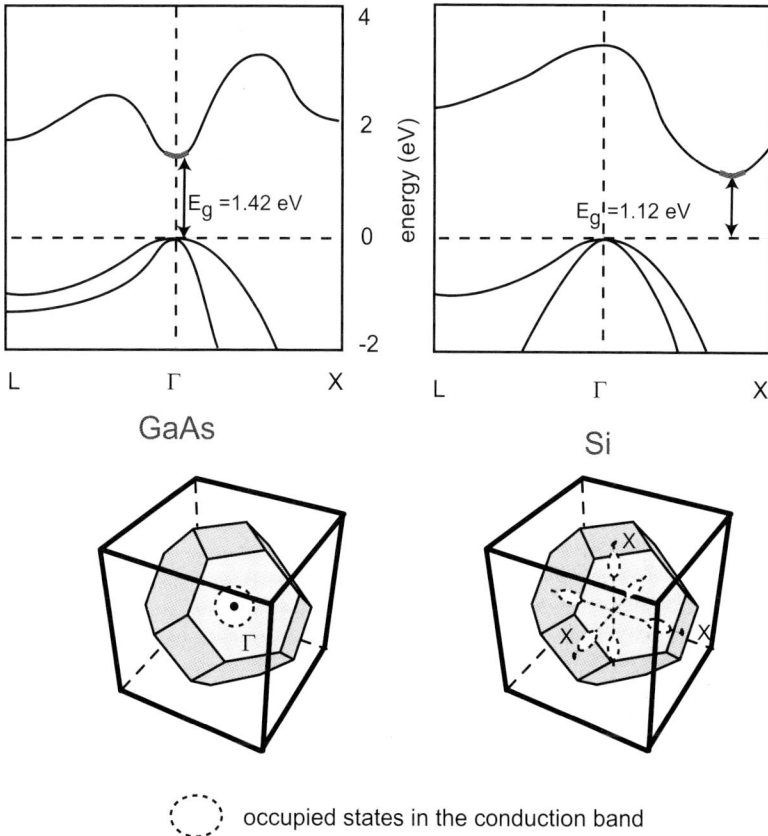

Figure 2.4: Top: electronic band structure of GaAs and Si close to the band gap. Bottom: schematic location and shape of the regions the electrons occupy at typical electron densities.

there are two (nearly) degenerate valence bands at the Γ-point. As in most semiconductors of interest, the valence band emerges from atomic p - states, which have a threefold orbital degeneracy and a spin degeneracy of 2. Typically, the corresponding σ - band formed by the atomic s-orbitals has its maximum well below the maximum of the p-bands and do not have to be taken into account for transport considerations. In the crystal, the degeneracy of the p - orbitals is removed, and 3 different, spin degenerate bands are obtained. Two of them are shown in Fig. 2.4, while the third one is split off and shifted to lower energies. This splitting has its origin in the spin-orbit interaction. The spin-orbit Hamiltonian is given by

$$H_{so} = \frac{\hbar}{4m^2c^2} \underline{\sigma} \cdot \vec{\nabla} V \times \vec{p} \qquad (2.9)$$

where V is the electrostatic potential, and $\underline{\sigma}$ are the Pauli matrices. This is a relativistic term, which means we have to replace the Schrödinger equation by the Dirac equation, and the wave function becomes a two-component spinor. In a spherical symmetric potential, the spin-orbit Hamiltonian becomes proportional to the scalar product of the angular momentum and the spin $\vec{L} \cdot \vec{S}$. To get an idea what the spin-orbit Hamiltonian does to the energies, we assume that the interaction in the solid can be approximated by that one in the individual atoms. It is then clear from atomic physics that this term separates the four-fold degenerate $j = 3/2$ states from the twofold degenerate $j = 1/2$ states, where j denotes the total angular momentum quantum number. The $j = 1/2$ state is lowered in energy, by an amount that essentially depends on the strength of the atomic Coulomb potential. The heavier the nucleus, the stronger is this spin-orbit splitting Δ_{so}. This tendency can be seen experimentally: $\Delta_{so}(graphite) \approx 6$ meV, $\Delta_{so}(Si) \approx 45$ meV, and $\Delta_{so}(GaAs) \approx 340$ meV.

The energy dispersions of the remaining four bands with $j = 3/2$ can be conveniently described within the $\vec{k} \cdot \vec{p}$ *approximation*, a method to model the dispersion around the extremal points of an energy band. We consider a semiconductor with a band maximum at $\vec{k} = 0$, as it is the case for the valence bands under study. Within the $\vec{k} \cdot \vec{p}$ model, the spatial derivatives in the Schrödinger equation of the crystal, eq. (2.2), are carried out only for the plane wave component of the Bloch function of the type (2.2). The equation

$$\left\{ \frac{p^2}{2m} + \frac{\hbar \vec{k} \cdot \vec{p}}{m} + \frac{\hbar \vec{k}^2}{2m} + V(\vec{r}) \right\} u_{n,\vec{k}}(\vec{r}) = E_{n,\vec{k}} u_{n,\vec{k}}(\vec{r}) \qquad (2.10)$$

emerges. Here, n denotes the band index. For $\vec{k} = 0$, it simplifies significantly, and we assume that an approximate solution can be found for all bands involved. A non-vanishing but small wave vector can then be treated as a perturbation.

First of all, the term $\propto \vec{k}^2$ produces an energy shift which depends on $\vec{k}$, but does not couple the bands. Technically, it can just be added to the crystal potential. The term containing $\vec{k} \cdot \vec{p}$, however, must be treated with degenerate perturbation theory. It turns out that the second-order term is the leading one, since the first-order term is linear in $\vec{k}$ and must thus vanish at a maximum. The matrix elements are given by

$$h_{ij}(\vec{k}) = \frac{\hbar^2}{m^2} \sum_{q=1; q \neq i,j}^{4} \frac{\langle n_i, 0 | \vec{k} \cdot \vec{p} | n_q, 0 \rangle \langle n_q, 0 | \vec{k} \cdot \vec{p} | n_j, 0 \rangle}{\epsilon_{n_i,0} - \epsilon_{n_q,0}} \qquad (2.11)$$

2.2 Electronic energy bands

The bands are labelled by $n_{i,j,q}$ here. This 4x4 matrix equation gives the energy eigenvalues of the type

$$E_{lh,hh} = \frac{\hbar^2}{2m}\left[\gamma_1 k^2 \pm \sqrt{4\gamma_2^2 k^4 + 12(\gamma_3^2 - \gamma_2^2)(k_x^2 k_y^2 + k_y^2 k_z^2 + k_z^2 k_x^2)}\right] \quad (2.12)$$

The γ_i are the Luttinger parameters, which depend on the material. For GaAs, $\gamma_1 = 6.95, \gamma_2 = 2.25$, and $\gamma_3 = 2.86$. In Si, $\gamma_1 = 4.29, \gamma_2 = 0.34$, and $\gamma_3 = 1.42$. The "+" energy dispersion

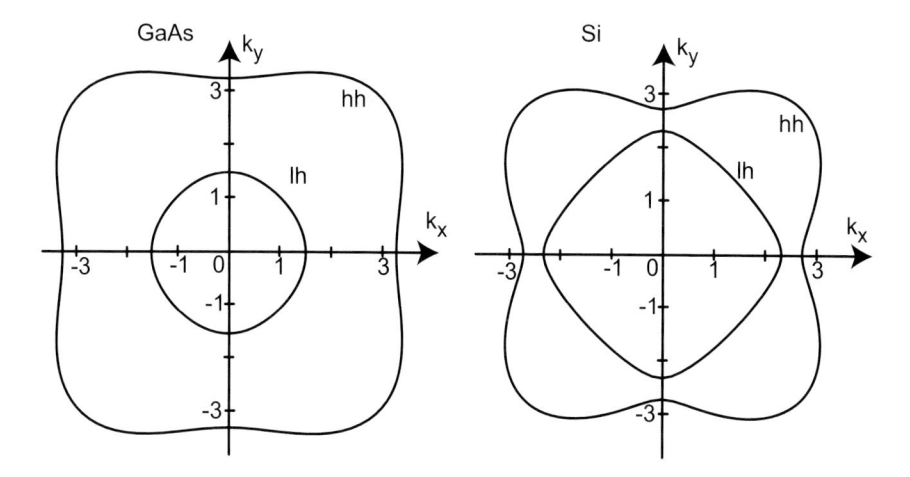

Figure 2.5: Warped surfaces of constant energy for heavy and light hole bands in GaAs (left) and Si (right). A cross section through the plane $k_z = 0$ is shown for a (typical) Fermi energy of 10 meV. The wave numbers are measured in units of 10^8 m^{-1}.

corresponds to a lighter effective mass for all directions in $\vec{k}$ - space. The band is therefore referred to as the *light hole (lh) band*. Correspondingly, the "-" - sign represents the energy dispersion for the *heavy hole (hh) band*. To get an idea of the shape of the hole bands, consider a surface of constant energy. The first term of the right hand side in eq. (2.12) describes a sphere, which is warped by the second term. The warping is $\vec{k}$- dependent and of opposite sign in the two bands for all directions. These surfaces are therefore known as *warped spheres*, see Fig. 2.5.

Note that both bands remain twofold degenerate in this treatment at $\vec{k} = 0$. This is known as the Kramers degeneracy. It is removed in polar crystals, such as GaAs or InP, due to the absence of an inversion center. The corresponding correction to the Hamiltonian is known as the Dresselhaus term. The resulting energy splitting, however, is small, i.e., in the range of μeV, although it causes measurable effects occur at very low temperatures.

Per definition, hole masses are negative. A hole is thus a quasiparticle with negative effective mass and a negative charge of $q = -e$. Since, as shown below, the ratio q/m^* enters in the conductivity, we can equivalently regard a hole as a particle with positive effective mass and a charge $q = +e$. Due to the warped structure of the valence bands, the hole masses are averaged over all directions in the valence bands. Typical literature values are $m_{hh}^*(Si) = 0.54m$, $m_{lh}^*(Si) = 0.15m$, $m_{hh}^*(GaAs) = 0.51m$, and $m_{lh}^*(Si) = 0.08m$.

GaAs has a direct band gap, meaning that the minimum in the conduction band is at the same location in k - space as the maximum of the valence band. The band gap of Si is indirect. A large momentum transfer is necessary for exciting electrons from the valence band maximum into the conduction band minimum.

Question 2.4: Compare the momentum of a photon with the energy of the Si band gap with the momentum difference between the Γ-point and the conduction band minimum in Si!

Therefore, Si can absorb photons with an energy close to the band gap only if phonons are absorbed/ emitted simultaneously. This is a rather unlikely process, which makes crystalline Si a poor material for optoelectronics.

Due to anharmonic contributions to the lattice vibrations, the crystals shrink as they get cooled down. As a consequence, the band gap increases with decreasing temperature. Empirically, one finds

$$E_{g,Si}(\Theta) = 1.17 \text{ eV} - \frac{4.73 \cdot 10^{-4} \Theta^2 \text{ K}^{-1}}{\Theta + 636 \text{ K}} \text{ eV}$$

$$E_{g,GaAs}(\Theta) = 1.52 \text{ eV} - \frac{5.4 \cdot 10^{-4} \Theta^2 \text{ K}^{-1}}{\Theta + 204 \text{ K}} \text{ eV}$$

For many applications, a fraction x of the Ga atoms in GaAs is replaced by Al atoms, and the ternary $Al_xGa_{1-x}As$ results. For $x \leq 0.38$, the band gap increases linearly with x, with a maximum at $E_g(\Gamma, x = 0.38) = 1.92$ eV, and can be tailored for a specific application (see Fig. 2.6). For $x \geq 0.38$, however, the local minimum close to the X-point becomes the

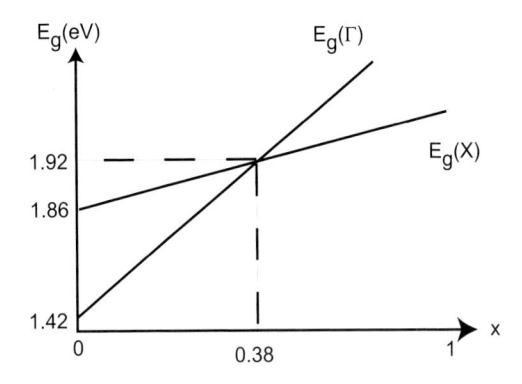

Figure 2.6: Sketch of the band gap in $Al_xGa_{1-x}As$ as a function of the Al concentration x.

global minimum of the conduction band, and the material becomes an indirect semiconductor. Pure AlAs as a band gap of $E_g = 2.16$ eV. Note that the positions of the Al atoms are random, which means that the ternary compound is *not* a crystal. Nevertheless, we can speak of band structures and effective masses, since such crystals can be treated within an an averaging procedure known as "virtual crystal approximation" (see [Bastard1989]).

2.3 Occupation of energy bands

In this section, we study how the electrons occupy the valence and conduction bands. The electron density n in a band is obtained by integrating over the spectral electron density $n(E)$, i.e. the density of electrons in the interval $[E, E+dE]$. The spectral electron density is given by the spectral density of electronic states $D_d(E)$ available, multiplied by their occupation probability. Here, the index d denotes the dimensionality of the system. We will briefly discuss these two quantities.

2.3.1 The electronic density of states

The electronic density of states $D_d(E)$ is the number of electronic states in $[E, E+dE]$ and per unit volume. It depends on the dimensionality d of the system and the energy dispersion $E(\vec{k})$ of the electronic band under consideration. The usual way to calculate $D_d(E)$ is to determine the electronic mode density in k - space $D_d(\vec{k})$ of a cavity of size L^d and transform it into the energy space via $E(\vec{k})$. We carry out this calculation for a two-dimensional system with a parabolic energy dispersion, since this is what we will encounter most frequently in the following.

Consider a two-dimensional crystal cube with a base length L, oriented along the x- and y-axes. We assume periodic boundary conditions, and use plane waves as base functions.[2] An electronic state Ψ exists at wave vector $\vec{k}$ if

$$\Psi(\vec{r} + (L, L)) = \Psi(\vec{r}) \Rightarrow \vec{k} = \frac{2\pi}{L}(n_x, n_y)$$

with n_i being an integer. The allowed wave vectors form a simple cubic lattice in k-space with a lattice constant of $\frac{2\pi}{L}$. Each state is g-fold degenerate due to spin and valley degeneracies. Hence, there are g states in the volume $(\frac{2\pi}{L})^2$. States of equal $|\vec{k}|$ are located at a circle. The number of states dN_2 in an annulus of radius $\vec{k}$ and widths $d\vec{k}$ is given by

$$dN_2 = g\frac{2\pi k}{(2\pi/L)^2}dk$$

with $k = |\vec{k}|$. This gives a density of states in k-space of

$$D_2(k) = \frac{1}{L^2}\frac{dN}{dk} = \frac{gk}{2\pi}$$

$D_2(E)$ is obtained from $D_2(k)$ by a coordinate transformation

$$D_2(E) = D_2(k)\frac{dk}{dE} = \frac{gm^*}{2\pi\hbar^2} \quad (2.13)$$

---
[2] It can be shown that the results do not depend on the boundary conditions.

Here, we have used the energy dispersion for electrons with an isotropic effective mass m^*,

$$E(\vec{k}) = \frac{\hbar^2 \vec{k}^2}{2m^*}$$

The density of states in 3, 2, and 1 dimensions are shown in Fig. 2.7.

Question 2.5: Calculate $D_3(E)$ and $D_1(E)$! Show that

$$D_3(E) = g \frac{(2m)^{3/2}}{4\pi^2 \hbar^3} \sqrt{E} \qquad (2.14)$$

and

$$D_1(E) = g \frac{\sqrt{2m}}{2\pi \hbar} \frac{1}{\sqrt{E}} \qquad (2.15)$$

How does the density of states look for a zero-dimensional system?

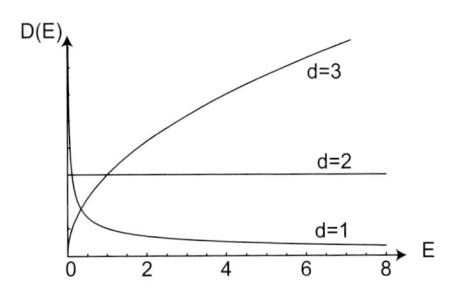

Figure 2.7: The electronic density of states within the effective mass approximation as a function of energy, in one, two, and three dimensions.

2.3.2 Occupation probability and chemical potential

In equilibrium, fermions occupy states of energy E with a probability given by the Fermi-Dirac distribution function

$$f(E, \Theta) = \frac{1}{e^{(E-\mu)/k_B \Theta} + 1} \qquad (2.16)$$

Here, μ denotes the chemical potential, i.e., the energy for which the density of occupied states with larger energies equals the density of empty states with lower energies. This definition, by the way, also holds if the occupation probability is not a Fermi-Dirac distribution. Furthermore, Θ is the temperature. The *Fermi energy* E_F is the energy at which $f(E, \Theta = 0)$ jumps

2.3 Occupation of energy bands

from 1 to 0. Clearly, $\mu = E_F$ at $\Theta = 0$. For $\Theta > 0$, μ may differ from E_F, depending on the energy dependence of the density of states.

In a metal, at least one band is per definition partly occupied at $\Theta = 0$. Therefore, E_F is located within an energy band. Semiconductors, on the other hand, are crystals where the conduction band is empty at $\Theta = 0$, and E_F thus resides in the band gap. The same is of course true for insulators. The electron density in a band ranging from E_{bottom} to E_{top} is obtained from

$$n = \int_{E_{bottom}}^{E_{top}} n(E)dE = \int_{E_{bottom}}^{E_{top}} D_d(E)f(E,\Theta)dE \qquad (2.17)$$

Note that the dimensionality d of $D_d(E)$ also determines the dimensionality of n. The Fermi function and the spectral electron density are sketched in Fig. 2.8.

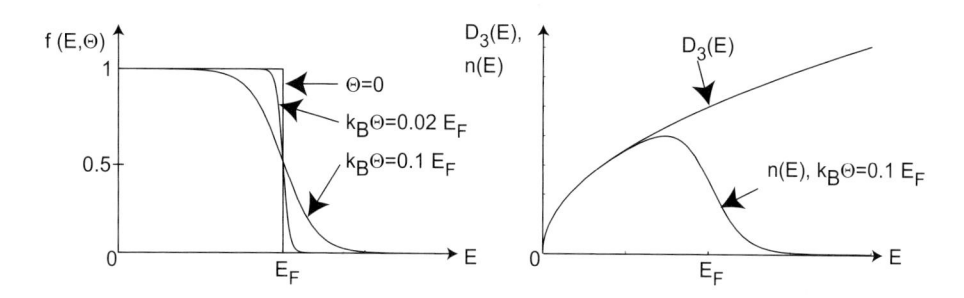

Figure 2.8: Thermal smearing of the Fermi function (left), and the density of states ($d = 3$) as well as the spectral carrier density n(E), right.

2.3.3 Intrinsic carrier concentration

The carrier concentration of a perfect, impurity-free crystal is called intrinsic. Here, the carriers are exclusively generated by thermal excitation of electrons from the valence band into the conduction band, which means that $n = p$ (p denotes the hole density). Within the effective mass approximations and for a spin degenerate system (g=2), the carrier densities are given by

$$n = \frac{\sqrt{2}m_e^{*3/2}}{\pi^2 \hbar^3} \int_{E_C}^{\infty} \sqrt{E - E_C} f(E,\Theta) dE;$$

$$p = \frac{\sqrt{2}m_h^{*3/2}}{\pi^2 \hbar^3} \int_{-\infty}^{E_V} \sqrt{E_V - E} f(E,\Theta) dE \qquad (2.18)$$

The chemical potential is close to the center of the band gap, slightly shifted towards the band with the lighter effective mass.[3] Therefore, it is safe to assume that $|E_{C,V} - \mu| \gg k_B\Theta$. This tells us that only the tails of the Fermi function, far away from the chemical potential, lie inside the bands, and can be well approximated by a Boltzmann distribution, i.e. $f(E,\Theta) = \exp[-(E-\mu)/k_B\Theta]$. A brief calculation gives

$$n = N_C e^{-(E_C-\mu)/k_B\Theta}; \qquad p = P_V e^{(E_V-\mu)/k_B\Theta} \qquad (2.19)$$

N_C and P_V are known as "effective density of states", given by

$$N_C = \frac{1}{4}\left(\frac{2m_e^* k_B\Theta}{\pi\hbar^2}\right)^{3/2}; \qquad P_V = \frac{1}{4}\left(\frac{2m_h^* k_B\Theta}{\pi\hbar^2}\right)^{3/2}$$

An immediate consequence is the "mass action law for charge carriers"

$$n \cdot p = N_C P_V e^{-E_g/k_B\Theta} \Rightarrow n = p = \sqrt{N_C P_V} e^{-E_g/2k_B\Theta}$$

Inserting p in eq. (2.19) leads to

$$\mu = E_V + \frac{1}{2}E_g + \frac{3}{4}k_B\Theta \ln\left(\frac{m_h^*}{m_e^*}\right) \qquad (2.20)$$

Fig. 2.9 summarizes the relations between density of states, Fermi function and intrinsic carrier densities in a semiconductor.

The exponential dependence of n and p on the temperature causes *carrier freezeout* as the temperature is reduced. At room temperature, we have an intrinsic electron density of $n_{\text{Si}} = 1.45 \cdot 10^{16} m^{-3}$ for silicon and $n_{\text{GaAs}} = 1.8 \cdot 10^{12} m^{-3}$ for GaAs, see exercise E 3.4.

At these typical, small carrier densities, the electron Fermi surface consists of six rotational ellipsoids in Si, and of a sphere in GaAs, as indicated in Fig. 2.4. In the valence band, the warped surfaces in the previous section represent Fermi spheres.

2.4 Envelope wave functions

So far, the materials have been homogeneous. The real crystal is certainly not perfect. Its translational symmetry can be perturbed, either by, e.g., unwanted lattice imperfections, or by intentionally built-in superpotentials. We will frequently see such superpotentials later on. How do the wave functions and energy levels in such a perturbed crystal look like?

Consider a lattice imperfection with the perturbation potential $V_p(\vec{r})$. For simplicity, we take only one electronic band into account. The Schrödinger equation for the imperfect crystal reads

$$\left[-\frac{\hbar^2}{2m}\Delta + V_{lattice}(\vec{r}) + V_p(\vec{r})\right]\Phi(\vec{r}) = E\Phi(\vec{r}) \qquad (2.21)$$

---

[3] This is qualitatively clear as μ is given by the condition $n = p$, and the density of states increases with increasing effective mass.

2.4 Envelope wave functions

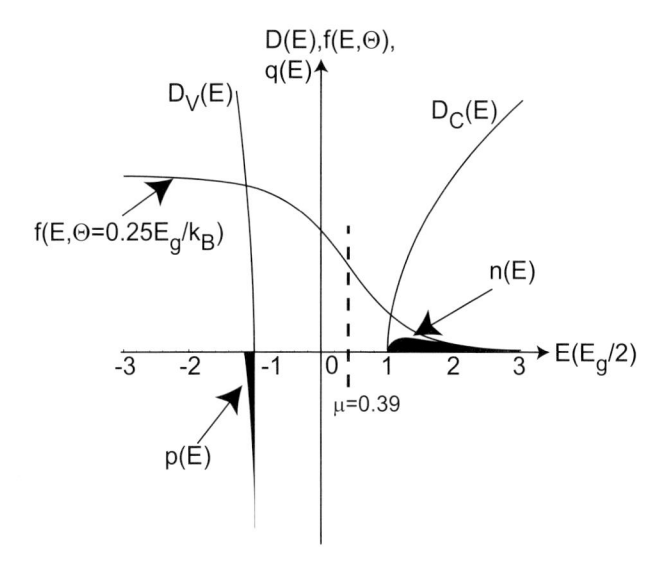

Figure 2.9: Fermi function, density of states and spectral carrier densities $q(E)$ (q=n,p) in an intrinsic semiconductor with $m_h^* = 2m_e^*$. The temperature is $\Theta = 0.25 E_g/k_B$, which corresponds to a chemical potential $\mu = 0.39 E_g/2$.

The solution $\Phi(\vec{r})$ is no longer a Bloch function, but it can be expanded in the Bloch wave functions of the unperturbed band

$$\Phi(\vec{r}) = \sum_{\vec{k}'} c_{\vec{k}'} \xi(\vec{k}', \vec{r}) \qquad (2.22)$$

Inserting this expansion into eq. (2.21), multiplying by $\xi^*(\vec{k}, \vec{r})$, integrating over the whole crystal gives

$$\epsilon(\vec{k}) c_{\vec{k}} + \sum_{\vec{k}'} c_{\vec{k}'} a(\vec{k}, \vec{k}') = E c_{\vec{k}} \qquad (2.23)$$

with the matrix elements

$$a(\vec{k}, \vec{k}') = \langle \xi(\vec{k}, \vec{r}) | V_p(\vec{r}) | \xi(\vec{k}', \vec{r}) \rangle \qquad (2.24)$$

We plan to rewrite eq. (2.23) in the form of a Schrödinger equation with a newly defined wave function, which will be the envelope function. This can be done by making two approximations, namely by (i) $V_p(\vec{r})$ varies "smoothly", meaning slowly on the scale of the lattice constant, and (ii) the effective mass approximation.

Our first task is finding an appropriate expression for $a(\vec{k},\vec{k}')$. We have assumed that $V_p(\vec{r})$ varies slowly on the scale of individual unit cells. This means that we can keep $V_p(\vec{r})$ constant within each cell, which is referred to by the corresponding lattice vector $\vec{R}$. In order to use this in eq. (2.24), we split the integral, which runs over the whole crystal, into integrals running over unit cells, and sum them up:

$$a(\vec{k},\vec{k}') = \sum_{\vec{R}} \int_{cell \vec{R}} \xi^*(\vec{k},\vec{r}) V_p(\vec{r}) \xi(\vec{k}',\vec{r}) d\vec{r} = \sum_{\vec{R}} V_p(\vec{R}) \int_{cell \vec{R}} \xi^*(\vec{k},\vec{r}) \xi(\vec{k}',\vec{r}) d\vec{r} \quad (2.25)$$

Since $\xi(\vec{k},\vec{r})$ is of the form given by eq.(2.2), i.e. $\xi(\vec{k},\vec{r}) = u_{\vec{k}}(\vec{r})e^{i\vec{k}\vec{r}}$, the cell integral can be written as

$$\int_{cell \vec{R}} u_{\vec{k}}^*(\vec{R}+\vec{r}) u_{\vec{k}'}(\vec{R}+\vec{r}) e^{i(\vec{k}'-\vec{k})(\vec{R}+\vec{r})} \quad (2.26)$$

The function $u_{\vec{k}}^*(\vec{r})u_{\vec{k}'}(\vec{r})$ has the periodicity of the lattice and can thus, according to the Fourier theorem, be expanded in harmonic functions with the same periodicity:

$$u_{\vec{k}}^*(\vec{r})u_{\vec{k}'}(\vec{r}) = \sum_{\vec{G}} \alpha(\vec{G}) e^{i\vec{G}\vec{r}}; \quad \alpha(\vec{G}) = \frac{1}{V} \int_V u_{\vec{k}}^*(\vec{r})u_{\vec{k}'}(\vec{r}) e^{-i\vec{G}\vec{r}} d\vec{r} \quad (2.27)$$

where $\vec{G}$ is a reciprocal lattice vector. With the Fourier expansion inserted in eq. (2.25), we obtain

$$a(\vec{k},\vec{k}') = \sum_{\vec{G},\vec{R}} \alpha(\vec{G}) V_p(\vec{R}) \int_{cell \vec{R}} e^{i(\vec{k}'-\vec{k})(\vec{R}+\vec{r})} e^{i\vec{G}(\vec{R}+\vec{r})} d\vec{r} \quad (2.28)$$

which can be simplified considerably. First of all, $e^{i\vec{G}\vec{R}} = 1$, Second, since $V_p(\vec{r})$ varies smoothly, only Bloch waves within a narrow interval of $\vec{k}$ vectors will contribute to $a(\vec{k},\vec{k}')$, and we can assume that $\vec{k}'-\vec{k}$ is small on the scale of the smallest reciprocal lattice vector. Therefore, it is justified to approximate $e^{i(\vec{k}'-\vec{k})(\vec{R}+\vec{r})} \approx e^{i(\vec{k}'-\vec{k})\vec{R}}$. After taking these considerations into account, eq. (2.28) reads

$$a(\vec{k},\vec{k}') \approx \sum_{\vec{G},\vec{R}} \alpha(\vec{G}) V_p(\vec{R}) e^{i(\vec{k}'-\vec{k})\vec{R}} \int_{cell \vec{R}} e^{i\vec{G}\vec{r}} d\vec{r}$$

In addition, Green's theorem for functions with the periodicity of the lattice [Ashcroft1985] tells us that

$$\int_{cell \vec{R}} e^{i\vec{G}\vec{r}} d\vec{r} = V_{cell} \delta_{\vec{G},0} \quad (2.29)$$

where V_{cell} is the volume of the unit cell.

2.4 Envelope wave functions

Question 2.6: Prove eq. (2.29) for a one-dimensional crystal.

With eq. (2.29), we obtain

$$a(\vec{k}, \vec{k}') = a(0) \sum_{\vec{R}} V_p(\vec{R}) V_{cell} e^{i(\vec{k}'-\vec{k})\vec{R}}$$

Summing up the contributions of all cells can now be replaced by an integration over the whole crystal, such that

$$a(\vec{k}, \vec{k}') = a(0) \int_V V_p(\vec{r}) e^{i(\vec{k}'-\vec{k})\vec{r}} d\vec{r}$$

It remains to determine $a(0)$, which we approximate by

$$a(0) = \frac{1}{V} \int_V u_{\vec{k}}^*(\vec{r}) u_{\vec{k}'}(\vec{r}) d\vec{r} \approx \frac{1}{V}$$

This is justified since $\vec{k}' \approx \vec{k}$, and the integral in the definition of $a(0)$ should give a value very close to 1 (recall that the functions $u_{\vec{k}}$ are normalized to 1). This finally leads to

$$a(\vec{k}, \vec{k}') = \frac{1}{V} \int_V V_p(\vec{r}) e^{i(\vec{k}'-\vec{k})\vec{r}} d\vec{r} \qquad (2.30)$$

Inserting eq. (2.30) in eq. (2.23) and by using the effective mass approximation for the unperturbed crystal,

$$\epsilon_{\vec{k}} = E_C + \hbar^2 \vec{k}^2 / 2m^* \qquad (2.31)$$

eq. (2.23) changes to

$$\frac{\hbar^2 \vec{k}^2}{2m^*} c_{\vec{k}} + [E_C - E] c_{\vec{k}} + \frac{1}{V} \sum_{\vec{k}'} c_{\vec{k}'} \int V_p(\vec{r}) e^{i(\vec{k}'-\vec{k})\vec{r}} d\vec{r} = 0 \qquad (2.32)$$

We proceed by defining the *envelope wave function* as

$$\psi(\vec{r}) = \frac{1}{\sqrt{V}} \sum_{\vec{k}'} c_{\vec{k}'} e^{i\vec{k}'\vec{r}} \qquad (2.33)$$

which we plan to insert the envelope wave function in eq. (2.32) by substituting $c_{\vec{k}}$ and $\vec{k}^2 c_{\vec{k}}$. This can be done via the relations

$$c_{\vec{k}} = \sum_{\vec{k}'} c_{\vec{k}'} \delta(\vec{k}-\vec{k}') = \frac{1}{V} \int \sum_{\vec{k}'} c_{\vec{k}'} e^{i\vec{k}'\vec{r}} e^{-i\vec{k}\vec{r}} d\vec{r} = \frac{1}{\sqrt{V}} \int \psi(\vec{r}) e^{-i\vec{k}\vec{r}} d\vec{r}$$

and

$$\vec{k}^2 c_{\vec{k}} = \sum_{\vec{k}'} \vec{k}'^2 c_{\vec{k}'} \delta(\vec{k} - \vec{k}') = \frac{1}{V} \int \sum_{\vec{k}'} \vec{k}'^2 c_{\vec{k}'} e^{i\vec{k}'\vec{r}} e^{-i\vec{k}\vec{r}} d\vec{r} =$$

$$\frac{1}{\sqrt{V}} \int (-\Delta \psi(\vec{r})) e^{-i\vec{k}\vec{r}} d\vec{r}$$

The equation

$$\int e^{-i\vec{k}\vec{r}} \left[-\frac{\hbar^2 \Delta}{2m^*} + E_C - E + V_p \right] \psi(\vec{r}) d\vec{r} = 0$$

is obtained, which is fulfilled for all $\vec{k}$ only if

$$\left[-\frac{\hbar^2 \Delta}{2m^*} + V_p(\vec{r}) \right] \psi(\vec{r}) = [E - E_C] \psi(\vec{r}) \qquad (2.34)$$

This is the *envelope wave equation*. For perturbation potentials that vary slowly on the scale of the crystal unit cell, the energy eigenvalues of the $V_p(\vec{r})$ in the crystal correspond to the energy eigenvalues of $V_p(\vec{r})$ in a homogeneous medium with the dielectric constant of the crystal, and for particles which have the effective mass of the corresponding electronic band. The energy eigenvalues obtained from the envelope wave equation are relative to E_C, the conduction band bottom, in our case. The envelope wave functions are thus just regular wave functions that solve eq. (2.34).

2.5 Doping

In many cases, it is desirable to have predominantly one type of mobile carrier, or to have a carrier density independent of temperature within a certain range. This can be achieved by implanting suitable impurities, also known as dopants, in the crystal. As an example, consider a Si atom replacing a Ga atom in a GaAs crystal (Fig. 2.10). Only three of the four valence electrons of Si can be placed in the covalent bonds with adjacent As atoms. The remaining electron will be bound to the attractive potential of the Si ion in the GaAs environment. This model will resemble a Coulomb potential in a medium with the dielectric constant of GaAs. It is straightforward to estimate the energy levels and the wave functions of this potential by using the effective mass approximation. The envelope wave equation for the electron in the donor potential reads

$$\left[-\frac{\hbar^2 \Delta}{2m_e^*} - \frac{e^2}{4\pi\epsilon\epsilon_0 r} \right] \psi(\vec{r}) = [E - E_C] \psi(\vec{r}) \qquad (2.35)$$

The hydrogen-like energy levels of the doping atom are given with respect to the conduction band bottom. Compared to hydrogen, the energy spectrum is compressed by the factor $\frac{1}{\epsilon^2} \frac{m_e^*}{m}$

$$E_D(n) = E_C + \frac{1}{\epsilon^2} \frac{m_e^*}{m} E_H = E_C - 13.6 \text{ eV} \cdot \frac{1}{j^2} \cdot \frac{1}{\epsilon^2} \frac{m_e^*}{m}.$$

2.5 Doping

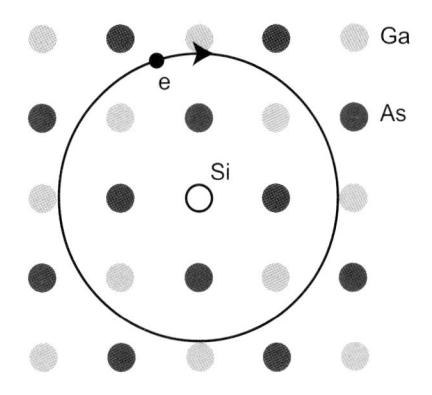

Figure 2.10: Schematic example of a donor atom in a semiconductor.

Since for semiconductors, $\epsilon \approx 10$ and $m_e^* \approx 0.1m$, the binding energy of a typical donor is reduced by a factor of ≈ 1000 as compared to the hydrogen atom, and is just a few meV. The effective Bohr radius becomes very large. For the ground state of the dopant (j=1), it is found

$$a_B^* = \epsilon^2 \frac{m}{m_e^*} a_B \approx 5 \text{ nm},$$

which is much larger than the lattice constant. This in retrospect justifies our assumption that the doping electrons actually see the average dielectric constant of the host crystal.

Such weakly bound electrons can be easily thermally excited into the conduction band. Impurities that generate such levels are called donors. Simultaneously, states just above the valence band can be occupied by electrons from the valence band by thermal excitation (suppose the Si atom would replace an As atom). Impurities that generate this kind of state are called acceptors. Equivalently, we can rephrase this process and say "the acceptors donates a hole in the valence band". In reality, the doping atoms usually do not replace the crystal atoms. Rather they are placed at interstitial sites, and it depends on the local potential whether the atom acts as a donor or as an acceptor. Typical n-dopants for Si are Sb and P, while B and Al are common p-dopants. In both cases, the binding energy of the electrons (holes) is in the range of 50 meV. Si and s are n-dopants for GaAs, with a binding enery of about 6 meV, while Be or Zn can be used for p-doping. here, the hole binding energy is of the order of 30 meV. Some dopants, such as oxygen or chromium, have deep doping levels, which mean that they lie somewhere around the center of the band gap. This cannot be explained with the envelope function model, where only the parameters of the semiconductor host enter. It remains to mention that there are also excited dopant levels, which are of no further interest to us.

How do the carrier densities change due to the doping process? The mass action law still holds, but all doping atoms have to be included in the effective density of states. This, together with the charge neutrality condition, determines the carrier densities. We denote by n_D and p_A

the total density of donors/ acceptors, by n_D^0 and p_A^0 the density of neutral donors/ acceptors, and by n_D^+ and p_A^- the density of ionized donors/ acceptors.

These quantities are related via $n_D = n_D^0 + n_D^+$ and $p_A = p_A^0 + p_A^-$. In addition, charge neutrality requires $n + p_A^- = p + n_D^+$.

Let us take n-doping as an example and calculate n. Now, the assumption made in the intrinsic case, $k_B\Theta \ll E_C - \mu$ is no longer justified. The full solution of this problem is beyond our scope. Instead, we look at a simplifying approximation, which catches the main points. Suppose that $p_A = 0$ and intrinsic carriers can be neglected. We further assume that $E_C - \mu > k_B\Theta$, but $E_D - \mu \approx k_B\Theta$. This means that the doping is so high that it pulls the chemical potential very close to the energy level of the dopant. It is important to note that for typical doping energies close to the valence or the conduction band, the occupation probability is no longer given by a Fermi-Dirac distribution. We discuss the origin qualitatively for a donor level. The derivation of eq. (2.16) is based on the assumption that each energy level can be occupied twice without an additional energy associated with the double occupancy. This is only true in the non-interacting case. In atoms, the Coulomb energy to be paid for sticking two electrons in the same orbital state typically exceeds the binding energy of the doping atom. If the donor ground state would be filled with two electrons, its energy increases above the conduction band edge, such that this state is unstable. Therefore, only three occupations have to be included in the quantum statistics leading to the probability distribution: the donor state is either empty, or occupied with an electron with spin up or spin down. The probability distribution

$$f(E_D, \Theta) = \frac{1}{\frac{1}{2}e^{(E_D-\mu)/k_B\Theta} + 1}$$

results. By a similar argument, it can be shown that for acceptor levels, the corresponding probability distribution reads

$$f(E_A, \Theta) = \frac{1}{\frac{1}{2}e^{(\mu-E_A)/k_B\Theta} + 1}$$

For a detailed discussion of this issue, see e.g. [Grosso2000]. Therefore, the density of occupied donor levels is given by

$$n_D^0 = n_D(1 + \frac{1}{2}e^{(E_D-\mu)/k_B\Theta})^{-1}$$

such that

$$n = n_D^+ = n_D - n_D^0 = \frac{n_D}{1 + 2e^{(\mu-E_D)/k_B\Theta}} \tag{2.36}$$

Also, within our approximation, we can write $n = N_C e^{-(E_C-\mu)/k_BT}$, similar to the intrinsic case. Inserting $e^{\mu/k_B\Theta} = \frac{n}{N_C}e^{E_C/k_B\Theta}$ results in a quadratic equation for n, the positive solution of which reads

$$n = \frac{N_C}{4}e^{(E_D-E_C)/k_B\Theta}\left[-1 + \sqrt{1 + \frac{8n_D}{N_C}e^{(E_C-E_D)/k_B\Theta}}\right] \tag{2.37}$$

Three regimes can be distinguished:

- $k_B\Theta \ll E_C - E_D$ (freezeout regime). In this limit, eq. (2.37) gives

$$n = \sqrt{\frac{N_C n_D}{2}} e^{-(E_C-E_D)/2k_B\Theta}$$

By comparing this expression with the intrinsic case, one finds that the energy levels of the donors plays the role of the valence band edge. In effect, the doping has reduced the band gap by three orders of magnitude. Due to the modified statistics, the effective donor density of states is half the doping density.

- $k_B\Theta \gg E_C - E_D$, but $k_B\Theta < E_g$ (saturation regime): expanding eq. (2.37) with respect to $(E_C - E_D)/k_B\Theta$ to first order gives $n \approx N_D$. In this regime, the carrier density is constant, as long as intrinsic carriers can be neglected.

- $k_B\Theta \approx E_g$ (intrinsic regime): This case is not included in eq. (2.37), but it is clear that now, the carrier concentration depends exponentially on the temperature, and the doping electrons can be neglected, since the doping density is per definition much lower than the density of crystal atoms. The three regimes are summarized in Fig. 2.11.

The chemical potential reacts accordingly to the temperature and can be easily calculated. For low temperatures, it resides close to the donor level, while at large temperatures, it approaches the middle of the band gap.

Question 2.7: Can you dope an insulator to make it conducting?

We have already mentioned that not all impurities generate shallow doping levels. Impurities with *deep levels* can be used for "undoping" samples. In some cases, it is desirable to have a semiconductor of extremely high resistivity at room temperature. Due to unavoidable, residual impurities which act as dopants, the resistivity of ultra-pure GaAs, for example, is not higher than about 1 Ωm. Chromium acts as an acceptor in GaAs with an energy level close to mid-gap. Hence, the residual doping electrons can be removed from the conduction band by a rather small density of Cr doping, which is, however, much higher than the residual n-doping (typical doping densities are of the order of $n_{Cr} \approx 2 \cdot 10^{23}$ m^{-3}). This way, the resistivities can be increased by more than three orders of magnitude. Therefore, such semiconductor materials are called *semi-insulating*. As a consequence, the Fermi level is typically fixed at the energy of the deep dopant, and one speaks of *pinning of the Fermi level*.

2.6 Diffusive transport and the Boltzmann equation

Before we discuss the basic components of diffusive transport theory, we briefly summarize some important results of solid state physics.

- Neither full nor empty bands carry current.

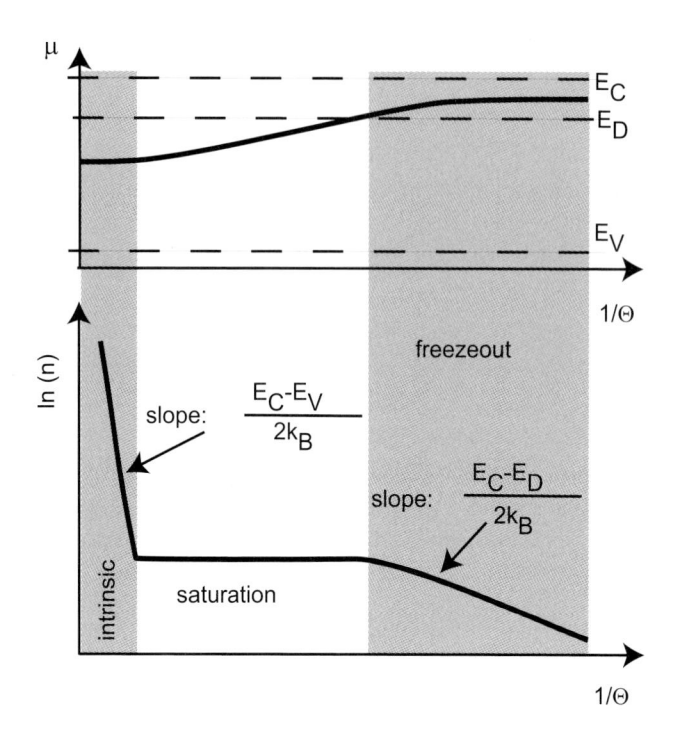

Figure 2.11: Electron density of a doped semiconductor as a function of the inverse temperature.

- The resistance of a perfect crystal with at least one partially filled electronic band vanishes. The electrons obey the semiclassical equations of motion

$$\vec{v}(\vec{k}) = \frac{1}{\hbar}\vec{\nabla}_{\vec{k}} E(\vec{k})$$

$$\frac{d\vec{k}}{dt} = -\frac{e}{\hbar}(\vec{E} + \vec{v}(\vec{k}) \times \vec{B})$$

Resistance is generated by deviations from the perfect lattice, such as phonons, impurities, lattice dislocations, but also by surfaces and interfaces.

- Electron-electron scattering changes the total momentum of the electron gas only in exceptional cases, and does therefore, to a good approximation, not generate resistance.

- For small electric fields applied, only a small fraction of the electrons around the Fermi energy contributes to the current.

In an introductory solid state physics course, transport usually means diffusive transport: a steady state is established between the external electromagnetic fields and the friction inside the solid, which on a microscopic scale is generated by various scattering events. The sample

2.6 Diffusive transport and the Boltzmann equation

size investigated is much larger that the mean free path, which is the distance an electron travels before it is scattered. This means we observe a homogeneous friction which stems from averaging over all microscopic scattering events.

The Boltzmann equation plays a central role in the theory of diffusive electronic transport. Even though electron-electron interactions and phase coherence are neglected already, the general version of the Boltzmann equation is a non-trivial integro-differential equation. Only after its linearization, the relaxation-time approximation and some further assumptions, the equation gives us a simple picture how an electric field acts on the carriers: essentially, within this approximation, the Fermi sphere is displaced in k-space without changing its shape. The relaxation time approximation introduces a phenomenological parameter known as "momentum relaxation time", frequently also referred to as the "Drude scattering time", τ. All important scattering mechanisms are contained in this parameter. We can take it and use it in the Drude model, which is convenient to include magnetic field effects. We will make use of the Drude model on several occasions later on.

Anything that disturbs the perfect lattice will lead to scattering of electrons. Lattice imperfections, that we describe by a perturbation Hamiltonian V_p, will scatter electronic waves from the initial state $|\vec{k}\rangle$ in a final state $|\vec{k}'\rangle$. The scattering matrix elements $W_{\vec{k},\vec{k}'}$ have to be calculated from

$$W_{\vec{k},\vec{k}'} \propto |\langle \vec{k}' | V_p | \vec{k} \rangle|^2 \tag{2.38}$$

A large subfield of transport theory is to calculate such matrix elements for all kinds of scatterers. We will mention some important scattering mechanisms below.

2.6.1 The Boltzmann equation

In general, both external fields as well as scattering will modify the Fermi distribution, which we write here as $f(\vec{k}) = [1 + e^{(E(\vec{k})-\mu)/k_B\Theta}]^{-1}$. The electron distribution function $\phi(\vec{k},\vec{r},t)$ is, in the most general case, not a Fermi function. It may depend on $\vec{r}$ and on the time t. Note that the points $\{\vec{k},\vec{r}\}$ constitute the phase space. Therefore, $\frac{2}{(2\pi)^3}\phi(\vec{k},\vec{r},t)d\vec{k}d\vec{r}$ is the number of electrons in $d\vec{k}d\vec{r}$ for systems with a spin degeneracy of 2.

We consider the evolution of $\phi(\vec{k},\vec{r},t)$ in the time interval dt after time t due to an external, static electric field $\vec{E}$. We could add the effect of a magnetic field, which is dealt with in a similar way, although this is somewhat more elaborate [Seeger1997]. Within dt, an electron located at $(\vec{k},\vec{r})$ in phase space at time t moves to $(\vec{k}+\delta\vec{k},\vec{r}+\delta\vec{r})$, which, according to the the semiclassical equations of motion, equals $(\vec{k} - \frac{e}{\hbar}\vec{E}dt, \vec{r} + \vec{v}(\vec{k})dt)$. This only holds if the electron is not scattered into a different region of the phase space. Also, not all electrons in $(\vec{k}+\delta\vec{k},\vec{r}+\delta\vec{r})$ at time $t+dt$ were at $(\vec{k},\vec{r})$ at time t: they could have been scattered into this volume within dt. These scattering events may cause a change $\delta\phi$ in the electron density, which we write as $\delta\phi = \left[\frac{\partial\phi(\vec{k},\vec{r},t)}{\partial t}\right]_{scatter} dt$. This results in

$$\phi(\vec{k} - \frac{e}{\hbar}\vec{E}dt, \vec{r} + \vec{v}(\vec{k})dt, t + dt)d\vec{k}d\vec{r} = \phi(\vec{k},\vec{r},t)d\vec{k}d\vec{r} + \left[\frac{\partial\phi(\vec{k},\vec{r},t)}{\partial t}\right]_{scatter} dt\, d\vec{k}d\vec{r}$$

The size of the volume element $d\vec{k}d\vec{r}$ cannot change, which is the statement of Liouville's theorem on the evolution of semiclassical systems in phase space. Now, the *general Boltzmann equation* is obtained by expanding the left hand side in a Taylor series in dt up to first order:

$$\vec{v}(\vec{k}) \cdot \vec{\nabla} \phi(\vec{k}, \vec{r}, t) - \frac{e\vec{E}}{\hbar} \cdot \vec{\nabla}_{\vec{k}} \phi(\vec{k}, \vec{r}, t) + \frac{\partial \phi(\vec{k}, \vec{r}, t)}{\partial t} = [\frac{\partial \phi(\vec{k}, \vec{r}, t)}{\partial t}]_{scatter} \quad (2.39)$$

In principle, the scattering term can be calculated from the scattering matrix elements for all scattering mechanisms of relevance (like, e.g., electron-phonon scattering or impurity scattering, see [Seeger1997] for a detailed discussion), each weighted by the corresponding occupation probabilities of the initial states and the probabilities for finding the final state empty. These probabilities, however, are just the distribution functions $\phi(\vec{k}, \vec{r}, t)$, and $1 - \phi(\vec{k}, \vec{r}, t)$, respectively. Therefore, the general Boltzmann equation is in fact a complicated integro-differential equation, and models as well as approximations are needed to evaluate the scattering term.

A rather crude approximation consists of putting all these scattering mechanisms together and assume they generate an average "relaxation time" τ, which we further assume to be independent of $\vec{k}$ and $\vec{r}$. It is based on the following picture. Let us further assume that the system is large enough to appear homogeneous in real space, such that we can drop the space coordinate. If we switch off the external field at time t_0, the distribution function will exponentially relax[4] to $f(\vec{k})$ with a decay time τ:

$$\phi(\vec{k}, t) = f(\vec{k}) + (\phi(\vec{k}, t_0) - f(\vec{k})) e^{-t/\tau}$$

This relaxation will take place exclusively via scattering, and hence

$$\frac{\partial \phi(\vec{k}, t)}{\partial t} = [\frac{\partial \phi(\vec{k}, t)}{\partial t}]_{scatter} = -\frac{\phi(\vec{k}, t) - f(\vec{k})}{\tau}$$

which simplifies the general Boltzmann equation considerably. In a stationary state (no time dependence), it now reads

$$-\frac{e\vec{E}}{\hbar} \cdot \nabla_{\vec{k}} \phi(\vec{k}) = -\frac{\phi(\vec{k}) - f(\vec{k})}{\tau} \quad (2.40)$$

eq.(2.40) can be further evaluated by considering small electric fields only. In this regime, the deviation of ϕ from the Fermi function should be roughly linear in $\vec{E}$, and we can thus write $\vec{\nabla}_{\vec{k}} \phi(\vec{k}) \approx \vec{\nabla}_{\vec{k}} f(\vec{k})$. Now, eq. (2.32) represents a Taylor expansion of $\phi(\vec{k})$ in $(e\tau \vec{E}/\hbar)$ up to first order:

$$\phi(\vec{k}) = f(\vec{k}) + \vec{\nabla}_{\vec{k}} f(\vec{k}) \frac{e\tau \vec{E}}{\hbar}$$

The right hand side is a good approximation for $f(\vec{k} + e\tau \vec{E}/\hbar)$, provided that $e\tau \vec{E}/\hbar \ll \vec{k}$. We thus finally find a *simplified Boltzmann equation* which states that under all the approximations made, small electric fields displace the Fermi surface in k-space by $e\tau \vec{E}/\hbar$:

$$\phi(\vec{k}) = f(\vec{k} + \frac{e\tau \vec{E}}{\hbar}) \quad (2.41)$$

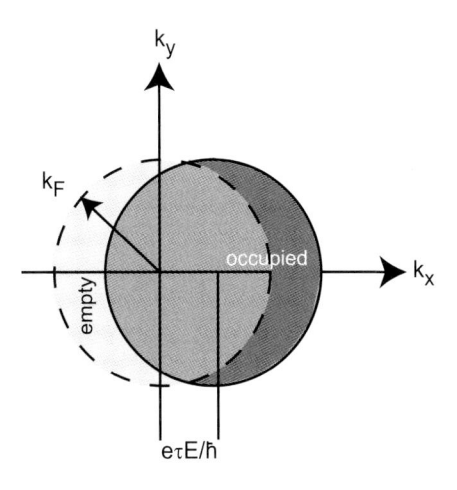

2.6 Diffusive transport and the Boltzmann equation

Figure 2.12: The displaced Fermi sphere as obtained from the Boltzmann equation.

Electrons get accelerated and scatter into empty states via elastic or inelastic processes, which emphasizes again the diffusive and dissipative character of the Boltzmann model. As a consequence, this displacement is quasi-static. In addition, we see that only electrons close to the edge of the Fermi sphere contribute to the current. For states deep inside the Fermi sphere, the partial current generated by an electron with momentum $\hbar\vec{k}$ is cancelled by the electron with momentum $-\hbar\vec{k}$.

2.6.2 The conductance predicted by the simplified Boltzmann equation

It remains to calculate the conductance σ predicted by the assumptions leading to eq.(2.41). In general, σ is a tensor defined by

$$\vec{j} = \sigma \vec{E}$$

However, it makes sense to assume $\vec{j} \parallel \vec{E}$, such that σ is actually a scalar. It is obtained from the current density via

$$\vec{j} = \sigma \vec{E} \rightarrow \sigma = \frac{\vec{j}\vec{E}}{\vec{E}^2} \tag{2.42}$$

In order to calculate $\vec{j}$, we have to integrate over the $\vec{k}$-space, weighing each state by its occupation probability. State $\vec{k}$ contributes a partial current of

$$\vec{j}(\vec{k}) = -e\phi(\vec{k})\vec{v}(\vec{k}) = -\frac{e\hbar}{m^*}\vec{k}\phi(\vec{k})$$

---

[4]Exponential relaxation is a consequence of Poisson processes, which are exemplified in chapter 8.

The total current density is obtained by summing up the contributions of all states. Since for a spin degeneracy of 2, each state occupies a volume of $4\pi^3$ in $\vec{k}$-space, this summation can be written as an integral

$$\vec{j} = \int \vec{j}(\vec{k})d\vec{k} = -\frac{e\hbar}{4\pi^3 m^*}\underbrace{\int \vec{k}f(\vec{k})d\vec{k}}_{=0} + \int \vec{k}\vec{\nabla}_{\vec{k}}f(\vec{k})\frac{e\tau \vec{E}}{\hbar}d\vec{k}$$

Since

$$\vec{\nabla}_{\vec{k}}f(\vec{k}) = \frac{\partial f(\vec{k})}{\partial E}\vec{\nabla}_{\vec{k}}E(\vec{k}) = \frac{\partial f(\vec{k})}{\partial E}\frac{\hbar^2 \vec{k}}{m^*}$$

the current density equals

$$\vec{j} = -\frac{e^2\tau\hbar^2}{4\pi^3 m^{*2}}\int \vec{k}\frac{\partial f(\vec{k})}{\partial E}[\vec{k}\vec{E}]d\vec{k}$$

With eq.(2.42), we can write

$$\sigma = -\frac{e^2\tau\hbar^2}{4\pi^3 m^{*2}}\int \frac{(\vec{k}\vec{E})^2}{\vec{E}^2}\frac{\partial f(\vec{k})}{\partial E}d\vec{k}$$

For sufficiently low temperatures,

$$-\frac{\partial f(E)}{\partial E} = \delta(E - E_F) = \delta(k - k_F)\frac{m^*}{\hbar^2 k}$$

which results in the surface integral

$$\sigma = \frac{e^2\tau}{4\pi^3 m^*}\int \frac{(\vec{k}\vec{E})^2}{\vec{E}^2}\delta(k - k_F)\frac{1}{k}d\vec{k} =$$

$$\frac{e^2\tau}{4\pi^3 m^*}\int_{\theta=0}^{2\pi}\int_{\varphi=0}^{\pi} k_F^3 \cos^2(\varphi)\sin(\varphi)d\varphi d\theta = \frac{e^2\tau k_F^3}{3\pi^2 m^*}$$

Since the electron density n is given by $n = 3\pi^2 k_F^3$, we find

$$\sigma = \frac{ne^2\tau_D}{m^*} = ne\mu \qquad (2.43)$$

Here, we have defined the electron mobility $\mu = e\tau_D/m^*$.

Question 2.8: Prove that $\sigma = ne\mu$ also holds in two dimensions.

2.6 Diffusive transport and the Boltzmann equation

Result (2.43) is at first sight quite strange: the conductivity is proportional to the total electron density, and it seems like all electrons would contribute equally to the current. However, we know that only the electrons at the Fermi surface carry current. The explanation is that a higher electron density increases the number of electrons and the electron velocity at the Fermi surface, which turns out to give a conductivity proportional to n.

We can use eq. (2.43) to define a useful quantity, the drift velocity $\vec{v}_d$ as

$$\vec{v}_d = -\frac{\vec{j}}{en} = -\mu\vec{E} \tag{2.44}$$

The drift velocity is thus an effective average velocity which leads to an equation for the current density that is formally identical to the Drude expression, which was derived by assuming that all electrons contribute equally to the current and move through the crystal with an average drift velocity.

Along similar lines, it can be shown that in the presence of additional magnetic fields, the current density can be written as

$$\vec{j} = \sigma(\vec{E} + \vec{v}_d \times \vec{B}) \tag{2.45}$$

This current density corresponds to the stationary solution of the classical equation of motion

$$m^*\frac{d^2\vec{r}}{dt^2} + \frac{m^*}{\tau}\vec{v}_d = -e(\vec{E} + \vec{v}_d \times \vec{B}) \tag{2.46}$$

for electrons moving at velocity $\vec{v}_d$ through the crystal, in the presence of electric and magnetic fields and with a Stokes-type friction term given by $m^*\vec{v}_d/\tau$.

2.6.3 The magneto-resistivity tensor

We proceed by studying the electron transport according to eq. (2.46) in weak magnetic fields, with "weak" being specified by

$$\omega_c = e\,|\vec{B}|\,/m_e^* \ll 1/\tau \tag{2.47}$$

This condition means that the distance the electrons travel before getting scattered (the *mean free path* $\ell_e = v_F\tau$) is small compared to the cyclotron radius r_c. We will see in chapters 6 and 7 what happens when the electrons can complete the cyclotron orbits without getting scattered. Suppose a magnetic field is applied in z direction, $\vec{B} = (0, 0, B)$. In such a case, we obtain

$$j_x = \sigma E_x + \sigma v_y B = \sigma E_x + \frac{ne^2\tau}{m_e^*}v_y B = \sigma E_x - j_y\omega_c\tau$$

$$j_y = \sigma E_y - \sigma v_x B = \sigma E_y - \frac{ne^2\tau}{m_e^*}v_x B = \sigma E_y + j_x\omega_c\tau$$

$$j_z = \sigma E_z$$

where v_i are the components of the drift velocity vector. Solving this system of equations for $\vec{j}$ gives $\vec{j} = \underline{\sigma}\vec{E}$ with

$$\underline{\sigma} = \frac{\sigma}{1+\omega_c^2\tau^2} \begin{pmatrix} 1 & -\omega_c\tau & 0 \\ \omega_c\tau & 1 & 0 \\ 0 & 0 & 1+\omega_c^2\tau^2 \end{pmatrix}$$

$\underline{\sigma}$ is known as the magneto-conductivity tensor. Its components can be experimentally determined by measuring four-probe resistances using "Hall bar" - shaped samples (Fig. 2.13). Voltage probes are attached to a rectangular thin film of the material, aligned parallel to the x- and y-direction, and perpendicular to the magnetic field direction. The transport in z direction remains unaffected by $\vec{B}$ is is of no further interest to us. We can determine the components ρ_{xx} and ρ_{xy} of the resistivity tensor by applying a current in x-direction and measuring the voltage drops V_x and V_y. Since

$$\begin{pmatrix} V_x \\ V_y \end{pmatrix} = \begin{pmatrix} \rho_{xx} & \rho_{xy} \\ -\rho_{xy} & \rho_{xx} \end{pmatrix} \cdot \begin{pmatrix} I_x \\ I_y \end{pmatrix}$$

and

$$\begin{pmatrix} I_x \\ I_y \end{pmatrix} = \begin{pmatrix} \sigma_{xx} & \sigma_{xy} \\ -\sigma_{xy} & \sigma_{xx} \end{pmatrix} \cdot \begin{pmatrix} V_x \\ V_y \end{pmatrix}$$

we can establish the relation between the components of the resistivity and the conductivity tensors:

$$\rho_{xx} = \frac{\sigma_{xx}}{\sigma_{xx}^2 + \sigma_{xy}^2}; \qquad \rho_{xy} = \frac{-\sigma_{xy}}{\sigma_{xx}^2 + \sigma_{xy}^2} \qquad (2.48)$$

We thus find that $\rho_{xx} = 1/\sigma$ does not depend on $\vec{B}$, and $\rho_{xy} = -\omega_c\tau/\sigma = -B/en$. ρ_{xy} is the Hall resistivity, and $R_H = -1/en$ is known as the Hall coefficient. Hall measurements are actually a standard tool to determine carrier densities. It may be counter-intuitive at first sight that for $\rho_{xx} = 0$, σ_{xx} becomes zero as well. Furthermore, the Onsager-Casimir symmetry relation should be mentioned, which states that the result of a measurement is exactly the same when all current- and voltage sources are exchanged, and the polarity of the magnetic field is reversed. One consequence is that two-probe measurements, i.e., for the voltage drop being measured at the source and drain contact, must be symmetric with respect to $B = 0$.

2.7 Scattering mechanisms

As mentioned in Section 2.6, many scattering mechanisms contribute to the average momentum relaxation time τ. Each process has its characteristic matrix element $W_{\vec{k},\vec{k}'}$, eq. (2.38). The relevance of a particular kind of scattering varies greatly and depends on the carrier density as well as on the temperature. How in detail the matrix elements are calculated is treated in several excellent books, e.g. by [Seeger1997], and by [Ridley1999]. Each scattering mechanism can be characterized by its contribution to the carrier mobility μ_i, that sum up to the

2.7 Scattering mechanisms

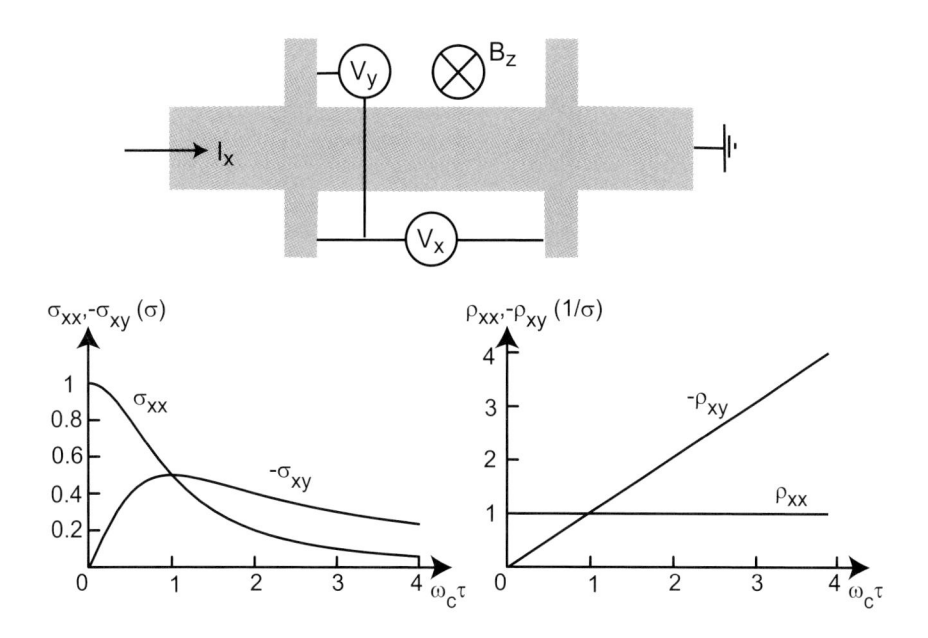

Figure 2.13: Top: top view of a Hall geometry. The magnetic field is applied perpendicular to the sheet. Bottom: the components of the conductivity and the resistivity tensors are shown to the left and to the right, respectively.

total mobility according to the Matthiesen rule, $1/\mu = \sum_i 1/\mu_i$. In pure crystals, the sole source of scattering are lattice vibrations. Electron-phonon scattering has several facets. In crystals with valley degeneracy, electrons may be scattered between valleys, which requires absorption or emission of a phonon. In polar and/or piezoelectric crystals, on the other hand, lattice vibrations go along with strong oscillating electric fields. In real crystals, charged impurities may dominate the scattering rates. We briefly present the most important scattering mechanisms below.

An impurity breaks the symmetry of the lattice and causes scattering. If the impurity is neutral, the scattering rates are usually negligible. Charged impurities, however, represent screened Coulomb scatterers, with peak potentials that can become comparable to the Fermi energy.[5] Clearly, an electron with a larger kinetic energy will get deflected by a smaller angle as it gets scattered, and we can expect that the mobility increases as the temperature, and with is the average electron kinetic energy, increases. In fact, an evaluation of the corresponding matrix element shows that for weak Coulomb potentials and within the Born approximation, the resulting mobility is $\propto \Theta^{3/2}$, multiplied by a logarithmic correction, i.e. a factor which depends logarithmically on Θ.

---

[5]The screened Coulomb potential is studied in exercise 2.3.

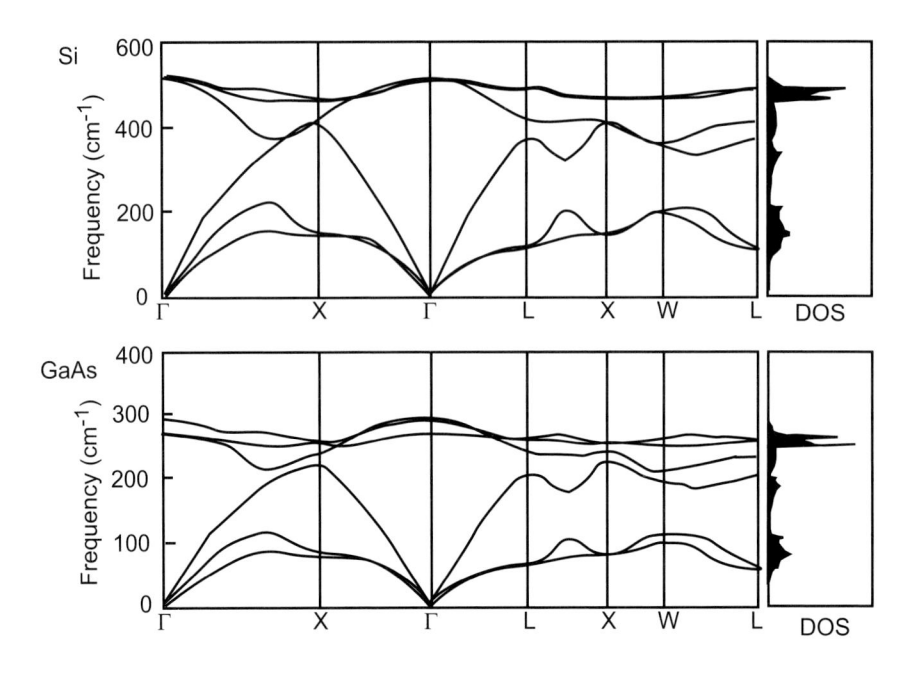

Figure 2.14: Phonon dispersions for Si (top) and GaAs (bottom). After [Giannozzi1991].

Electron-phonon scattering can be divided into deformation potential scattering and scattering of electrons at the corresponding electric fields. By deformation potential scattering, we mean scattering at the lattice deformations caused by the phonons. Here, scattering at acoustic phonons is the most important mechanism. Since the energy transfers are small in electron-acoustic phonon scattering, it can be treated as quasi-elastic. A simple argument gives the correct temperature dependence. The density of acoustic phonons n_{ac} is proportional to the Bose-Einstein distribution which, for large temperatures as compared to the phonon energy, varies as $1/\Theta$. Since the mobility is proportional to the $n_{ac}/\bar{v}$ ($\bar{v}$ is the average electron velocity, which is $\propto \sqrt{\Theta}$), we expect that the mobility due electron-acoustic phonon scattering is $\propto \Theta^{-3/2}$. This is in fact observed experimentally.

Furthermore, both optical and acoustic phonons can assist the electron in scattering between the valleys in a crystal with valley degeneracy, such as Si. The corresponding momentum transfers are quite large, since the separation of the valley in reciprocal space is of the order of the size of the Brillouin zone.

This completes the list of the scattering mechanisms relevant in Si. In this material, ionized impurity dominates the mobility at low temperatures, while quasi-elastic acoustic phonon scattering is the most important mechanism at intermediate temperatures. For $\Theta > 200$ K, inter-valley scattering becomes significant as well. Consequently, the mobility in Si shows a maximum as a function of temperature. Its position depends on both the impurity density and the carrier density. Electron mobilities up to $1 \text{ m}^2/\text{Vs}$ have been achieved in Si.

GaAs is a polar material, and consequently, lattice vibrations are always accompanied by oscillating electric fields. They are particularly strong for optical phonons. The resulting scattering mechanism is called polar scattering. Optical phonons vanish for temperatures below ≈ 60 K, and consequently, polar scattering is relevant only above this temperature. In the limit $k_B\Theta \gg \hbar\omega_{op}$ (ω_{op} denotes the optical phonon frequency, which for GaAs is of the order of 5 meV, see Fig. 2.14), it can be shown that the resulting mobility varies as $\Theta^{-1/2}$. If the crystal is piezoelectric like GaAs, a crystal deformation generates a polarization field as well, which is another source of scattering, which is called piezoelectric. As the polar scattering, the mobility due to piezoelectric scattering is $\propto \Theta^{-1/2}$, although this temperature dependence holds for a larger range of temperatures.

Fig.2.15 summarizes the contributions of different scattering mechanisms to the electron total mobility of GaAs. A comparison to measurements reveals that at low temperatures, ionized impurity scattering dominates, while at larger temperatures, the mobility is entirely determined by polar scattering. In a small temperature range around the emerging maximum of the mobility, piezoelectric scattering is very significant. Furthermore, is is seen that acoustic phonon scattering plays no role, in contrast to the scattering in Si.

2.8 Screening

The conduction electrons react to external perturbations. They collect in the potential valleys and avoid the peaks. As a consequence, the external potential is reduced to an effective potential in the crystal; the electrons "screen" the perturbation. The goal of this paragraph is to present a qualitative picture how the electron density is modified by perturbations. For a more detailed discussion of screening and electron-electron interactions, the reader is refereed to textbooks on solid state physics ([Ashcroft1985], [Grosso2000],[Ziman1995]).

By time dependent perturbation theory, it can be shown that the screening in a free electron gas depends on the wave vector $\vec{q}$ and the frequency ω of the perturbation. It can be expressed by a dielectric function $\epsilon(\vec{q},\omega)$ of the type

$$\epsilon(\vec{q},\omega) = 1 + \epsilon_{lattice} + \frac{e^2}{\epsilon_0 q^2} \sum_{\vec{k}} \frac{f(\vec{k}) - f(\vec{k}+\vec{q})}{E(\vec{k}+\vec{q}) - E(\vec{k}) + i\alpha} \quad (2.49)$$

Here, $\epsilon_{lattice}$ means the dielectric function of the lattice ([Ashcroft1985]), and $\alpha \to 0$ is the (small) convergence parameter that can be related to a scattering time (see the exercises). The dielectric function describes how the Fourier components of the external potential energy $V_{ext}(\vec{q},\omega)$ are screened and result in a effective potential energy $V_{eff}(\vec{q},\omega)$, namely

$$V_{eff}(\vec{q},\omega) = \frac{V_{ext}(\vec{q},\omega)}{\epsilon(\vec{q},\omega)} \quad (2.50)$$

In the static limit ($\omega \to 0$), within the effective mass approximation, and for low temperatures $k_B\Theta \ll E_F$, the sum in eq. (2.49) can be calculated analytically, and one finds

$$\epsilon(\vec{q}) = 1 + \frac{k_{TF}^2}{\epsilon_0 q^2} F(\frac{q}{2k_F});$$

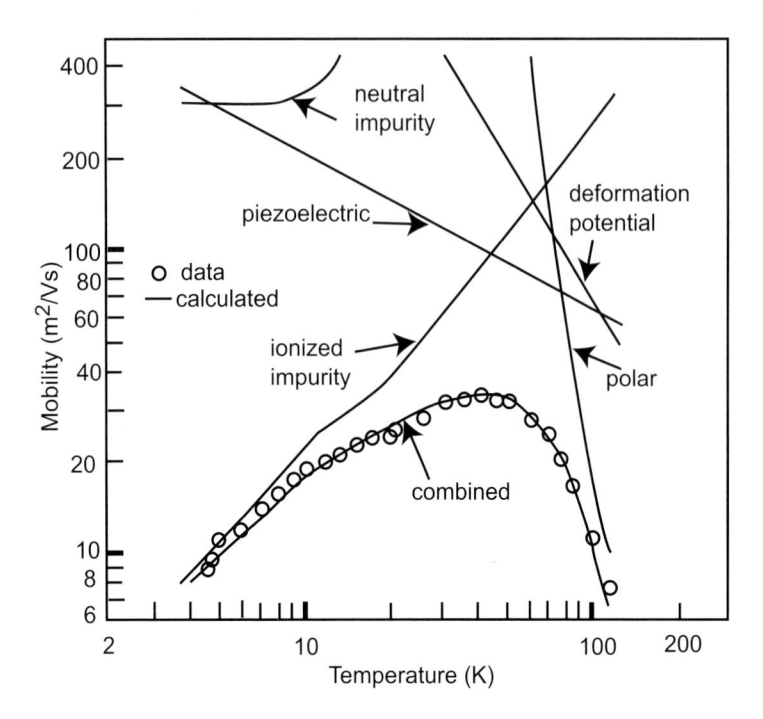

Figure 2.15: Measured electron mobility in GaAs (circles) as function of temperature, including the theoretical contributions of relevant scattering mechanisms (full lines). The sample contained a donor density of $n_D = 4.8 \cdot 10^{19}$ m^{-3} and an acceptor density of $n_A = 2.1 \cdot 10^{19}$ m^{-3}. After [Stillman1976].

$$F(s) = \frac{1}{2} + \frac{1-s^2}{4s} ln\left|\frac{1+s}{1-s}\right| \qquad (2.51)$$

Here, we have defined the Thomas-Fermi screening vector

$$k_{TF} = e\sqrt{\frac{D(E_F)}{\epsilon_0}} \qquad (2.52)$$

The functions $F(s)$ and $\epsilon(\vec{q})$ are shown in Fig. 2.16. Most notably, $\epsilon(\vec{q})$ drops significantly as $\vec{q}$ increases. Above $2k_F$, it rapidly approaches 1. The reason is simply that for low temperatures, the wave vector of occupied states differs by no more than $2k_F$. The term $f(\vec{k}) - f(\vec{k}+\vec{q})$ means that only states contribute where the occupation of the two states characterized by $\vec{k}$ and $\vec{k}+\vec{q}$ is different. The number of contributing states thus increases as $0 \leq |\vec{q}| \leq 2k_F$, but remains constant for $|\vec{q}| > 2k_F$ in eq. (2.49), which means that $\epsilon(\vec{q})$ drops significantly at $|\vec{q}| = 2k_F$. This point in fact represents a logarithmic singularity, which has important consequences for the screening. As an example, consider the potential of a point charge with

2.8 Screening

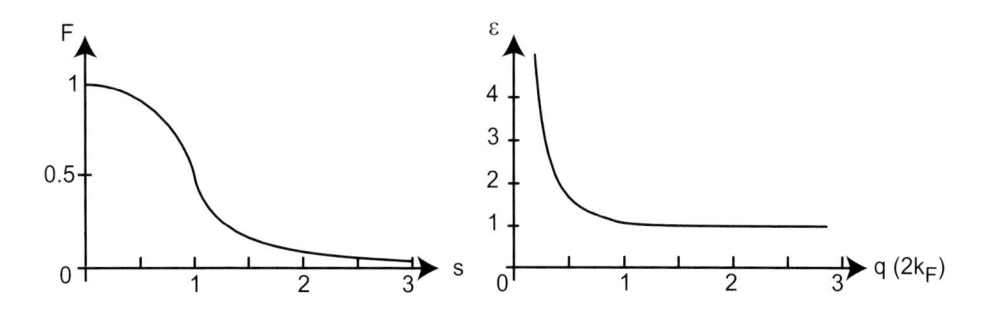

Figure 2.16: The function F(s), left, and the static dielectric function $\epsilon(\vec{q})$ for free electrons with typical experimental parameters in doped semiconductors ($k_F = 10^8$ m^{-1}).

an external potential energy given by

$$V_{ext}(r) = -Ze/r = -\frac{Ze^2}{(2\pi)^3}\int \frac{4\pi}{q^2} e^{i\vec{q}\vec{r}} d\vec{q}$$

Correspondingly, the effective potential energy can be written as

$$V_{eff}(r) = -Ze/r = -\frac{Ze^2}{(2\pi)^3}\int \frac{1}{\epsilon(\vec{q})}\frac{4\pi}{q^2} e^{i\vec{q}\vec{r}} d\vec{q}$$

The induced charge density $\rho_{ind}(\vec{r})$ is then given by the Poisson equation $\Delta V_{ind}(\vec{r}) = \Delta[V_{eff}(\vec{r}) - V_{ext}(\vec{r})] = -e\frac{\rho_{ind}(\vec{r})}{\epsilon_0}$, such that

$$\rho_{ind}(\vec{r}) = \frac{Ze}{(2\pi)^3}\int \frac{1}{\epsilon(\vec{q})-1} e^{i\vec{q}\vec{r}} d\vec{q}$$

Evaluating this integral shows that only terms with the argument $|\vec{q}| \approx 2k_F$ contribute significantly, and that for large r,

$$\rho_{ind}(\vec{r}) = \frac{Ze}{\pi}\frac{k_{TF}^2}{k_F^2(4+k_{TF}^2/2k_F^2)}\frac{\cos(2k_F r)}{r^3} \qquad (2.53)$$

The charge density thus develops a periodic component, with a period of half the Fermi wavelength. This can be understood in terms of a standing wave due to a superposition of the incoming waves and the waves reflected at the perturbation potential. These oscillations are known as Friedel oscillations. [6]

---

[6]Friedel oscillations are not a special property of screened Coulomb potentials. It can be shown [Gruner1977] that for large distances form the perturbing potential, $\rho_{ind}(\vec{r}) = \frac{A}{r^3}\frac{\cos(2k_F r+\phi)}{r^3}$, where the phase ϕ and the constant A depend on the potential.

Papers and Exercises

E 2.1: A microchip factory processes "3-inch wafers", i.e., mono-crystalline, cylindrically shaped semiconductor disks, with a diameter of 3 inches and a height of 0.5 mm. Unfortunately, the silicon wafer badge and the GaAs wafer badge have not been labelled. Someone suggests to determine the material by weighing the wafers. Is this realistic?
Si has an atomic mass of 28.09 amu and crystallizes with a lattice constant of 0.543 nm. For Ga and As, the atomic mass is 69.72 amu, and 74.92 amu, respectively.; GaAs has a lattice constant of 0.565 nm. Calculate the weight of the two wafer types!

E 2.2: Suppose certain macromolecules form two-dimensional crystals. For each unit cell, $2\pi/3$ electrons are available for electronic bands. The unit cell is defined by the lattice vectors $\vec{a_1} = (4 \text{ nm}, 0)$ and $\vec{a_2} = (1 \text{ nm}, 3 \text{ nm})$.

(a) Calculate the reciprocal lattice vectors $\vec{b_1}, \vec{b_2}$. Draw the reciprocal lattice and construct the first and the second Brillouin zone.

(b) Suppose the energy dispersion can be approximated by that one of free electrons. What is the radius of the Fermi sphere? Draw the Fermi surface in the reciprocal lattice!

(c) Which period(s) would you expect in a de Haas -van Alphen experiment? The magnetic field is applied perpendicular to the crystal plane.

E 2.3: Study the dielectric function in the limit $|\vec{q}| \ll 2k_F$. Show that the potential of a screened point charge can be written as

$$V_{eff}(\vec{r}) = -\frac{Ze^2}{r} e^{-k_{TF}r}$$

Screening in this limit is known as Thomas-Fermi screening.

E 2.4: Consider a periodic potential composed of δ-functions

$$V(x) = -V_0 \sum_{n=-\infty}^{\infty} \delta(x + na)$$

with $V_0 > 0$ and n integer.

(a) Determine the eigenvalue E_0 and the eigenfunction $\Phi_0(x)$ of a single δ-function, $V_{single}(x) = -V_0 \delta(x)$.

(b) Show that the tight-binding wave functions of the crystal

$$\Psi_k(x) = \sum_{j=-\infty}^{\infty} \Phi_0(x - ja) \cdot e^{ikja}$$

satisfy Bloch's theorem.

(c) Use the wave function and show by the method sketched in the text, that the dispersion relation takes the form

$$E(k) = E_0 + \frac{\beta + \sum_{n=1}^{\infty} \gamma_n \cos(kna)}{1 + \sum_{n=1}^{\infty} \alpha_n \cos(kna)}$$

Note that life gets easier if the term with $n = 0$ is treated separately.

(d) What is the effective mass around $k = 0$?

E 2.5: Use the band gap energies and the effective masses given in the text to calculate the effective densities of states N_c and P_v, as well as the intrinsic carrier concentrations, for Si and GaAs at room temperature! What happens to the carrier concentrations as the materials are cooled to liquid nitrogen temperature?

Further Reading

There are several excellent books on solid state physics available, e.g., [Ashcroft1985],[Ziman1995], or [Grosso2000]. For particular properties of semiconductors, see [Seeger1997], [Yu1999]. The material parameters of the important semiconductors, are listed in a condensed, yet informative way in [Grahn1999].

3 Surfaces, Interfaces, and Layered Devices

Infinite crystals are convenient for introducing solid state physics. Quite often, surface effects can be neglected in real crystals, as the number of surface atoms is vanishingly small as compared to the number of bulk atoms. Surfaces play a very important role, though. First of all, they are the interface between the crystal and the outside world. Across surfaces, the energy bands get related to the vacuum level, which is the energy of an electron at rest outside the crystal. It takes the energy Φ_A, also known as *work function*, to transfer an electron at the chemical potential in the crystal into the vacuum. In pure semiconductors and insulators, this is an impossible process, since there are no states at the chemical potential. The *electron affinity* χ_e is therefore introduced in addition. It measures the energy difference between the vacuum level and the bottom of the conduction band. Their numeric values depend on both the bulk band structure, as well as on surface-specific properties [Ashcroft1985].

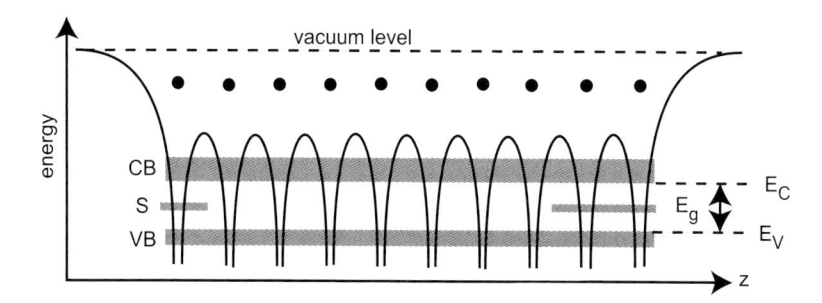

Figure 3.1: Schematic representation of the potential landscape in a finite crystal, which gets modified close to the surface. Surface states (S) may result, with typical energies inside the gap between the valence band (VB) and the conduction band (CB).

The regions of interest in most mesoscopic samples, as well as in the majority of commercial microchips, are very close to surfaces and/or to crystal interfaces, and their influence on the active region is highly relevant. In fact, crystal interfaces are frequently tailored to provide useful properties. The most elementary interface is that one between a crystal and the vacuum, which is the topic of section 3.1. We will see that at a surface, electronic states can exist which are absent in the bulk, with typical energies in the band gap of the bulk material, as sketched in Fig. 3.1. These states are not additional states, though. Rather, they emerge from valence and conduction band states. This is evident if we recall that the number of electronic states in the crystal equals the number of the states in all atoms the crystal contains. In order to appreciate the mechanism leading to surface states, a model calculation for a one-dimensional crystal within the tight-binding model is presented in section 3.1.1. Generalizing the results to three dimensions is conceptually simple, and results in surface bands which are two-dimensional in character (see Section 3.1.2). Typically, surface bands are partly filled. The chemical potential at the surface will usually be located somewhere inside the surface band, i.e., inside the band gap of the bulk. In equilibrium, the surface chemical potential will align with the that one in the bulk, which in general requires a charge transfer between bulk states and surface states. The consequences of this mechanism are *band bending* and *Fermi level pinning*, introduced in section 3.1.3. Both effects are of utmost importance in semiconductor nanostructures.

Generalizing the properties of crystal surfaces to other types of interfaces is straightforward, once the concepts are available. Similar to surfaces, charge rearrangements will align the two chemical potentials. Nevertheless, interface states have a somewhat different character than surface states. Two types of interfaces are relevant for us. Metal-semiconductor interface are studied in section 3.2. They come in two flavors, Schottky contacts and Ohmic contacts. Equally important is the semiconductor heterointerface, the topic of section 3.3. As we shall see, it is quite common to combine layers of different semiconductors and take advantage of the resulting band alignment.

After these preparations, we are ready to look at examples of devices that rely on interface effects. The most important structures for our purposes are the Si-MOSFET and the Ga[Al]As-HEMT, which will be introduced in section 3.4. It will become clear that due to band bending at interfaces, carrier gases are formed which can be *two-dimensional*. Such systems will be the workhorse for most of the experiments discussed in subsequent chapters. There are, however, many more interesting ways to combine semiconductors layers, and we will briefly present some examples in that section as well.

This chapter is concluded by some remarks related to further modifications the electrons experience close to interfaces.

3.1 Electronic surface states

3.1.1 Surface states in one dimension

A surface breaks the translational symmetry of the crystal. How this fact modifies the electronic structure can be studied in various models. Their line of arguing, however, is similar: Bloch's theorem allows Bloch functions with complex wave vectors. However, they correspond to evanescent waves, which is unphysical in an infinite crystal. This is no longer necessarily true at surfaces. Here, wave functions may be obtained which are strongly localized at the surface and have real eigenvalues.

As a simple example, we use the tight-binding model to study how surface states emerge from a σ energy band of the bulk, e.g. a band which is formed from atomic s orbitals. This model goes back to [Goodwin1939b] and [Shockley1939]. We start from the Schrödinger equation of a finite one-dimensional crystal composed of N atoms with a lattice constant a.

$$\frac{\hbar^2}{2m}\Delta\Psi(z) + \left[E - V(z)\right]\Psi(z) = 0 \qquad (3.1)$$

3.1 Electronic surface states

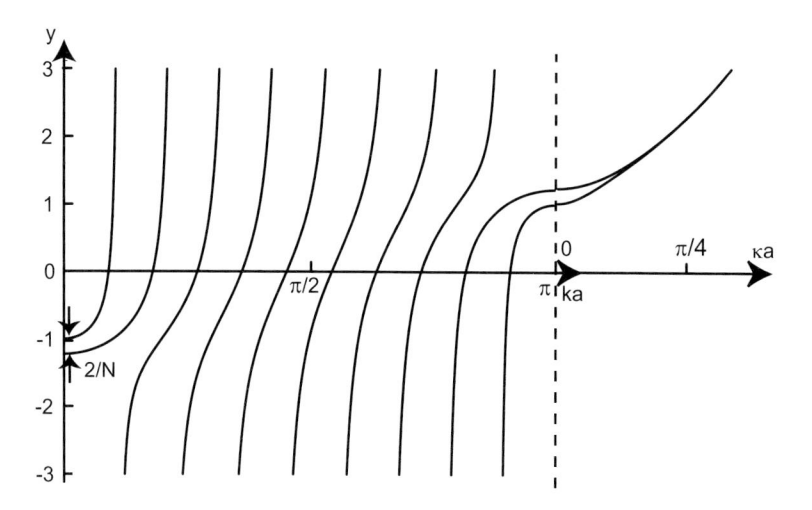

Figure 3.2: Plot of $y(k)$ for a crystal consisting of $N = 10$ atoms. For $-1 \leq y \leq 1$, 10 real wave numbers k are obtained. For $y > 1$, two values of k have non-vanishing imaginary components, $k = \pi/a + i\kappa$. In this regime, 8 real and 2 complex solutions are obtained.

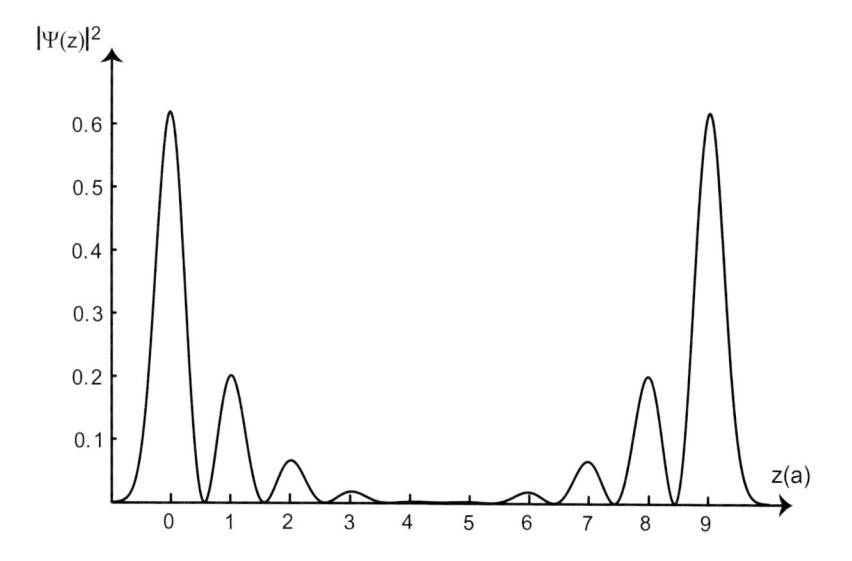

Figure 3.3: Model wave function for a surface state in a 10-atom crystal. The atomic wave functions $\Phi_n(z) = e^{-4(z-na)^2}$, and $\kappa a = 0.5$, corresponding to $y \approx 1.65$, have been used.

where $V(z)$ is the periodic lattice potential. The individual atoms are described by

$$\frac{\hbar^2}{2m}\Delta\Phi(z - z_n) + \left[E_0 - U(z - z_n)\right]\Phi(z - z_n) = 0 \tag{3.2}$$

Here, E_0 is the energy eigenvalue of the atom for the s state under consideration. We insert the ansatz

$$\Psi = \sum_n c_n \Phi(z - z_n) \tag{3.3}$$

in eq.(3.1), multiply with $\Phi^*(z - z_m)$ and integrate over space. If only nearest neighbor coupling is considered, the matrix elements

$$\langle \Phi_m^* | \Phi_n \rangle = \delta_{mn} + \beta \delta_{m,n\pm 1}$$

$$\langle \Phi_m^* | V - U | \Phi_n \rangle = \begin{cases} -\alpha & n = m \notin \{1, N\} \\ -\alpha' & n = m \in \{1, N\} \\ -\gamma & m = n \pm 1 \\ 0 & else \end{cases} \tag{3.4}$$

are obtained. Since both $(E - E_0)$ and β are small, their product can be neglected. With the definitions $E - E_0 + \alpha = \epsilon$ and $\alpha - \alpha' = \epsilon_0$, the coefficients c_n obey the relations

$$\gamma c_{n-1} + \epsilon c_n + \gamma c_{n+1} = 0; n \in \{2, N-1\} \tag{3.5}$$

$$(\epsilon - \epsilon_0) c_1 + \gamma c_2 = 0 \tag{3.6}$$

$$\gamma c_{N-1} + (\epsilon - \epsilon_0) c_N = 0 \tag{3.7}$$

We write c_n in the form

$$c_n = A e^{ikna} + B e^{-ikna} \tag{3.8}$$

For the bulk, eq.(3.5) gives the well-known dispersion relation

$$\epsilon = 2\gamma \cos(ka) \tag{3.9}$$

which we use in eqns. (3.6) and (3.7) to calculate the allowed wave numbers in the finite crystal. After some algebra and by using the relations $\sin x \sin y = \frac{1}{2}\cos(x-y) - \frac{1}{2}\cos(x+y)$ as well as $\cos(2x) = \cos^2 x - \sin^2 x$, one finds the condition

$$y = -\frac{\epsilon_0}{\gamma} = \frac{-\sin(Nka) \pm \sin(ka)}{\sin((N-1)ka)} \tag{3.10}$$

This function is plotted in Fig. 3.2 for $N = 10$. Clearly, a N-atom crystal contains N electronic states that emerge from the atomic s states. In fact, for $-1 \leq y \leq 1$, condition (3.10) delivers N real wave numbers. For $|y| > 1$, however, only $N - 2$ real solutions are found![1] It can be shown that $\gamma < 0$ for s bands [Goodwin1939b]. We thus focus on the region $y > 1$. Apparently, two states have formed with an imaginary component of the wave number, which corresponds to states outside the σ band. Inserting $k = \pi/a + i\kappa$ into (3.10) gives

$$y = \frac{\sinh(Nka) \pm \sinh(ka)}{\sinh((N-1)ka)} \approx e^{\kappa a} \tag{3.11}$$

---

[1] There is a small interval $1 \leq |y| \leq 1 + 2/N$ for which $N - 1$ real solutions are obtained. As N is usually large, this region is irrelevant.

3.1 Electronic surface states

where the approximation holds in the limit of large N. In this case, the states have the energy dispersion

$$E = E_0 - \alpha + 2\gamma \cosh(\kappa a) \qquad (3.12)$$

Since $y > 1$ and $\cosh[ln(y)] > 1$, the energies of these states ar *larger* than those of the bulk states, see Fig. 3.2.

How do the wave functions of these split-off states look like? The complex wave numbers describe evanescent wave functions, localized at the crystal surface. This is easily established by calculating the coefficients c_n. Inserting the dispersion relation (3.12) in eqns.(3.6) and (3.7) gives

$$c_2 = -c_1 e^{-\kappa a}; \qquad c_{N-1} = -c_N e^{-\kappa a} \qquad (3.13)$$

The remaining coefficients are obtained by recursively applying eq.(3.5). Since for symmetry reasons $c_1 = c_N$, one obtains

$$c_{n+1} = c_1(-1)^n [e^{-n\kappa a} + (-1)^N e^{-(N-n)\kappa a}] \qquad (3.14)$$

Such a wave function is sketched in Fig.3.3 for our 10-atom model crystal. Even for our tiny 10 - atom crystal, it is strongly localized at the crystal surfaces and extends only a few lattice constants into the bulk! These are the surface states, which have emerged form the bulk states of the σ - band.

Question 3.1: Determine the wave functions of the bulk states (with the dispersion given by eq. (3.9))!

Shockley included the effect of band crossings in an extended version of this model [Shockley1939], and studied the energy of the surface states as a function of the lattice constant. His famous results are summarized in Fig. 3.4. He found that for lattice constants where the bands have crossed, a common case in real crystals, both the upper and the lower band contribute one surface state inside the band gap. The energies of the surface states are typically close to mid-gap.

Surface states of one-dimensional crystals are sometimes categorized in "Maue-Shockley"-states (following the models of [Maue1935] and [Shockley1939], where the potential at the surface is not modified, except that the periodic potential is interrupted) and "Tamm-Goodwin"-states, where the surface states occur due to modifications of the surface potential as compared to the bulk ([Tamm1932],[Goodwin1939b]). From a more general point of view, these different types of surface states are special cases of a symmetry condition given by [Zak1985], which is the topic of [P 3.1].

It is furthermore instructive to see how surface states are formed within the nearly free electron model. Within this standard model of solid state physics, the periodic potential $V(z)$ is assumed to be weak as compared to the kinetic energy of the electrons. It can thus be well approximated by its Fourier expansion up to first order,

$$V(z) = V_0 + V_g e^{igz} \qquad (3.15)$$

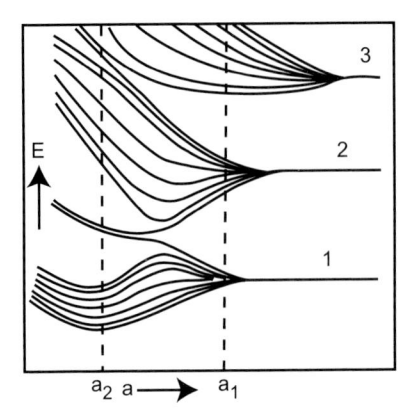

Figure 3.4: Energy of surface states in the one-dimensional Shockley model, shown as a function of the lattice constant a. At e.g. a_2, both a donor-like and an acceptor-like surface states are present. After [Shockley 1939].

Here, V_0 is the constant part of the potential inside the crystal with respect to the vacuum level. and g denotes the smallest reciprocal lattice vector, $g = 2\pi/a$. Because $V(z)$ is weak, it can be treated as a perturbation. The states belonging to wave numbers k and $k - 2\pi/a$ with k very close to the edge of the first Brillouin zone are nearly degenerate.[2] The effect of the weak periodic potential on the energies of such states can be studied by degenerate perturbation theory. The result of such a calculation for $q = \frac{\pi}{a} - k \ll \frac{\pi}{a}$ is the energy dispersion

$$E(q) = V_0 + \epsilon_{\pi/a} + \epsilon_q \pm \sqrt{4\epsilon_{\pi/a}\epsilon_q + |V_g|^2} \qquad (3.16)$$

with

$$\epsilon_p = \frac{\hbar^2 p^2}{2m}; \qquad p = \{\frac{\pi}{a}, q\}$$

At $q = 0$, a band gap of $2|V_g|$ results, [Ashcroft 1985, Ziman 1995] which corresponds to the range of energies where the wave number has an imaginary part. Due to the non-vanishing imaginary part of the wave vector, the corresponding wave functions decay exponentially in space, which is unphysical in a crystal with perfect periodicity. Therefore, no states exist within the band gap.

Note that $E(q)$ is a continuous function in the complex plane. For energies in $[\epsilon_{\pi/a} - |V_g|, \epsilon_{\pi/a} + |V_g|]$, q becomes imaginary, and by substituting $q = -i\kappa_c$ (κ_c is a real number), eq. (3.16) can be rewritten as

$$E(\kappa_c) = V_0 + \epsilon_{\pi/a} - \epsilon_{\kappa_c} \pm \sqrt{|V_g|^2 - 4\epsilon_{\pi/a}\epsilon_{\kappa_c}} \qquad (3.17)$$

---
[2] Of course, similar degeneracies exist at the boundaries of higher Brillouin zones.

3.1 Electronic surface states

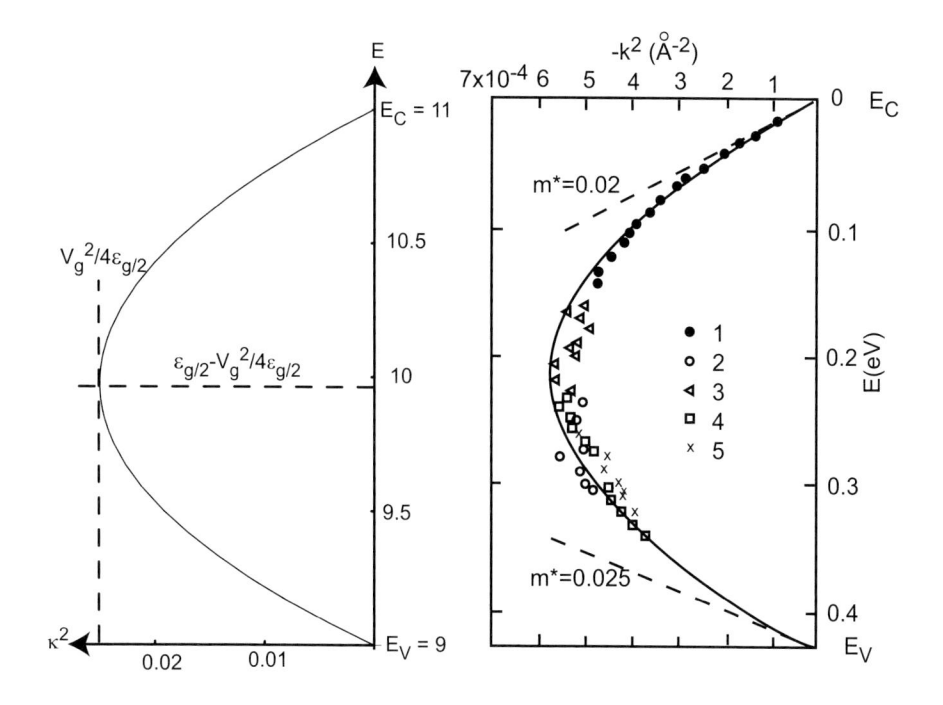

Figure 3.5: Left: imaginary energy dispersion inside the band gap as obtained within the approximation of a weak periodic potential, according to eq. (3.1). $\epsilon_{g/2} = 10$ and $V_g = 1$ were used as model parameters. Right: measurement of the imaginary energy dispersion in InAs by surface-barrier tunneling experiments, after [Parker1968]. Here, the symbols correspond to different samples. Note that the energy is plotted vs. κ^2.

This energy dispersion is shown in Fig. 3.5. It can be measured across the whole band gap in certain materials, for example in InAs, by tunneling experiments [Parker1968].

$E(\kappa)$ resembles a semicircle that connects E_V and E_C. How do the surface states emerge from these considerations? Well, eigenstate close to the surface can exist, provided its wave functions can be properly matched to a wave function that decays exponentially into the vacuum. This scenario is schematically depicted in 3.6. Consider a crystal with a surface at $z = 0$. For weak periodic potentials, the total potential is given by

$$V(z) = \begin{cases} 0 & z \leq 0 \\ V_0 + 2V_g \cos(gz) & z > 0 \end{cases} \quad (3.18)$$

The wave function $\Phi_s(z)$ of a surface state is composed out of two components,

$$\Phi_s(z) = \begin{cases} \Phi_v(z) \propto e^{\kappa_v z} & z \leq 0 \\ \Phi_c(z) \propto e^{-\kappa_c z} \cos\left(2\frac{\pi}{a}z + \phi_0\right) & z > 0 \end{cases}$$

where ϕ_0 denotes a possible phase factor. From the continuity conditions for $\Phi_s(z)$ and its

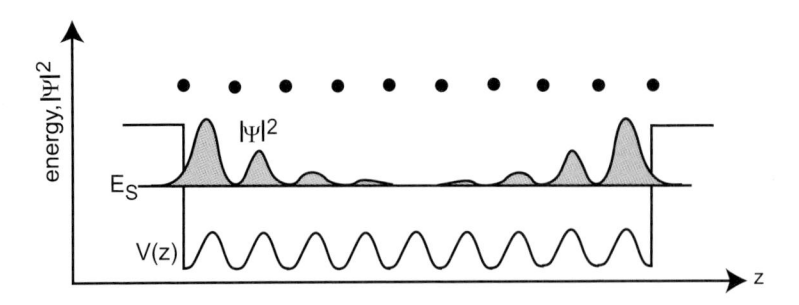

Figure 3.6: Sketch of a surface state probability amplitude $|\Psi(z)|^2$ in a ten-atom crystal within the nearly free electron model. The potential is approximated by a harmonic function. The wave functions decay exponentially into the vacuum. Inside the crystal, the wave function oscillates with the period of the lattice, while the amplitude drops exponentially as the distance to the surface increases. Note that there may be a phase shift of $\Psi(z)$ with respect to the position of the atoms.

first derivative with respect to z at $z = 0$, the condition

$$\cos^{-2}\phi = \frac{V_0 + V_g}{\epsilon_{\pi/a}} \tag{3.19}$$

can be derived after some lengthy algebra. If a solution for ϕ exists, the finite crystal has a surface state at energy

$$E_s = -\frac{\hbar^2 \kappa_v^2}{2m} \tag{3.20}$$

with

$$\kappa_v = \frac{\pi}{a}\tan\phi - \kappa_c; \quad \kappa_c = \frac{mV_g a}{\pi\hbar^2 \sin^2\phi}$$

The details of these calculations can be found in the references given at the end of this chapter.

Thus, the model of nearly free electrons gives qualitatively the same result as the tight-binding model, namely that localized states may exist at the surface, with typical energies inside the band gaps.

3.1.2 Surfaces of 3-dimensional crystals

Extending the previous results to three dimensions is straightforward, although a quantitative treatment can be a formidable task. It requires not only symmetry considerations, but also inclusion of surface recombinations. The periodic pattern of chemical bonds is interrupted at the surface, and the unsaturated bonds (so-called dangling bonds) rearrange themselves. This usually goes along with a change of both the surface crystal structure and the allowed energies for electrons. A famous example for such a surface recombination is the 7×7 - reconstruction of Si(111) [Binnig1983]. Another common scenario is chemical binding to a monolayer of

3.1 Electronic surface states

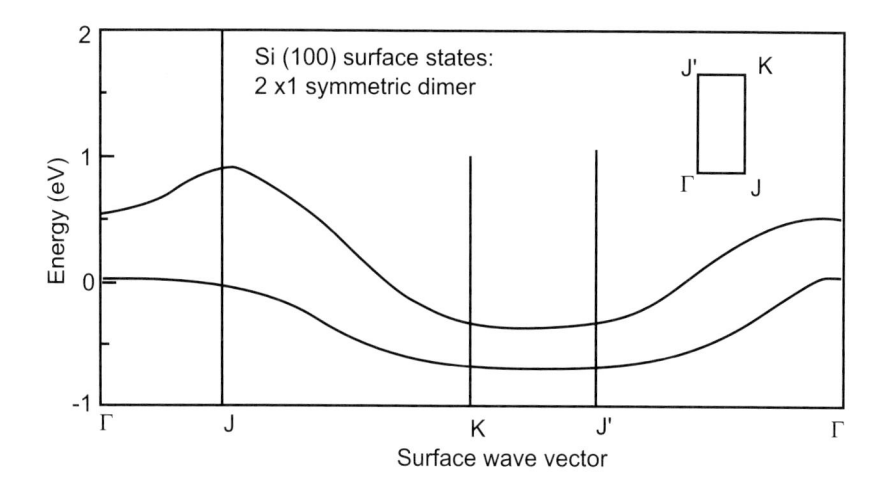

Figure 3.7: Calculated energy dispersion of the two-dimensional bands of a Si(100) surface. After [Chadi1979].

adatoms which saturate the dangling bonds. In either case, it is clear that the electronic surface structure has little to do with the bulk structure.

A perfect crystal surface is certainly periodic within the surface plane, and it is reasonable to set up a tight-binding model for the two-dimensional surface layer. Each surface state obtained within a one-dimensional model now corresponds to a two-dimensional band, with a band width given by the transfer and overlap integrals between surface atoms. The number of states per unit area in a surface band corresponds to the density of surface atoms, n_s, which is of the order of $5 \cdot 10^{18} m^{-2}$, see., e.g. [Louie1977]. Measurements of the surface density of states give typical results of the order of $4 \cdot 10^{18} m^{-2} eV^{-1}$, such that the band width of the surface bands is of the order of the band gap E_g of the bulk material. More quantitative calculations of surface band energy dispersions (an example is given in Fig. 3.7) essentially confirm these simple considerations. We will represent surface bands in graphical representations as shown in Fig. 3.8.

Independent of such issues, charge neutrality must be maintained at the surface. For a semiconductor with a neutral surface, this means that the number of occupied surface states must equal the number of states that have been removed from the valence band due to the formation of the surface band. Since the surface band with valence character can overlap with the surface band that has emerged from conduction band states, both types of surface bands can be partially filled (see Fig. 3.8). Surface states with valence band character can be regarded as donor-like. Likewise, we can call surface states with predominantly conduction band character as acceptor-like.[3]

---

[3]In general, surface states do not have pure valence band (or conduction band, respectively) character. Rather, they are an admixture of both types of band states. Surface states will be more valence-like the closer their energies are to the valence band, and vice versa.

An important quantity is the "charge neutrality level" μ_{CN}. At this energy, the character of the surface states changes from predominantly donor-like to predominantly acceptor-like. Typically, μ_{CN} has an energy close to the center of the band gap. For a neutral surface, the surface states are filled up to μ_{CN}. In general, the surface can be charged, however, and the chemical potential at the surface, μ_S, may differ from μ_{CN}.

3.1.3 Band bending and Fermi level pinning

So far, we have considered intrinsic materials, where the energy of the electrons in the donor-like surface states increases as compared to their bulk value, but the surface remains neutral. This changes in doped semiconductors. Let us take an n-doped semiconductor as an example (Fig. 3.8). If only donor-like surface states existed, nothing would happen. However, usually

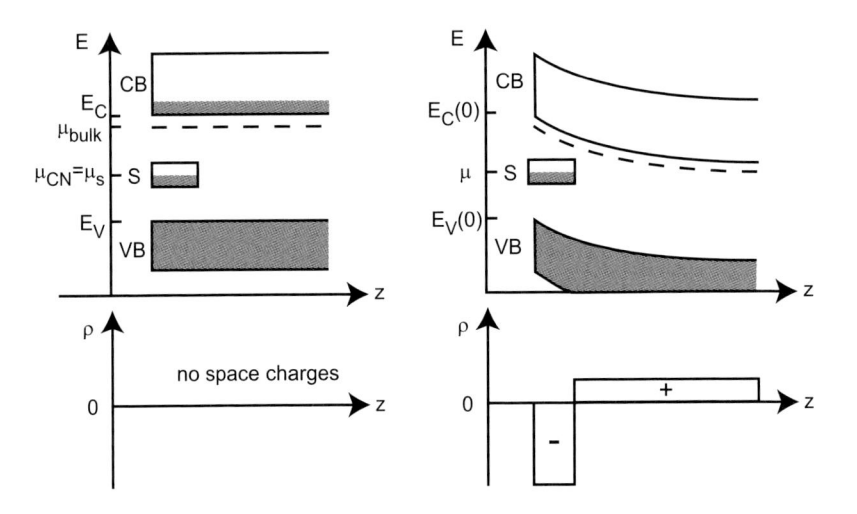

Figure 3.8: Band bending at the surface of a n-doped semiconductor before equilibration. Gray areas indicate occupied states. Left: schematic representation of the band structure close to the surface before equilibration. The sketched surface bands have a with of roughly $0.2 E_g$. Right: after equilibration, the surface gets charged, an upward band bending results, and the Fermi level gets pinned close to the charge neutrality level μ_{CN}.

both donor- and acceptor-like surface states are present. In that case, some donor electrons in the conduction band will reduce their energy by occupying the acceptor-like surface states. A negative surface charge is generated, counterbalanced by a positive space charge that originates from ionized donors within a depletion length z_{dep} away from the surface, such that overall charge neutrality is maintained. Consequently, an electric field and the corresponding electrostatic potential will build up, and the energy bands bend upwards as they approach the surface. It is self-evident that in a p-doped semiconductor, the band bending will occur towards lower electron energies, since holes accumulate at the surface. In equilibrium, the chemical potential is constant throughout the crystal. To be somewhat more quantitative, consider a n-doped semiconductor within a distance z_{dep} from the surface. In this region, the complete ionization of the donors leads to a space charge density of $\rho = eN_D$. The missing electrons occupy the surface states. The Poisson equation gives us the z-dependence of the potential $V(z)$ via

$$\frac{d^2V}{dz^2} = -\frac{eN_D}{\epsilon\epsilon_0} \Rightarrow V(z) = -\frac{eN_D}{\epsilon\epsilon_0}(z - z_{dep})^2 \quad (3.21)$$

for $z \in [0, z_{dep}]$. Note that this is the potential an electron feels as it approaches the surface from $z \geq z_{dep}$ *without* exiting the crystal. Therefore, the potential maximum of the conduction band is reached at $z = 0$ and equals

$$V(0) = -\frac{eN_D z_{dep}^2}{2\epsilon\epsilon_0}$$

since the surface electron density is given by $n_s = -N_D z_{dep}$, we can write

$$z_{dep} = \sqrt{\frac{2\epsilon\epsilon_0 V(0)}{eN_D}} \quad (3.22)$$

Assuming that the surface states that get filled lie close to the center of the band gap, we can estimate the depletion length for typical parameters, say $E_g = 1.4eV$, $N_D = 10^{24} m^{-3}$, and $\epsilon = 12$, and find a depletion length of $z_{dep} \approx 30nm$. The band bending thus extends across many lattice constants, and the depleted region is much larger than the spatial extension of the surface states.

For our model parameters, a surface charge density of $n_s \approx 3 \cdot 10^{16} m^{-2}$ results, which is much smaller than the integrated density of surface states. Since $D_s(E) \approx 5 \cdot 10^{18} m^{-2}$, this means that the chemical potential at the surface μ_S changes only by a few meV due to the charge transfer. Hence, to a good approximation, μ_S does not depend on the doping density. This property is often coined by the statement that the Fermi level is *pinned* by the surface states at μ_{CN}. In that respect, surface states act similarly to deep dopants used to generate semi-insulating materials, see chapter 2.

3.2 Semiconductor-metal interfaces

The ideas of wave function matching and modified transfer integrals at surfaces are also applicable to the important metal-semiconductor interface. Before the metal and the semiconductor get in contact, their common energy scale is the vacuum level, and the relative position of the bands in both materials is trivial. As the interface is formed, however,[4] the local lattice structure at the interface changes, which can give rise to interface states. We distinguish between *Schottky barriers*, where charge carriers have to tunnel through a barrier as they move across the interface, and *Ohmic contacts*, where such a barrier is absent or highly penetrable.

---

[4] We assume that the surfaces are clean, and contain no oxides. This is in fact more or less the case in real devices, as possible oxide layers can be etched away, and and the metal is usually deposited on top of the semiconductor in a high vacuum environment. Furthermore, the interfaces considered in theory are usually atomically flat. This is quite hard to achieve experimentally.

3.2.1 Band alignment and Schottky barriers

The character of the previously mentioned interface states depends on the energy. At an energy that lies inside the gap between full bands of both materials, localized states may form, with a character very similar to that one of surface states. For energies inside a full band of both materials, the semi-extended wave functions have to be matched. Both situations have no further consequences, as all the states involved are filled anyway. A new scenario is obtained for energies that lie in an energy band of one component, but inside a gap of the second one. The most relevant case is a semiconductor band gap with energies inside the conduction band of the metal, and we focus on this scenario. Heine has shown that the metallic wave

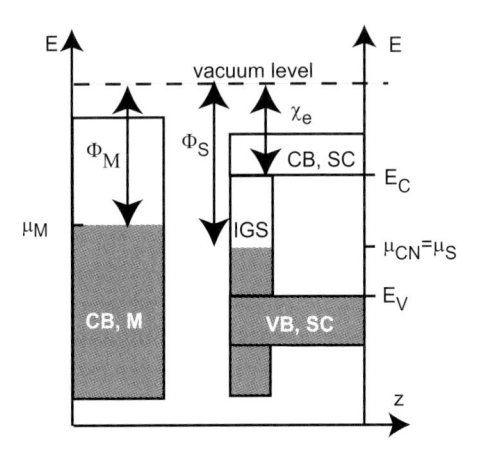

Figure 3.9: Typical energy band alignment between a metal (left) and a semiconductor (right) before charge transfer across the interface is allowed. Extended states in the metal induce gap states in the semiconductor at all energies inside the band gaps. The induced gap states (IGS) are filled up to μ_{CN}, and the common energy scale of both band structures is the vacuum level.

function can be matched to the evanescent wave functions in the semiconductor for all energies [Heine1965]. Hence the metal induces a continuum of interface states in the semiconductor band gap, so-called *induced gap states - IGS*. Close to the interface, the semiconductor thus develops a non-zero density of states at energies inside the metallic bands, Fig. 3.9. Suppose that at the beginning, both materials are well separated and do interact. Clearly, their common energy scale is the vacuum level, and the energy difference between the chemical potential in the metal μ_m and the electron affinity in the semiconductor $\chi_{e,sc}$ equals the energy difference between semiconductor conduction band bottom and μ_m, Fig. 3.7). In many, but not all, cases this means that μ_m is somewhere in between the top of the valence band and conduction band bottom of the semiconductor. Next, we assume that the surfaces are so close to each other that IGS are formed in the semiconductor band gap. Fig. 3.10 shows a calculation of the density of states across a GaAs - Al interface. Within the first two GaAs double layers away from the interface, a significant density of IGS is generated, while the Al density of states remains essentially unchanged even for the Al layer at the interface.

3.2 Semiconductor-metal interfaces

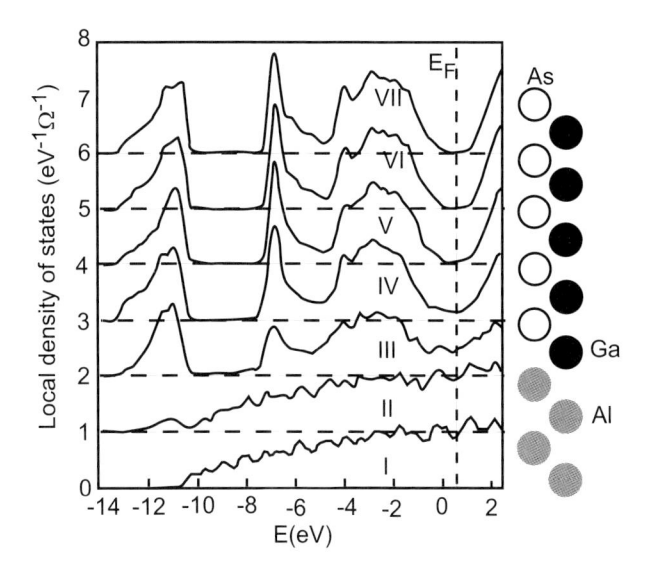

Figure 3.10: Calculated density of states across a Al-GaAs interface. After [Berthod1998].

Since the IGS inside the semiconductor are built from the virtual gap states of the semiconductor, their properties are semiconductor specific. Like a surface state, such an interface state is predominantly acceptor-like or donor-like. A charge neutrality level and the interface work function of the semiconductor Φ_S can be defined, and again the integrated density of states must remain constant. How do the band structures of the metal and the semiconductor align with respect to each other? In general, the charge neutrality level in the semiconductor μ_{CN} will be different from μ_m, and a charge transfer will take place. We discuss the alignment using the interface between a metal and an n-doped semiconductor as an example (Fig. 3.11a). In a gedanken experiment, we assume that for now, the donor electrons are not allowed to occupy surface states. Charge transfer across the interface forms a dipole and aligns, via the dipole potential that obeys the Poisson equation, μ_M with μ_S. This dipole is strongly localized, since the IGS only extend over a few lattice constants into the semiconductor (Fig. 3.11b). The semiconductor bands have been bent upwards in the case drawn here, and this energy difference adds to the difference between Φ_M and χ_e. The barrier at the interface is known as the Schottky barrier V_S. In a n-doped semiconductor, the IGS will get occupied by donor electrons as well. As in the previous section, a space charge layer builds up in the semiconductor close to the interface and generates the band bending sketched in Fig. 3.11c. Since the region of depleted donor electrons z_{dep} is much larger than the width of the IGS, the band bending due to the interface dipole is often drawn as a step function, resulting in a band diagram as shown in Fig. 3.11d. Similarly, of course, an opposite bending can occur for a p-doped semiconductor.

Within this picture, we would expect V_S to depend upon Φ_M, and this is in fact the case. Fig. 3.12 shows that an approximately linear relation exists between V_S and Φ_M, with a slope characteristic for each semiconductor ($d(eV_S)/d\Phi_M \approx 0.05$ for n-GaAs and ≈ 0.25 for Si).

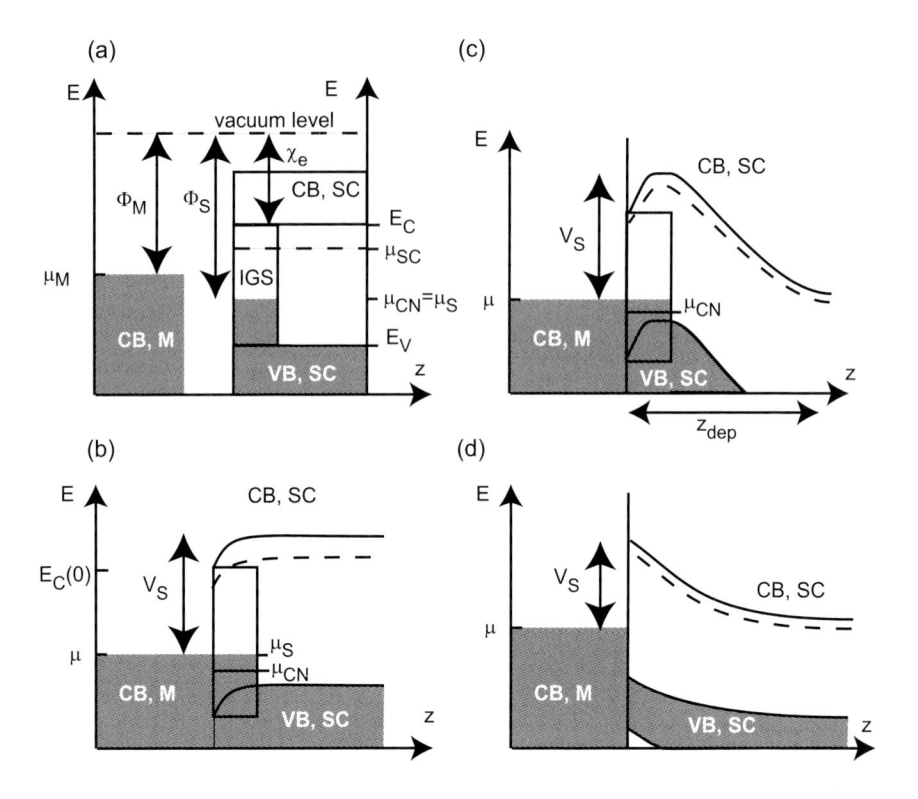

Figure 3.11: Metal-induced gap states and band alignment at a metal-semiconductor interface. (a): relative band energies of a metal and an n-doped semiconductor after formation of IGS, but before charge is transferred. (b): charge transfer across the interface generates an interfacial dipole, aligns μ_M with μ_S, and generates a highly localized band bending. (c): The donor electrons will occupy the IGS as well, generating a depletion layer of width z_{dep} and an additional band bending. Since z_{dep} is much larger than the width of the IGS, the interface band bending is usually drawn as a sharp step, as indicated in (d).

The Schottky model

The Schottky model neglects the consequences of interface states and models the Schottky barrier formation entirely by using bulk parameters [Schottky1938]. It is frequently used, and we thus briefly sketch it here, using again an n-doped semiconductor as an example. In a consideration similar to the previous one, we start with the two materials separated by an impenetrable tunnel barrier, such that charge transfer is impossible. As pointed out above, the difference between the metal Fermi level and the conduction band bottom of the semiconductor is $\Phi_M - \chi_{e,sc}$. Bringing the materials closer together will at some point allow for charge transfer. In the case depicted here, the donor electrons get transferred into the metal until the chemical potentials μ_M and μ_{SC} are aligned (Fig. 3.13). The Schottky model thus predicts

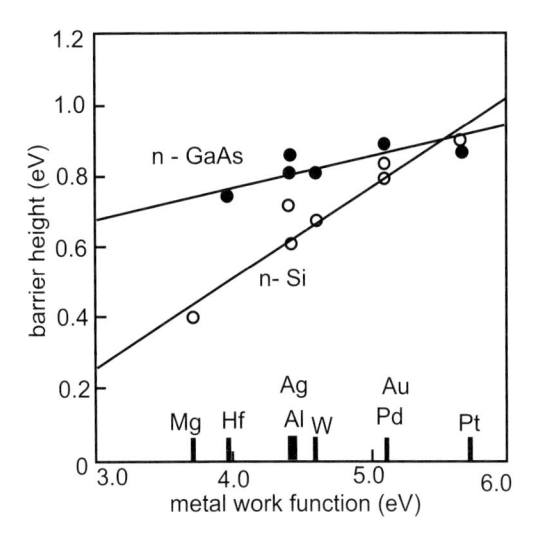

Figure 3.12: Schottky barriers of Si and GaAs in contact with different metals, plotted as a function of the metal work function. After [Sze1985].

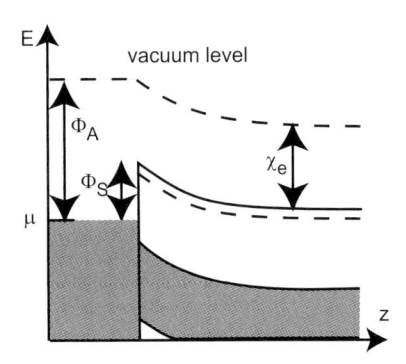

Figure 3.13: Positions of the Fermi levels of a metal and a n-doped semiconductor in equilibrium as obtained within the Schottky model.

that $V_S = \Phi_M - \chi_{e,sc}$, which can be taken as a coarse approximation to the observed barrier heights shown in Fig. 3.12.

The Schottky diode

Semiconductor-metal interfaces with a Schottky barrier act as a diode. The resistance of such a barrier is dominated by the depletion zone in the semiconductor, as well as by the Schottky barrier height. Applying a positive voltage to the metal with respect to the grounded semiconductor pulls the electrons in the semiconductor towards the interface, thus reducing the width of the space charge layer. Therefore, the tunnel barrier the electrons have to penetrate during their trip into the metal gets reduced. Reversing the bias voltage increases the space charge region and therefore reduce the conductance (Fig. 3.14). Since the transmission of a tunnel barrier depends exponentially on both its height and width, we expect a current that depends exponentially on the applied voltage. Although we will not make use of Schottky barriers as diodes in subsequent chapters,[5] it is important to keep in mind how the band bending in semiconductors close to metal can be modified by a voltage applied across the Schottky barrier.

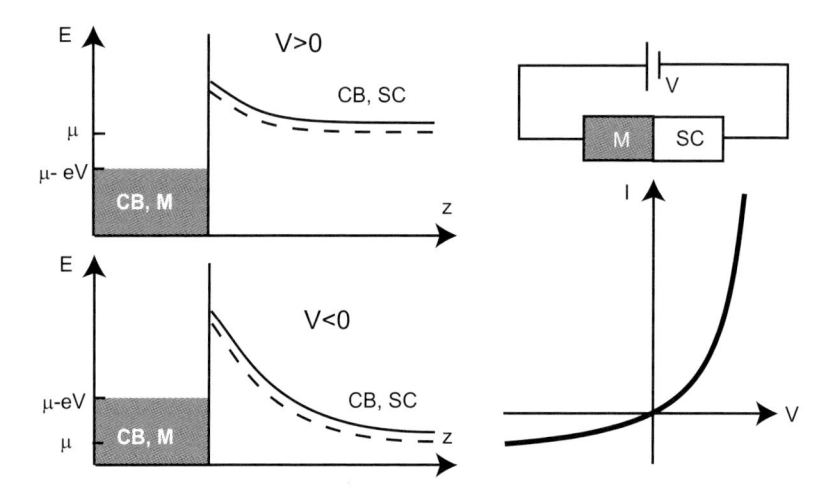

Figure 3.14: The tunnel barrier formed by the depletion layer at a Schottky contact depends on the bias voltage (left), which results in a diodic current-voltage characteristics (lower right). The upper right scheme shows the direction of the applied voltage.

3.2.2 Ohmic contacts

So far, we have assumed that the chemical potential at the interface in equilibrium is at some energy inside the semiconductor band gap. This is not what we always need. Operating a semiconductor device requires to feed current carried by a metal wire into the semiconductor at some point. This should be done with the lowest resistance possible, and a diodic current-voltage characteristics would certainly be an annoyance. The interface should behave according to Ohm's law.

---

[5] In some experimental setups, the currents across Schottky contacts actually disturb the experiment. Such currents are referred to as gate leakage, for reasons that will soon become clear.

3.3 Semiconductor heterointerfaces

Such Ohmic contacts can be formed naturally at some metal-semiconductor interfaces where the equilibrium Fermi level lies above above $E_{C,sc}$, see Fig. 3.15(a). No tunnel barrier is formed in that case, and charge can flow freely across the interface. An example is the interface between InAs and a metal. Most material combinations of importance, however, form Schottky barriers, and we have to design some sort of Ohmic contact. This is done by reducing the Schottky barrier to insignificant heights and widths by two means. First of all, a metal is used for Ohmic contact formation, such that V_S is reduced as much as possible. Second, the semiconductor is heavily doped, which reduces the width of the tunnel barrier, according to eq. (3.21), see Fig. 3.15(b). Since the resistance across the Schottky barrier depends exponentially on both the width and the height of the tunnel barrier, a current voltage characteristics results which is Ohmic for practical purposes.

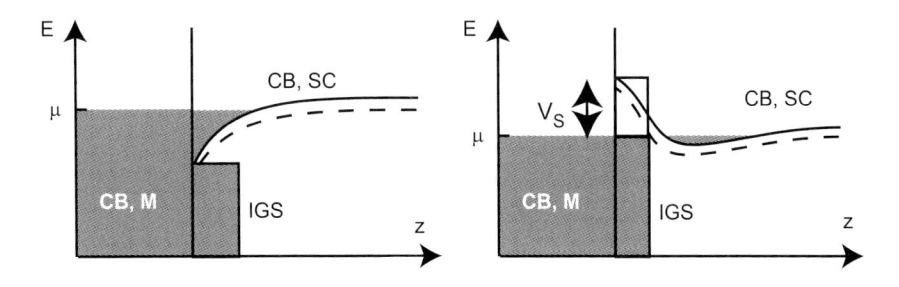

Figure 3.15: (a): an Ohmic contact without a Schottky barrier between the metal and the semiconductor. (b): scheme of a metal-semiconductor interface with a Schottky barrier that works as Ohmic contact.

3.3 Semiconductor heterointerfaces

The probably best known example of a semiconductor interface is the p-n junction, where a p-doped region meets a n-doped region of the same semiconductor host crystal. Close to the junction, donor electrons recombine with acceptors and generate a space charge region, which can be tuned by applying a voltage across the junction. The p-n- junction is the basis of bipolar devices and plays no role in subsequent chapters. The interested reader is referred to the standard textbooks on solid state physics and semiconductor physics [Ashcroft1985],[Seeger1997]. Instead, the semiconductor devices we will study frequently contain *heterointerfaces*, i.e. interfaces between two different semiconductor crystals. As in the metal-semiconductor interface, an important question is how the band structures align with respect to each other. We can get a qualitative understanding of the alignment mechanism by modifying the above considerations accordingly, and consider an interface between semiconductor 1 (n-doped) and semiconductor 2 (p-doped) as an example, Fig. 3.16.[6] Again we begin with the relative band

---
[6]Doping is not necessary for the alignment mechanisms to work, but simplifies the discussion somewhat. In intrinsic materials, polarization charges contribute significantly to the dipole at the interface.

energies of the two semiconductors before charge transfer has taken place, Fig. 3.16(a). For energies inside the band gap of one semiconductor and inside an energy band of the second semiconductor, IGS are generated. Additional states may arise for energies inside the band gap of both materials, which we neglect for simplicity. In any case, surface chemical potentials are defined for both materials, which may differ from the chemical potentials of the bulk as well as from each other. Correspondingly, a charge transfer will take place across the interface. Fig. 3.16(b) shows the resulting band alignment after electrons have been transferred from the valence band in semiconductor 2 into the IGS of semiconductor 1. Finally, the bulk chemical potentials will be aligned by electron transfer from the n-doped semiconductor into the p-doped semiconductor, and a space charge dipole layer, with a typical extension of several tens of nanometers into both materials, is obtained. This is shown in Fig. 3.16(c), where the dipole potential due to the occupation of IGS has been included already in the band offsets, similar to Fig. 3.11(d).

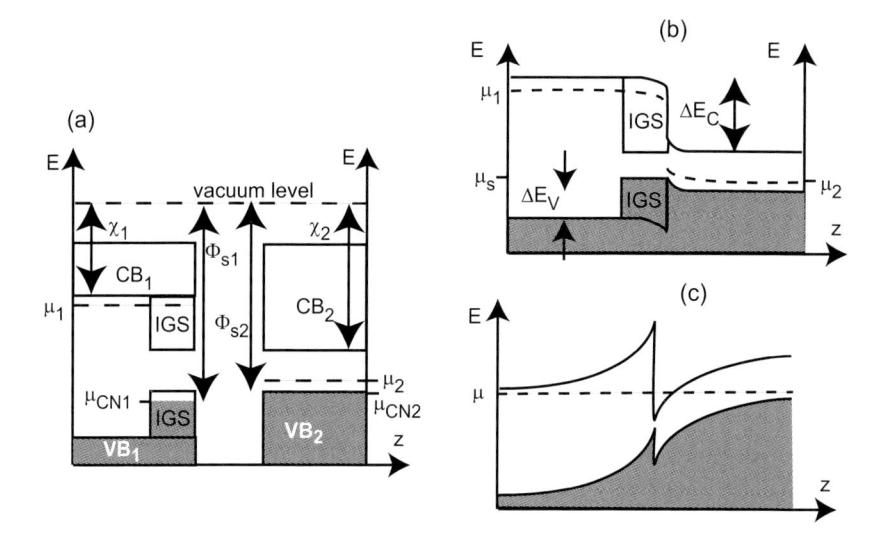

Figure 3.16: Band alignment at a semiconductor heterointerface. (a) Both band structures are drawn with respect to the vacuum level before charge is transferred. Gap states are induced, along with possible localized states in the band gap of both materials (not shown). (b) The band alignment is modified by a charge dipole which aligns the surface chemical potentials. Here, we have kept the doping charges immobile, such that the resulting band alignment is the same as for the intrinsic case. (c) In case of doped semiconductors, the bulk chemical potentials will align by transfer of doping charges across the heterointerface and the corresponding formation of depletion layers which are much larger than the spatial extension of the IGS. Here, the band offsets are drawn as sharp steps, neglecting the spatial extension of the IGS.

Question 3.2: Discuss the band alignment between two semiconductors when surface effects are neglected, i.e. a model similar to the Schottky model for the metal-semiconductor

3.3 Semiconductor heterointerfaces

interface!

Phenomenologically, one distinguishes between several types of alignment (see Fig. 3.17):

1. Type I: the smaller band gap lies completely inside the larger band gap. An important example is the $GaAs - Al_xGa_{1-x}As$ interface discussed below in detail.

2. Type II: here, both bands of crystal 1 lie above the corresponding bands of crystal 2. The $InAs - AlSb$ interface has such a structure. In the "staggered alignment", one of the band edges of material 1 resides inside the band gap of material 2, while in the misaligned type, the top of the valence band of material 1 lies above the conduction band bottom of material 2. The most prominent example for this type of alignment is the $InAs - GaSb$ interface.

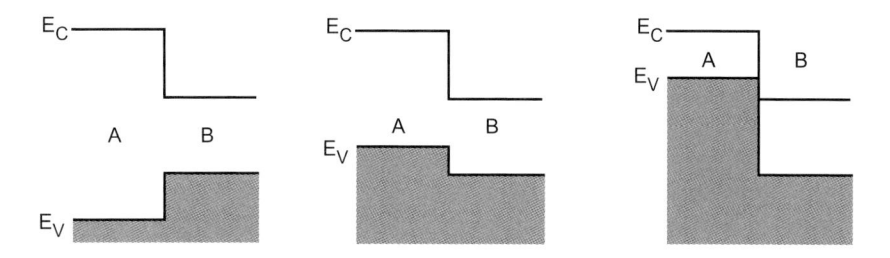

Figure 3.17: Different types of band alignments at semiconductor heterointerfaces. Left: type I, center: type II staggered, and type II misaligned (right).

To finish this section, it should be mentioned that very similar considerations give the band alignment between a semiconductor and an insulator.

Several theoretical treatments of band alignments have been developed. The concept of electronegativity, introduced by [Pauling1960], has proved highly useful in molecular physics and can be applied to crystal interfaces as well. An instructive evaluation of this approach is given in [P 3.1]. Furthermore, various extensive tight-binding models for interfaces have turned out to be highly successful. These concepts are developed in [Harrison1999], and references therein.

It should be remarked that agreement between theory and experiment is often hampered by imperfect interfaces containing defects or impurity atoms. A problem we have excluded is interface strain due to different lattice constants. With state-of-the art technology, however, close-to perfect interfaces can be grown, and the measured band alignments agree well with more sophisticated theoretical considerations, see, e.g. [Monch2001].

3.4 Field effect transistors and quantum wells

The properties of interfaces can be used to construct both useful devices as well as fascinating nanostructures. Field effect transistors are very important in both respects. Many mesoscopic samples are, in one way or another, some sort of field effect transistor, which are frequently denoted by the acronym FET. These devices heavily rely on interface effects. The two most important FETs in our context are the Si-MOSFET and the GaAs-HEMT. These are by no means the only systems though. Particularly in research, a wide variety of heterostructure devices is used. Some examples are given at the end of this section.

3.4.1 The silicon metal-oxide-semiconductor FET (Si-MOSFET)

This type of FET is the basic building block of the vast majority of present-day integrated circuits. A scheme of the Si-MOSFET is shown in Fig. 3.18(a). A silicon chip is, say, p-doped and electrically contacted with two Ohmic contacts that act as source and drain. A metal electrode resides in between the Ohmic contacts, separated by a SiO_2-layer from the Si. This M-O-S layer sequence can be thought of a Schottky diode of a very high resistance. Currents between the metal electrode and the semiconductor are neglected in the following. With no voltage applied, the resulting band structure across the interface is shown in Fig. 3.18(b). The p-doping is typically rather weak, say $N_D \approx 10^{21} \mathrm{m}^{-3}$, such that the resistivity of the Si is high. By applying a voltage to the metal electrode with respect to drain, a band bending is induced in the Si, and a corresponding charge accumulation at the $Si - SiO_2$-interface is generated, as depicted for the case of a positive voltage in Fig. 3.18(c). Here, E_c of the Si has dropped below the Fermi level, and electrons collect at the interface *in the conduction band*. Hence, an electron gas is generated which is confined in z-direction, but free in the directions parallel to the surface. For sufficiently high electron densities in this free electron gas, its conductance is much higher as compared to the p-doped bulk. We speak of *inversion* if the free carrier gas has the opposite sign than the carriers in the bulk due to doping. For appropriate doping densities, we can generate a free hole gas at the O - S interface by applying negative voltages to the metal electrode. This situation is refereed to as *accumulation*. Devices which offer the possibility of generating both electron and hole gases are known as *ambipolar*.

The current that flows between source and drain can thus be controlled by the voltage applied to the metal electrode, which is therefore known as the *gate*. The oxide prevents current flow between the gate and the silicon, which would reduce the performance of the switch. This three-terminal device thus represents a transistor that relies on the electrostatic field effect. However, we are not so much interested in the technological applications of MOSFETs in our context. Rather, we focus our attention at the electron gas that has formed at the O-S interface in Fig. 3.18(c). Apparently, its spatial extension in z direction is very small, as we have seen already above. Typically one finds that E_c is below the Fermi level for about 20 nm. Furthermore, the electron densities in such interface layers are much smaller than metallic densities, and the Fermi wavelength is larger. A crude estimation gives $\lambda_F \approx 20$ nm. Therefore, we expect size quantization effects in the electron gas. Fig. 3.19 shows a zoom-in of the conduction band structure at the oxide-semiconductor interface. The potential is roughly triangular. By applying an appropriate gate voltage, a situation can be established in which only the energy of the first quantized state is below the Fermi level. Since parallel to

3.4 Field effect transistors and quantum wells

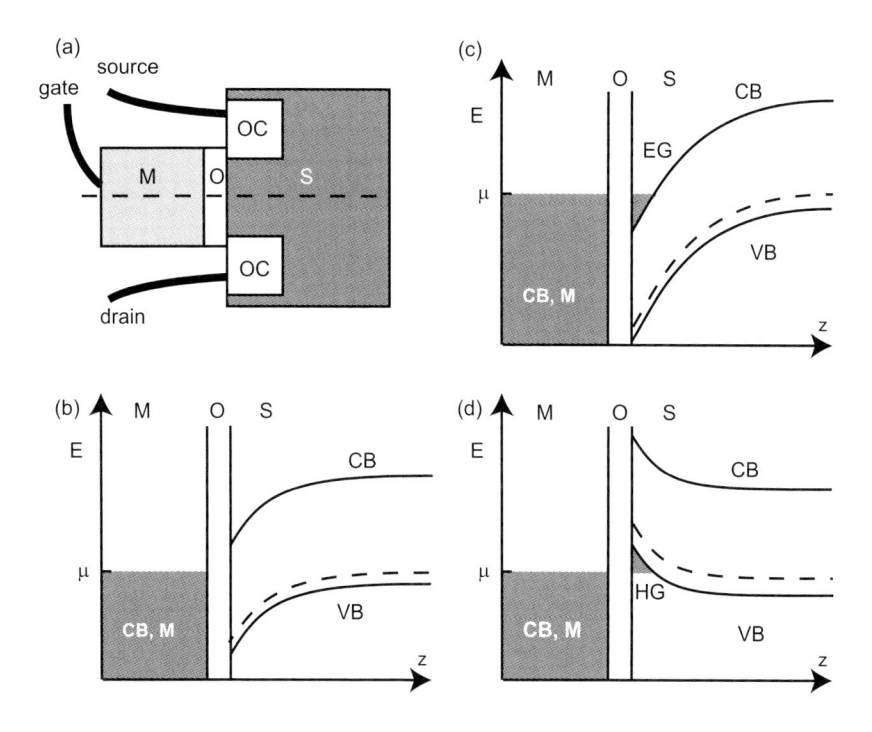

Figure 3.18: (a) Schematic illustration of a silicon MOSFET. A source-drain voltage is applied to a p-doped silicon wafer at two Ohmic contacts (OC). A metal electrode M ("gate") in between the Ohmic contacts is separated from the silicon by a SiO_2 layer. (b) Band alignment across the M-O-S interface (dashed line in (a) for $V_g = 0$). Applying a positive voltage to the gate increases the band bending. Above a threshold gate voltage, the conduction band bottom drops below μ at the O-S interface, and an electron gas (EG) is induced (inversion). A sufficiently large negative gate voltage pulls the top of the valence band above μ, and a hole has (HG) is generated at the surface (accumulation, (d)).

the interface, the electrons are not confined, a *two-dimensional electron gas* (2DEG) results. The conduction band bottom of this 2DEG is at E_0 in Fig. 3.19. We sometimes speak a two-dimensional subband. If more than one subband is occupied, the electron gas is said to be quasi-two-dimensional.

Like in three dimensions, this electron gas can be described by an effective mass and by the two-dimensional density of states. However, some care is required in adopting the bulk parameters to a two-dimensional carrier gas at an interface. We will meet some of the related issues later on. For now, we just look at the effective electron mass of the 2DEG in the Si-MOSFET. Suppose the Si crystal plane at the interface is a (100) plane, a very common case. The electrons move freely parallel to this plane only. Therefore, it is self-evident to project the valley-degenerated Fermi ellipsoids into that plane, see Fig. 3.20, which results in

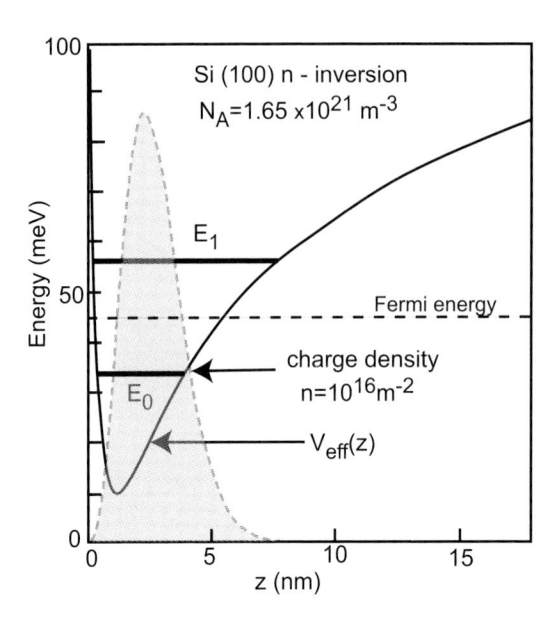

[Figure: plot of Energy (meV) vs z (nm) with Si(100) n-inversion, $N_A = 1.65 \times 10^{21}$ m^{-3}, showing E_0, E_1, Fermi energy, charge density $n = 10^{16}$ m^{-2}, and $V_{\text{eff}}(z)$]

Figure 3.19: Energy diagram of the conduction band in a Si-MOSFET close to the O-S interface, as obtained from a self-consistent calculation that includes electron-electron interactions and screening. In a potential well, quantized states are formed. The resulting electron gas is effectively two-dimensional as long as only the first quantized state lies below the Fermi level. Also indicated is the electronic wave function. After [Ando1976].

4 spin degenerate ellipses and a twofold valley-degenerated and spin degenerated circle at the center. Due to interface effects, however, the degeneracy between the ellipses and the circles gets removed, and the conduction band at the circle is about 20 meV below the conduction band minimum in the ellipses [Ando1982]. At room temperatures, both types of minima are occupied. At low temperatures, however, i.e. for $\Theta < 4.2$ K the electrons have a single effective mass of $m* = 0.19 m_e$ parallel to the surface, and the valley degeneracy is reduced to 2.

The two-dimensionality of this interface electron gas has some most surprising consequences, as will be seen in Chapter 6. But this is not the sole interesting property of such electron gases. Furthermore, the electron densities are much smaller that in conventional metals, and can be tuned. The Fermi wavelength is comparatively large, and size quantization effects can be expected also laterally, provided the MOSFET is patterned accordingly. Also, low density means that electron-electron interactions are more important, due to reduced screening.

Since the electrons are to some degree spatially separated from the ionized donors, impurity scattering is reduced and the electron mobility increases. In fact, the mobility of an electron gas at a O-S interface can be two orders of magnitude larger than the mobility of bulk Si. The mobility is typically dominated by scattering at impurities embedded in the ox-

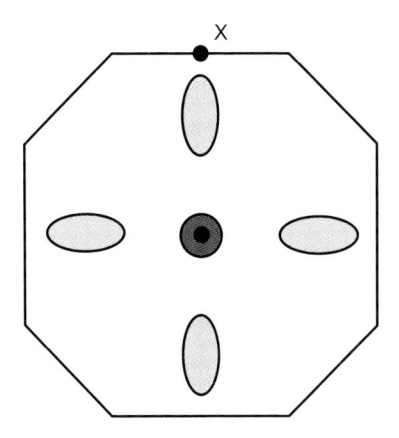

3.4 Field effect transistors and quantum wells 81

Figure 3.20: Projection of the Si Fermi surface for typical electron densities onto a (100) plane. Two ellipsoids get projected onto the Γ point. Their energy is reduced as compared to the 4 projected ellipses due to interface effects.

ide. Furthermore, the oxide is amorphous. The oxide atoms are by no means periodically arranged, which will cause additional electron scattering. However, due to size quantization, the probability of finding electrons right at the O - S interface is reduced, see the wave function in Fig. 3.19. The maximum of the probability density is several nanometers away from the interface.

3.4.2 The Ga[Al]As high electron mobility transistor (GaAs-HEMT)

In this system, the two-dimensional electron gas is generated inside the GaAs, at the interface formed between $Al_xGa_{1-x}As$ and GaAs. The band alignment of this interface is of type I. The band offsets depend on x, see Fig. 2.6. A typical choice is x=0.3. In that case, the conduction band of $Al_{0.3}Ga_{0.7}As$ is 300 meV above that one of GaAs. The top of the $Al_{0.3}Ga_{0.7}As$ valence band is located about 160 meV below that one of GaAs. This is of no further interest here, as we are going to consider again an electron gas.

In contrast to Si, the GaAs remains undoped. Instead, the electrons are provided by a doping layer inside the $Al_{0.3}Ga_{0.7}As$. Usually, Si is used as a donor. The doping layer can be spatially separated from the $Al_{0.3}Ga_{0.7}As$ by several tens of nanometers, see Fig. 3.21 (a). While most of the doping electrons that get thermally excited into the conduction band occupy the nearby surface states, some of them (typically about 10 %) reduce their energy by falling across the interface into the GaAs conduction band. This doping technique is called modulation doping; it was first demonstrated by [Dingle1978]. An accurate doping density is essential in designing a good HEMT structure: only a few percent deviation from the correct doping density can either cause mobile electrons in the doping layer (a "bypass"), or the triangular potential at the heterointerface remains empty. While the doping density and the thickness of the spacer layer determine the density of the 2DEG, it can be tuned with a top

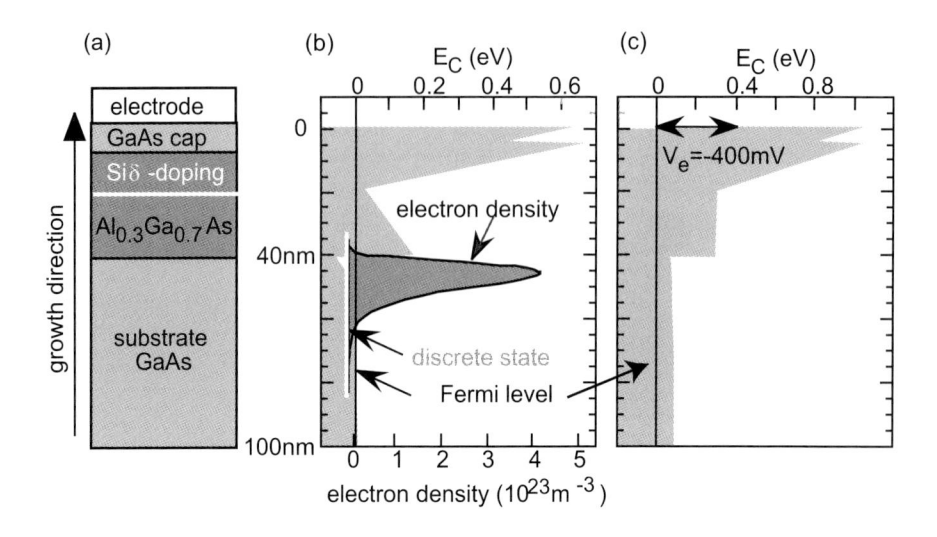

Figure 3.21: (a): band alignment at a modulation doped $GaAs - Al_xGa_{1-x}As$ interface. (b): schematic structure of a GaAs HEMT with the gate electrode grounded. (c) For gate voltages below -400mV, E_C at the interface moves above the chemical potential, and the electron gas is depleted.

gate over wide ranges.

Consequently, two charge dipoles build up, one between the surface and the doping layer, and a second one between the GaAs - $Al_{0.3}Ga_{0.7}As$ heterointerface.[7] This results in the band structure sketched in 3.21 (b). As in the Si-MOSFET, the resulting electron gas can be two-dimensional, and its carrier density can be tuned by applying voltages to a gate on top of the heterostructure, see Fig. 3.21 (c). Thus, the electron gas is present in this structure if no gate voltage is applied, or if there is no gate at all. Modulation doping of GaAs heterostructures caused a big progress in the electron mobilities that could be achieved (Fig. 3.22). The reason is twofold. First of all, $Al_xGa_{1-x}As$ is quasi-crystalline, in contrast to the SiO_2 - layer in a Si-MOSFET. Although the Al atoms replace the Ga atoms at random sites, this ternary compound is a somewhat distorted zincblend crystal structure with a well-defined lattice constant. The lattice mismatch between GaAs and $Al_{0.3}Ga_{0.7}As$ is only 0.4 %. Hence, the electrons in the 2DEG see an almost perfectly periodic environment, and the interface causes much less scattering as compared to the O - S interface in a Si-MOSFET. Second, the ionized donors, which are a strong source of scattering, are spatially separated from the electron gas. Consequently, the screened Coulomb potentials the electrons see are much weaker and generate predominantly small-angle scattering. In the years 1978 to 1985, the layer sequences and the compositions of Ga[Al]As-HEMTs had been optimized, and the increase in low-temperature electron mobilities achieved in this period is truly remarkable, see Fig.3.22. Today, the world

---

[7] Note the thin GaAs cap layer at the surface. Its purpose is to avoid oxidation of the $Al_{0.3}Ga_{0.7}As$ layer when exposed to air.

record is $\mu = 1440\text{m}^2/\text{Vs}$, held by [Umansky1997]. This corresponds to a mean free path of 120 µm. Although very similar devices can be built of several materials, like, e.g. Ga[Al]N, the Ga[Al]As heterostructure has remained unsurpassed in terms of electron mobility.

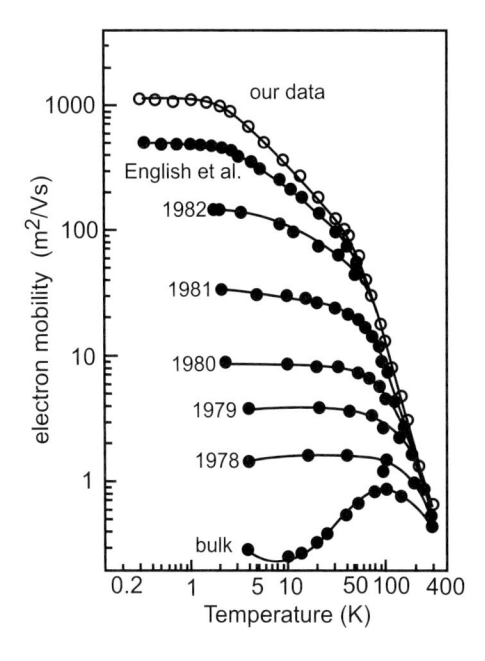

Figure 3.22: Evolution of electron mobilities over time, after modulation doping was introduced. After [Pfeiffer1989].

Another advantage of the Ga[Al]As system is the possibility to design the spatial variation of the band structure by controlling the Al content during sample growth. For example, quantum wells can be grown by embedding a thin layer of GaAs in two $\text{Al}_{0.3}\text{Ga}_{0.7}\text{As}$ layers. Varying the Al content parabolically during growth, i.e. $x \propto (z - z_0)^2$, results in a parabolic quantum well in growth direction, see 3.23. Hence, quantum mechanical model potentials can be experimentally realized this way, as long as the envelope function approximation is reasonable. We will occasionally meet such structures later on.

3.4.3 Other types of layered devices

We conclude this section with a selection of further interesting heterostructures. In particular, the Si[Ge] quantum well and the InAs/AlSb quantum wells are presented. Also, we will have a look at organic FETs. The materials cannot be combined arbitrarily, though. The lattice constants of the two components that form the interface should differ as little as possible. Differences in the lattice constants will inevitably lead to strained layers, which generates lattice dislocations and thus additional scattering. If the strain gets larger than $\approx 1\%$ homogeneous film growth is no longer possible, and strained islands of one material form instead. While

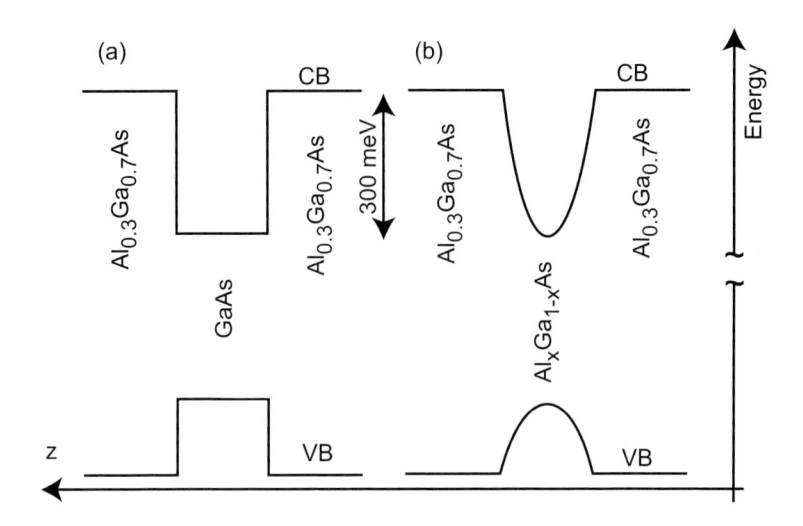

Figure 3.23: Sketch of a square (a) and parabolic (b) quantum well, fabricated in the Ga[Al]As system. In (b), the Al concentration x is varied $\propto z^2$ around the center of the well, up to $x = 0.3$.

these islands have fascinating properties (see the following chapter), they are of course unacceptable when a clean and homogeneous interface is required. A plot of the band gap of different semiconductors vs. their lattice constant is known as the band gap engineer's map. It reveals what kind of materials can possibly be combined, see Fig. 3.24.

1. The AlSb/InAs/AlSb quantum well. The band alignment of this material system is type II misaligned. Fig. 3.25 shows the band structure of a InAs quantum well, sandwiched in between two layers of AlSb. The surface is again capped with a binary crystal that does not contain Al, GaSb in this case. Modulation doping of this system is a tricky business. The standard donor for III-V systems is Si, which unfortunately acts as acceptor in GaSb and AlSb. Te is known to generate n-doping in AlSb, but this element causes technological problems during sample growth. Although there are ways to circumvent this problem, see e.g. [Bennett1998], doping is not necessary to fill the quantum well with electrons. The reason is that in this system, the surface chemical potential in the GaSb layer is just 180 meV below the GaSb conduction band, well above the conduction band bottom of InAs. Therefore, electrons get transferred from the surface band into the quantum well. Rather large electron densities up to $n \approx 1.2 \times 10^{16}$ m^{-2} can be achieved in such undoped quantum wells, by keeping the distance of the well from the surface sufficiently small, e.g. of the order of 30 nm. In intentionally doped systems, electron densities up to $n \approx 5.6 \times 10^{16}$ m^{-2} have been reported [Bennett1998]. At room temperature, mobilities can be of the order of 3 m^2/Vs [Bolognesi1996], which increase to $\mu \geq 30$ m^2/Vs at liquid helium temperatures. This material is also interesting because its large effective electronic g-factor of $g^*(\text{InAs}) = -14$, which is about 32 times larger than in GaAs ($g^*(\text{GaAs}) = -0.44$). The Zeeman splitting of the energy levels in external

3.4 Field effect transistors and quantum wells

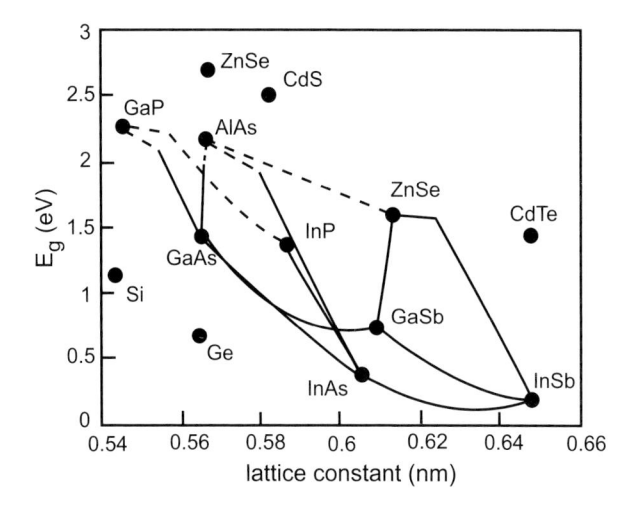

Figure 3.24: The band gap engineer's map shows what materials can be combined such that the lattice mismatch remains tolerable. Apparently, GaAs and AlAs match very well, while, e.g., the combination of Si with Ge will be accompanied by strain effects.

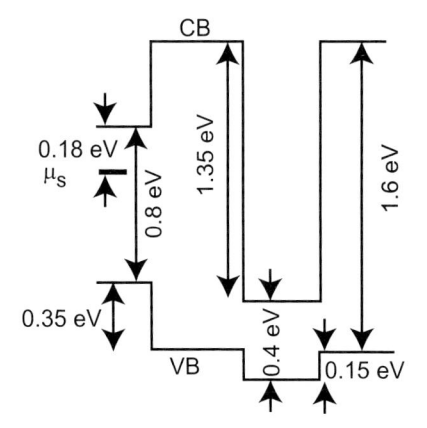

Figure 3.25: Schematic sketch of the band structure in a InAs/AlSb quantum well. The surface is capped by a GaSb layer with a surface chemical potential 180 rm meV below the bottom of the conduction band.

magnetic fields is thus very strong.

2. Hole gas in $Si/Si_{1-x}Ge_x/Si$ quantum wells. As an example of a hole has, we consider the $Si/Si_{0.85}Ge_{0.15}/Si$ quantum well depicted in Fig. 3.26. The interface is type II staggered. The band offset depends on x and occurs to a large fraction in the valence band. In our example, a boron-doped layer was grown in between the quantum well

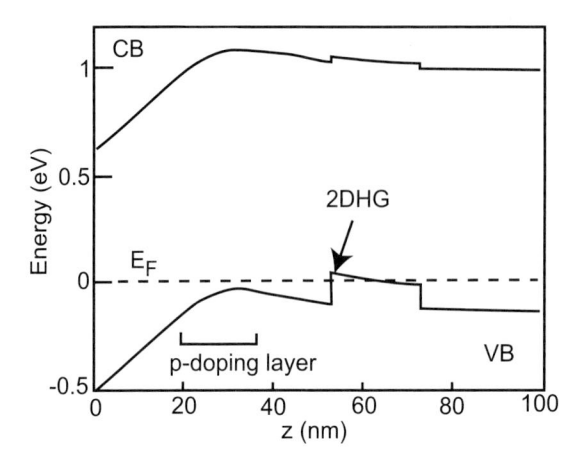

Figure 3.26: Band alignment of a modulation-doped $Si/Si_{0.85}Ge_{0.15}/Si$ quantum well. A two-dimensional hole gas (2DHG) is formed in the $Si_{0.85}Ge_{0.15}$ valence band. The square bracket denotes the Si layer that was doped with B, which acts as acceptor. The doping density was $N_A = 6.5 \times 10^{25}$ m^{-3} over a height of 15 nm in z - direction. Adapted from [Senz2002].

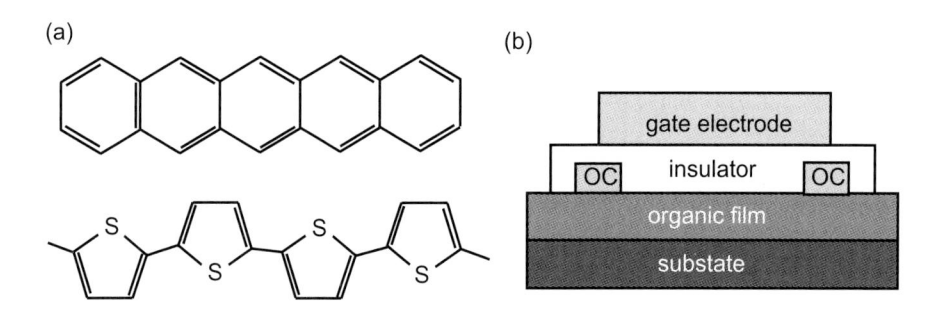

Figure 3.27: (a) Organic oligomers and polymers, like pentacene (top) of polythiophene (bottom) are examples for plastic FET materials. A typical schematic layout of such a transistor is shown in (b).

and the surface. Similar to the Ga[Al]As heterostructure discussed previously, the holes partly fill the surface band, and are partly transferred into the quantum well. Although hole mobilities in this system are rather low, namely of the order of $\mu \approx 1$ m^2/Vs even at reduced temperatures, this system represents a way to generate modulation doping in a Si-based material, which is of technological importance. From a fundamental point of view, this material is interesting, for example, because hole gases with greatly reduced screening properties can be generated. Hence, it is a system suited to study effects based on strong electron-electron interactions. [8]

3. As a final example, we mention field effect transistors that use an organic semiconductor , such as polythiophene - based materials or pentacene, see Fig.3.27, as host for the electron gas. Recently, such *plastic transistors* have received broad attention. They are potentially much less expensive than Si or GaAs based transistors. In addition, they are mechanically soft, such that novel types of applications can be thought of. Their electronic performance does not match those of established transistors so far. Room temperature mobilities of $\mu = 7 \times 10^{-6}$ m^2/Vs have been reported by [Horowitz1999], for example. For many applications, though, the high quality of established materials is not really needed. The investigation of low-temperature transport properties in nanostructured organic semiconductors is just at its beginning.

3.4.4 Quantum confined carriers in comparison to bulk carriers

It has been mentioned above already that the effective mass of electrons in a Si-MOSEFT differs from their effective mass in bulk silicon. This raises the question to what extent the bulk properties are relevant for mobile carriers in heterostructures.

First of all, quantum confinement changes the band structure and the effective masses. Consider, for example, a 2DEG in a quantum well of finite height. Clearly, the tails of the wave function extends into the barrier material, where the electrons have a different effective mass. Nevertheless, the conditions of the wave function and its derivative being continuous at the interface remain valid. This implies that the energy dispersion of the electrons is changed. See [Einevoll1994] for a discussion of such boundary conditions. Holes experience more dramatic modifications. Since the quantized energies of a quantum well depend on the effective mass, it is intuitively clear that the degeneracy of heavy holes and light holes at the Γ - point is removed by the quantum confinement. It can be shown that in fact the effective masses are reversed close to $\vec{k} = 0$, i.e. the light holes in the bulk material become the heavy holes in the quantum well. Furthermore, the confinement causes a mixing of the two bands, which leads to further strong modifications of the hole energy dispersion. As a result, it is often quite misleading to speak of heavy and light holes in quantum confined structures. For a quantitative discussion of these issues, the reader is referred to [Bastard1989].

All these descriptions implicitly assume that the envelope function approximation and the concept of effective masses remain valid in heterostructures. This requires that the superposed potential varies slowly on the scale of the lattice constant, which is clearly not the case at a heterointerface, or a narrow quantum well. It has been shown, though, that the envelope wave equation can also be derived for abrupt interfaces, as long as the envelope *function* still varies slowly. This issue is addressed in [P 3.2].

Second, the screening properties are modified in two dimensions, which is intuitively easy to see, since the scattering potential is still three-dimensional, but can be screened only in two dimensions be the electrons, while only polarization charges can screen in the third direction. It can be shown that for a strictly two-dimensional carrier system the static dielectric constant

---

[8]Two-dimensional electron gases can be generated in this material system as well. Here, the electrons collect in a Si quantum well that forms between two layers of Si$_{1-x}$Ge$_x$. In contrast to the hole gases, the mobility of such electron gases can be very high, e.g., values of $\mu \approx 50$ m^2/Vs have been reported by [Ismail1995].

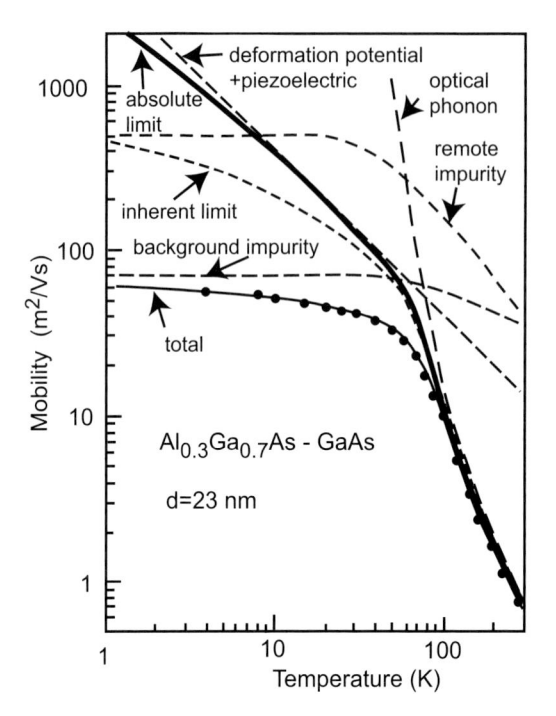

Figure 3.28: Significance of various scattering processes in a Ga[Al]As-HEMT. Black dots denote experimental results for a typical structure, with an electron density of n = 2.2×10^{15} m^{-2}, and a spacer thickness of d=23 nm. The density of the modulation doping was 8.6×10^{22} m^{-3}. This doping, which causes the remote impurities, was present within a 20 nm - layer between the surface and the spacer. In addition, a homogeneous density of background impurities of 9×10^{19} m^{-3} was assumed, which is a typical number for high-quality GaAs. After [Walukiewicz1984].

in the limit of low temperatures (see eq. (2.51) for the three-dimensional case) is given by

$$\epsilon(\vec{q}) = \begin{cases} 1 + \frac{k_{TF}}{q} & q \leq 2k_F \\ 1 + \frac{k_{TF}}{q}\sqrt{1 - \left(\frac{2k_F}{q}\right)^2} & q > 2k_F \end{cases} \quad (3.23)$$

and the resulting charge density induced by a Coulomb potential reads at large distances from the scattering center

$$V_{eff}(\vec{r}) = \frac{Ze}{\epsilon \epsilon_0} \frac{4k_{TF}k_F^2}{(2k_F + k_{TF})^2} \frac{sin(2k_F r)}{(2k_F r)^2} \quad (3.24)$$

Thus, the screened potential drops with r^{-2} as compared to r^{-3} in three dimensions.

Furthermore, additional scattering mechanisms, which are absent in bulk materials, are possible in quantum confined systems. The scattering of electrons on ionized impurities has

3.4 Field effect transistors and quantum wells

a somewhat different character in modulation doped systems as compared to bulk materials, since they are spatially separated from the electrons by a spacer layer. The residual and usually small density of ionized impurities inside the electron gas is comparatively small in high quality systems. One may be tempted to guess that the broader the spacer layer, the higher the mobility. This is not the case, though, since as the spacer thickness becomes larger, the carrier density gets smaller, and screening becomes less effective. Hence, a maximum in the mobility as a function of the spacer thickness is observed. In Ga[Al]As-HEMTs, the optimum spacer thickness depends on the cleanliness of the material and the doping density. It varies between $\approx$ 20 nm and $\approx$ 60 nm. Another scattering mechanism in FET structures is interface roughness scattering. The interface clearly constitutes a deviation from perfect periodicity and consequently generates scattering. In case of a Ga[Al]As HEMT, this is of minor importance. In narrow quantum wells, however, where fluctuations at both interfaces are important, this mechanism may become important. In Si-MOSFETs, on the other hand, the oxide is amorphous, and interface roughness scattering is not negligible.

Alloy scattering occurs in compound materials such as $Al_xGa_{1-x}As$. The replacement of Ga atoms by Al atoms takes place at random positions, and a non-periodic potential results. This kind of scattering usually plays no significant role, as long as the carriers reside in a crystalline material, such as GaAs, with a barrier made of a ternary compound, since only the evanescent tails of the wave function feel this kind of disorder. In Fig. 3.28 a model

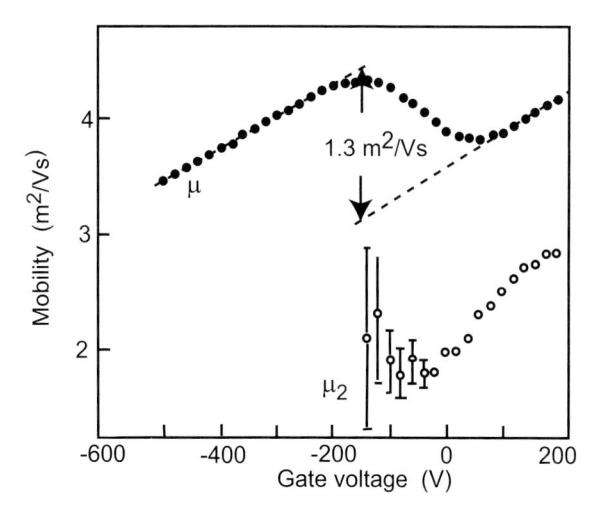

Figure 3.29: Electron mobilities in a GaAs-HEMT as a function of the gate voltage. Above a threshold electron density, the second two-dimensional subband gets occupied, and the mobility drops due to additional intersubband scattering. After [Stormer1982].

calculation adopted to some typical data is shown, which surveys the relevance of various scattering mechanisms in a Ga[Al]As-HEMT. While alloy scattering and interface roughness scattering are irrelevant except at very low temperatures and in extremely clean samples, the ionized impurities are split into two components, namely a density of homogeneously distributed background impurity, which can be reduced in principle by fabricating cleaner samples, and a density of remote impurities, which is necessary, since they are the donors that provide the electrons. The *inherent limit* in this figure indicates the mobility that would be obtained in a sample in the absence of background impurities, but with the remote donors still in the sample. The absolute limit represents a situation where the remote donors do not influence the mobility, which is then given by processes intrinsic to a perfect GaAs - $Al_xGa_{1-x}As$ interface. It is worth comparing this behavior with the temperature dependence of bulk GaAs (Fig. 2.15) and Fig. 3.22. Most strikingly, the reduction of the mobility as the temperature is lowered in bulk GaAs is absent. This is the effect of the modulation doping, which separates the conduction electrons from the ionized donors, and thus breaks the $\mu \propto \Theta^{3/2}$ - law. Second, it becomes apparent that the best samples shown in Fig. 3.22 hit the absolute mobility limit at intermediate and high temperatures ($4\ K \leq \Theta \leq 400\ K$) set by deformation potential scattering, piezoelectric scattering and optical phonon scattering. The saturation at very low temperatures is due the residual impurities, but it occurs not very much below the absolute limit. We can thus not expect to see another huge increase in electron mobilities in this material.

Finally, intersubband scattering should be mentioned, which denotes scattering events for which the incoming carrier is scattered into another subband of the confined potential. For elastic scattering events, this is only possible if at least two subbands are already occupied, and the system is thus not strictly two-dimensional.

Papers and Exercises

P 3.1: [Frensley1977] predicted band alignments between semiconductor heterostructures within the concept of electronegativities. Discuss the basic idea of this approach!

P 3.2: A highly instructive article on the validity of the envelope function approximation in semiconductor heterostructures has been written by [Burt1992]. Work out the author's line of arguing!

E 3.1: The wave functions of the ground state in a triangular potential can be approximated by the Fang-Howard function

$$\Phi(z) = \sqrt{\frac{b^3}{2}} z e^{-bz/2}$$

Calculate the expectation value of the electron location. Discuss the consequences of this result for, e.g., a Si-MOSEFT.

Further Reading

Extensive and comprehensive reviews on the physics of surface states are [Desjonqueres1998] and [Monch2001].

An introduction to the theory of crystal surfaces can be found in [Davidson1992]. For the quantum mechanical properties of layered devices, the book by [Bastard1989] is a valuable source of information.

4 Experimental Techniques

This chapter introduces the experimental tools and techniques involved in fabricating nanostructures and measuring their transport properties. The majority of the mesoscopic devices is made of semiconductor heterostructures. Section 4.1 describes how they are fabricated. This process usually includes single crystal growth, followed by some sort of lateral patterning of the crystal slices. The lateral patterning is done by various lithographic techniques, while occasionally, self-assembly is used as well. Already in the introduction, is has become clear that many of the experiments are performed at liquid helium temperatures, which is the regime below $\Theta = 4.2$ K, the boiling temperature of ^{4}He at 1 bar. Therefore, the concepts and techniques of generating such a low temperature environment are discussed in section 4.2. This includes the relevant properties of liquid helium as well as the essentials of helium cryostats. Finally, some basic understanding of electronics is very helpful for the discussion of the transport experiments. This is the topic of section 4.3.

The present chapter cannot replace a thorough treatment of these issues, which would require a bulky textbook for each section. Rather, our goal is to provide the knowledge needed to appreciate the constraints of the experiments set by the technology.

4.1 Sample fabrication

Fabricating nanostructures for mesoscopic transport experiments is a major technological challenge. The requirements concerning material purity, lithographic resolution and process control are at the edge of present-day technology. We will exemplify the technology using a Si and GaAs as examples, and mention occasionally some special properties of other materials. As a rule, we refrain from specifying process parameters and experimental recipes. The reader is referred to the specialized literature at the end of this chapter.

Silicon dominates in industry, while GaAs and other materials are essentially only used where silicon devices are significantly less useful, like in optoelectronics or in ultra-high speed, ultra-low noise applications. From a technological point of view, Si has several major advantages. First of all, the raw material is quartz sand, easily available and cheap. Second, its high mechanical stability simplifies all process steps. A very important point is the fact that Si can be easily oxidized into SiO_2, which has excellent mechanical and electronic properties, like high breakdown electric fields and large resistivities. GaAs oxides, on the other hand, have poor electronic properties and are more or less useless for electronic applications, like in capacitors.

During fabrication, we have to prevent that dust particles hit the crucial area, since they may cause defects in the lithographic patterns, which get transferred into the nanostructure

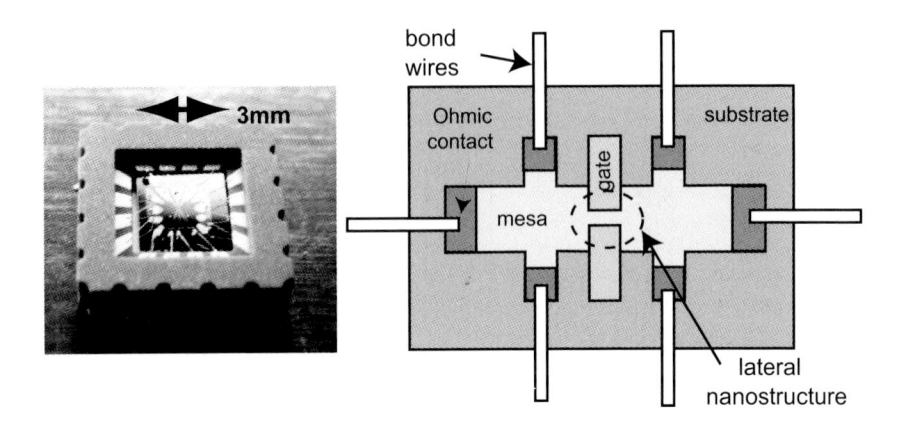

Figure 4.1: Left: photograph of an assembled Ga[Al]As-HEMT structure. The chip resides inside a ceramic chip carrier. Electric contacts between the carrier and the chip is made by bonded wires. The chip-carrier has a size of 5 mm x 5 mm. Right: schematic components of a typical microchip designed for mesoscopic transport measurements.

in subsequent process steps. Therefore, processing is usually done in a *clean room* in which the air is heavily filtered. Clean rooms are classified by the number N of dust particles per cubic foot, sized larger than 500 nm. For example, a class 100 clean room contains N=100 of such particles in one cubic foot. It is true that feature sizes have been scaled down to well below 500 nm, and along with this, the size of the relevant dust particles has decreased accordingly. Nevertheless, the clean room class as defined above is a useful quantity, as the particle size distributions in air are well known. Mesoscopic researchers typically fabricate a small number of individual samples, with an active area below a square millimeter. Of course, it is much harder to keep a 4 inch wafer free of dust particles, and the demands posed on an industrial clean room are much higher. Furthermore, although a high sample yield is desirable in research as well, a research lab has no problems living with yields below 100%.

Fig. 4.1 shows a processed Ga[Al]As-HEMT structure. It becomes immediately apparent that a whole bunch of fabrication steps is required. To begin with, the semiconductor heterostructure must be grown. This involves single crystal growth, plus some sort of epitaxy to add the layers of different materials. Next, a piece of the wafer has to be processed by lateral lithography. The two-dimensional electron gas is contained in the *mesa*, which should have a suitable geometry. Ohmic contacts and gate electrodes have to be patterned. Finally, wires must be connected to the electrodes.

We will discuss these fabrication steps below. Although the details of the processes depend on the material used, the technological concepts are similar.

4.1 Sample fabrication 95

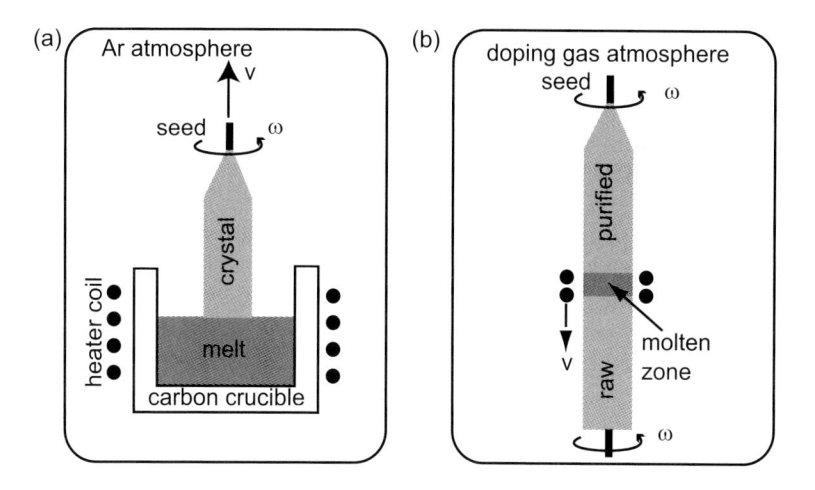

Figure 4.2: The Czochralski scheme for crystal growth (a) used to grow Si single crystals. The angular velocity and the velocity are ω and v, respectively. (b): the zone pulling technique.

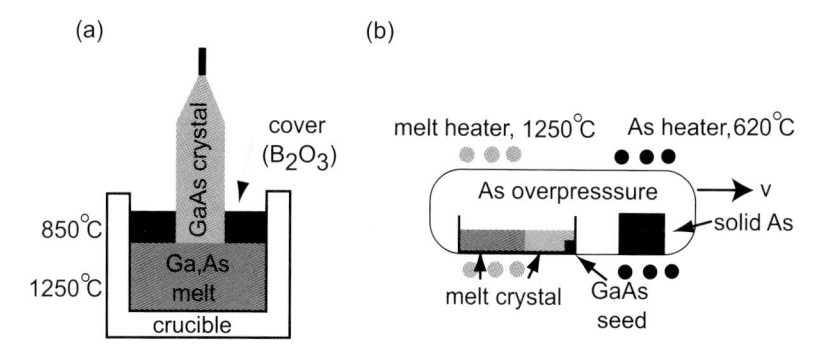

Figure 4.3: Schemes for growing of two-component crystals, such as GaAs. (a): in the LEC technique, the melt is covered by an impenetrable liquid. The crystal is pulled through this cover layer. In the Bridgman technique (b), a closed tube contains a crucible that hosts the melt, a seed and the freshly grown crystal, as well as a piece of solid As. The temperature of the solid As is kept at a temperature that corresponds to an As overpressure of 0.9 bars. The melt solidifies at the location where the spatially varying temperature reaches 1238 °C. Crystal growth is established by moving the tube along the temperature profile.

4.1.1 Single crystal growth

A standard method to grow silicon single crystals is the so-called *Czochralski method*, Fig. 4.2(a). A small single crystal (the *seed*, which also determines the crystal direction) is immersed in a purified silicon melt (Si has a melting point of $1412\,°C$ under ambient conditions). The atmosphere should be inert; an argon atmosphere is often used. The seed is rotated ($\omega \approx 2\pi s^{-1}$) and slowly (with a speed of, say, 1 mm/s) pulled out of the melt. The pulling speed determines the diameter of the crystal cylinder, which may be as large as 10 inches. A typical length is 1 m.

Another widely used technique is zone pulling, Fig. 4.2(b). Here, the raw crystal, which may still be poly-crystraline or a Czochralski grown crystal, is molten locally with via Eddy current heating with an RF coil. The temperature is about $1450\,°C$. The setup resides inside a high vacuum chamber. For impurity atoms, it is energetically favorable to be in the melt. They collect in the molten zone, evaporate there and can be pumped away. With this technique, impurity concentrations below 10^{13} cm^{-3} can be obtained. Undoped Si crystals can have resistivities above 10 Ωm, which reflect their high purity. If the crystal has to be doped, a doping gas atmosphere is established in the growth chamber, e.g. a B_2H_6 atmosphere for boron doping, or a PH_3 atmosphere for phosphorus doping.

GaAs is a binary material and as such more difficult to grow. A general problem with multi-component melts is the different vapor pressure of the components. In order to avoid compositional changes of the melt over time, the vapor pressures must be controlled. In case of GaAs, As has an overpressure of 0.9 bar, while e.g. in InP, the P overpressure is 60 bar. Two methods are common for compensating the overpressures. In the *liquid encapsulated Chochralski (LEC)* technique, the melt is covered with a fluid that does not intermix, Fig. 4.3 (a). As a result, no gas can escape from the melt. In the Bridgman technique, the melt resides in a closed quartz tube, in which the correct As overpressure is established by heating solid As outside the melt to the corresponding temperature (Fig. 4.3 (b)). Here, the crystal is grown by moving the melt along a suitable temperature gradient and at an appropriate speed.

4.1.2 Growth of layered structures

Pulling a crystal out of a melt is perfect for fabricating substrates, which are usually obtained by cutting the crystal cylinder into thin disks called *wafers*. A typical substrate thickness is 300 microns, while surfaces roughnesses of a few nanometers can be achieved by mechanical polishing. These techniques are not suited to grow layered structures, such a s a Ga[Al]As heterostructure. Here, we need something that ideally provides ultra-clean growth of individual monolayers, and the material composition must be controllable. Several techniques are established for fabricating layered semiconductor structures, and we will briefly discuss two of them.

MOCVD stands for *metal organic chemical vapor deposition*. The substrate is mounted in a vacuum chamber, see Fig. 4.4(a). The atoms of the semiconductor components to be grown are introduced via suitable molecular gas flows. The gas molecules crack at the surface and deposit the semiconductor atom on the substrate. In case of GaAs, a possible chemical reaction with $Ga(CH_3)_3$ and AsH_3 as input gases is,

$$Ga(CH_3)_3 + AsH_3 \rightarrow GaAs + 3CH_4$$

taking place around a temperature of $1120\,°C$. The advantage of MOCVD is the relatively low cost. A disadvantage is the high toxicity of the gases involved. In addition, the material grown

4.1 Sample fabrication

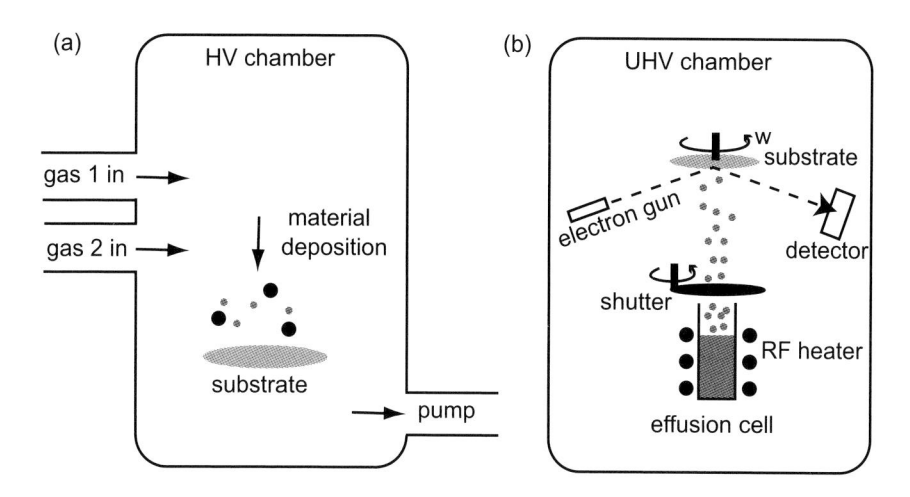

Figure 4.4: Schematic view into a MOCVD chamber (a), and into a UHV chamber for molecular beam epitaxy (b).

is not as clean as that one obtained with the second technique we look at, known as *molecular beam epitaxy (MBE)*. Here, atomic layers are grown in a ultra-high vacuum chamber, with pressures of the order of 10^{-11} mbars. A substrate is inserted in the UHV chamber, heated and slowly rotated (Fig. 4.4). The components of the semiconductor are supplied by effusion cells, which can be individually heated to provide the flux needed, as well as opened and closed. For growing standard GaAs HEMTs, for example, Ga, As, Al, and Si (for n-doping) effusion cells are needed. This way, monolayer by monolayer of the crystal can be grown, which can be selectively doped: the modulation doping encountered in chapter 3 can be easily incorporated. Typical growth rates are 0.1 nm/s. The growth can be calibrated and monitored with RHEED, which is an acronym for *reflection of high-energy electron diffraction*. Here, an electron beam with an energy of about 10 keV hits the surface under a very small angle (a degree or so), and gets reflected at the surface. Its penetration depth is a few monolayers only, such that the reflected interference pattern is highly sensitive to the roughness and the crystal structure of the surface. The reflected intensity shows a minimum if the coverage of the monolayer is 50%, which corresponds to maximum roughness. When a monolayer has just been completed, the scattering has the highest specularity, and the reflected intensity shows a maximum.

Although MBE is very expensive and time-consuming, it is widely used to grow heterostructures for mesoscopic transport experiments, since the quality of the samples is unsurpassed by any other methods. The high pressure is needed to make the residual gas monolayer formation time sufficiently large, see exercise [E 4.1].

Ga[Al]As heterostructures are frequently grown by MBE. After some surface cleaning,

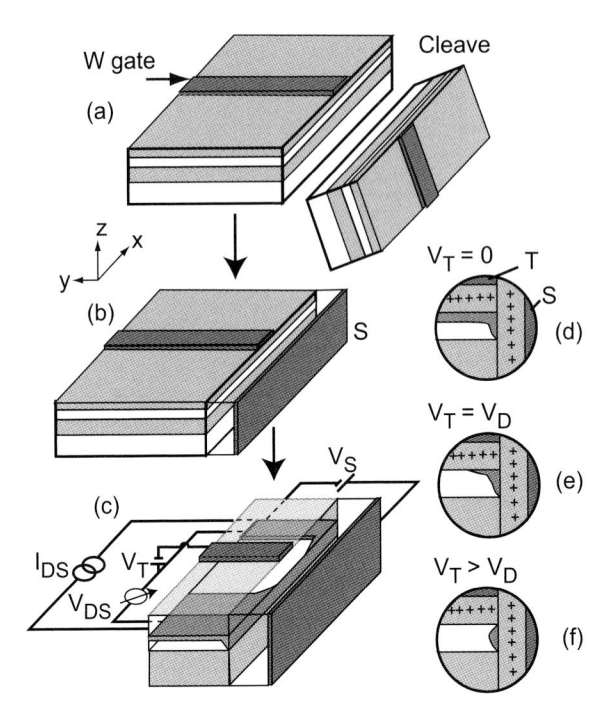

Figure 4.5: Quantum wire production by CEO. In (a), a thin GaAs layer (white), embedded in $Al_xGa_{1-x}As$ layers (gray) is shown. The sample is partly covered with a tungsten gate stripe. This structure has been obtained by MBE growth in z direction; the $Al_xGa_{1-x}As$ layer on top of the $GaAs$ well is modulation doped with Si. After the growth has been completed, the wafer is cleaved perpendicular to the stripe. (b) MBE growth is continued on the freshly cleaved surface, i.e., in y-direction. A modulation doped $Al_xGa_{1-x}As$ layer is covered by another tungsten layer. In (c), the electrical connections to the different elements are shown. Voltages can be applied to the two gates T and S, and a current is applied between the two areas separated by gate T. In Figs (d) to (f), it is sketched how different gate voltage combinations shape the electron gas. In particular, a one-dimensional wire is formed in (f), which extends along the quantum well at the cleaved and regrown interface. After [Yacoby1996].

the growth begins with a buffer layer, consisting of a series on GaAs -AlAs superlattice with a short period (a *short period superlattice - SPS*). The purpose of the SPS is twofold. First of all, the mechanically polished GaAs substrate is not atomically flat. It has been found that a SPS reduces the roughness due to polishing to nearly atomic flatness [Petroff1984]. Second, the superlattice tends to trap impurities which may diffuse from the substrate into the electronically active layers grown on top.

MBE can also be used to fabricate more complicated structures than sequences of layers with translation invariance in two dimensions. These technologies are in the focus of

4.1 Sample fabrication

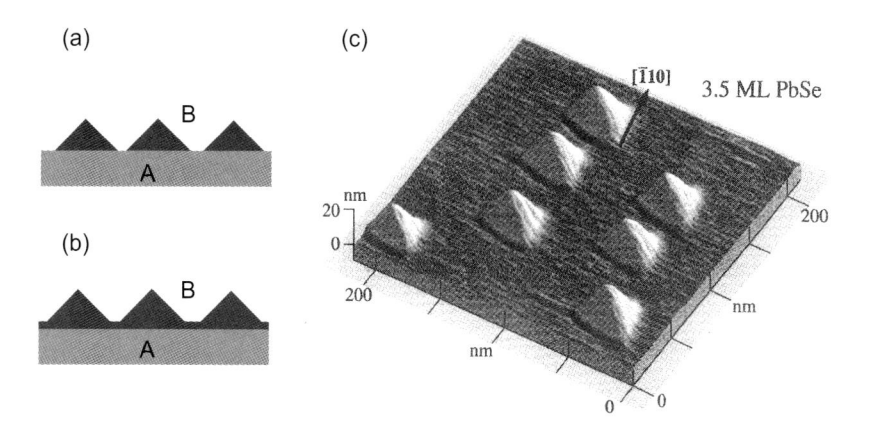

Figure 4.6: Growth of material B on top of material A in the Volmer-Weber (a) and in the Stranski-Krastanov (b) mode. (c): atomic force microscope picture of PbSe islands on a PbTe (111) substrate. Note the homogeneous size distribution and the orientation of the pyramids. Fig. (c) has been reproduced from [Pinczolits1998].

present-day research. As a first example, we consider a technique called *cleaved edge overgrowth (CEO)*, Fig. 4.5. Here, the layer growth is interrupted at the right point, and the wafer is cleaved inside the MBE chamber, such that the grown sequence of layers appears on an atomically flat surface. *Cleaving* includes scratching the wafer at its edge and subsequently breaking it by mechanical pressure. The GaAs wafer breaks at the scratched position along a single crystal plane. Subsequently, MBE growth is continued on top of this freshly cleaved surface.

Extremely small nanostructures of effectively one- and even zero-dimensional character, with atomically flat interfaces, have been produced this this way. For a review, see [Wegscheider1998]. We will discuss some properties of such nanostructures in subsequent chapters. Self-assembled quantum dots are a second example of ongoing MBE research activities. Growing a semiconductor (B) on top of a appropriate substrate (A) does not necessarily lead to atomically flat films that build up monolayer by monolayer. Rather, one distinguishes three growth modes, the one we just encountered is known as *Frank - van der Merve* growth. [Frank1949]. Alternatively, growth of a new material can take place in terms of isolated islands (the *Volmer-Weber* mode [Volmer1926]), or in terms of islands connected via a thin layer of the same material [Stranski1939], the so-called wetting layer, see Fig. 4.6. This type of growth is called *Stranski-Krastanov*. Which kind of growth takes place depends on an interplay of different energy scales. In a strongly simplified view, we can assume that there are surface energies per area E_A and E_B related to the surfaces of material A and B, respectively. In addition, there is an interface energy per area E_{AB}. Let S be the fraction of the surface of A covered by B. The total energy is then given by

$$E = (1 - S)E_A + SE_B + SE_{AB}$$

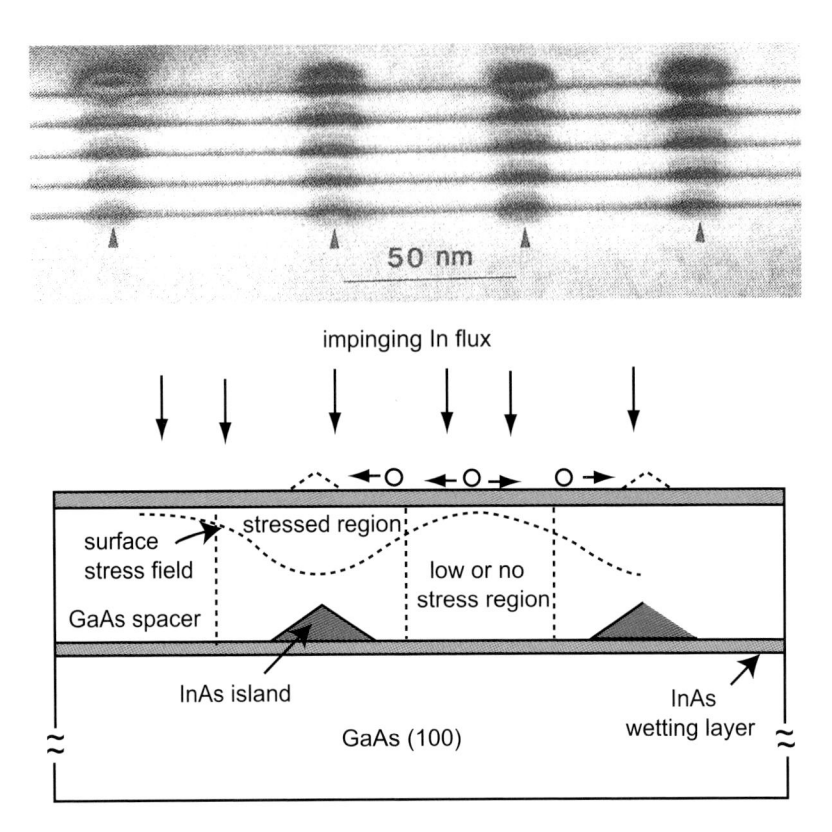

Figure 4.7: Top: transmission electron microscope picture of InAs layers, spaced by 36 monolayers of GaAs. The vertical direction is the growth direction. The InAs SAQDs, seen a dark spots, align on top of each other. The bottom figure shows a schematic sketch of the SAQDs and the wetting layer. From [Xie1995].

Here, we have assumed that the surface of B remains flat, and that none of the materials get strained, i.e. that A and B have identical lattice constants. This energy will get minimized. It follows that for $E_{AB} + E_B < E_A$, S will get maximized, and Frank - van der Merve growth takes place. On the other hand, for $E_{AB} + E_B > E_A$, a minimized S minimizes the energy, and we have Volmer-Weber growth. In order to establish Stranski-Krastanov growth, we need a lattice mismatch between A and B, which gives an elastic strain energy per area in B in addition, given by $E_{str}d$, where d denotes the thickness of the layer. Minimizing the total energy again predicts homogenous film formation for

$$d < \frac{E_A - E_B}{E_{str}}$$

provided we can neglect the interface energy and material A suffers no strain. This simple picture shows that after the wetting layer of thickness d (typically two to four monolayers) is

4.1 Sample fabrication

completed, is is energetically more favorable to continue growth with island formation. Experimental studies as well as theoretical considerations have demonstrated that strain does not cause, as one might assume, dislocations inside the dots; rather, the substrate gets elastically strained as well. [Eaglesham1990]. Such islands, e.g. InAs grown on GaAs are typically of pyramidal shape and have very homogeneous size distributions; the size variances are of the order of 10% only [Leonard1993]. Since their sizes are in the range of 20 nm in width and a few nanometers in height[1], strong quantization effects can be expected inside, and have in fact been observed in many experiments, some of which we will discuss later on. Therefore, they are referred to as *self-assembled quantum dots (SAQDs)*. In some systems, the SAQDs even align with each other and form lattices of various dimensions. A three-dimensional SAQD superlattice is the topic of [P 4.2]. Here, we look at growth of linear chains of InAs SAQD islands, embedded in GaAs, Fig. 4.7. In this example, the strain in the GaAs induced by the buried InAs islands modulates the GaAs surface energy, and thus the freshly offered InAs will preferentially form dots at locations where the lattice mismatch is minimum, which is above the locations of the SAQDs next to the surface. Clearly, the spacing between adjacent InAs layers can be optimized for maximum probability of SAQD alignment. For large distances, the strain modulation at the surface becomes too weak, while for very small spacings, neighboring points of extremal strain begin to overlap.

A very different process for fabricating a layered structure is thermal oxidation of Si. For growing oxides used in electronic applications, the technique of choice is usually *dry oxidation*. The Si wafer is placed in a furnace at a temperature of about $1000\,°C$ and exposed to oxygen. Via the reaction $Si + O_2 \rightarrow SiO_2$, the wafer oxidizes. The oxygen diffuses through the already grown oxide layer and reacts with the Si at the Si/SiO_2 interface. The oxide growth rate therefore drops as its thickness increases. Furthermore, the oxide penetrates into the Si. About 50% of the oxide layer is located below the original wafer surface. Breakdown electric fields for oxides grown with this technique can be of the order of $5 \cdot 10^8$ V/m, and are thus well suited for electronic applications.

4.1.3 Lateral patterning

The special MBE techniques CEO and SAQD growth are by no means standard technology. It is more usual to grow a heterostructure with two-dimensional translation invariance parallel to the surface, and pattern the nanostructure subsequently by some sort of lateral processing. Examples for typical sequences of process steps are given in Fig. 4.8. Column (a) is typical for many Si fabrication steps. The substrate is covered with a homogeneous metal layer, which is subsequently coated with a suitable resist. Illumination and development of the resist through a mask exposes some areas of the metal layer, while others are protected by the resist. The illumination is usually carried out with ultraviolet light or with electrons. An etch step follows, which selectively removes the free metal surfaces. here, the resist acts as an etch mask. Finally, the resist gets removed, and a patterned metal layer on the substrate results. This technique is rarely ever used in GaAs processing, since essentially all suitable metal etchants attack GaAs as well. Therefore, fabrication scheme (b) is typically used. Here, the substrate is first covered by resist, which gets illuminated and developed. Now, the metal is

---
[1] This is below the resolution limit of lithographic techniques, as we will see shortly.

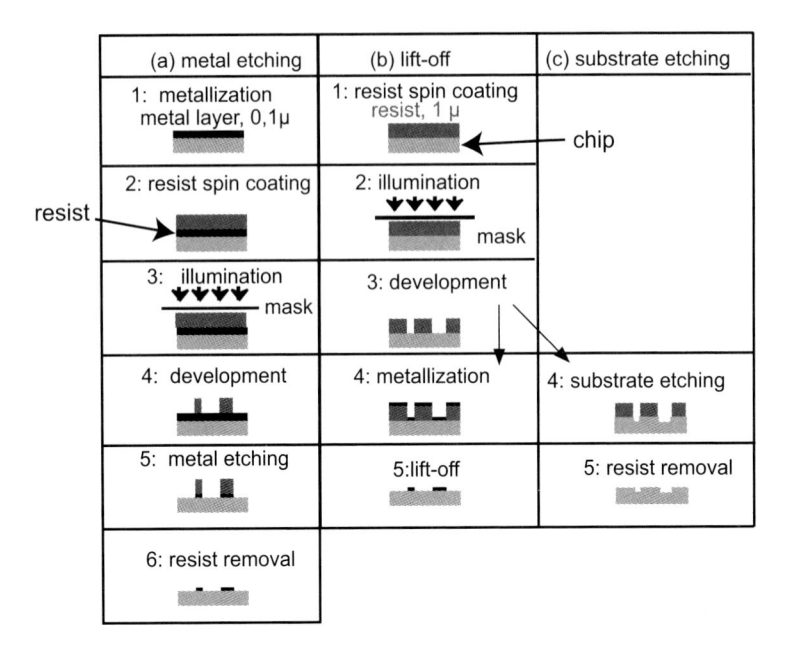

	(a) metal etching	(b) lift-off	(c) substrate etching
1	metallization metal layer, 0,1μ	resist spin coating resist, 1 μ	chip
2	resist spin coating	illumination ▼▼▼▼ mask	
3	illumination ▼▼▼▼ mask	development	
4	development	metallization	substrate etching
5	metal etching	lift-off	resist removal
6	resist removal		

Figure 4.8: Comparison of different lateral patterning schemes for semiconductors.

evaporated on the substrate, with the patterned resist acting as evaporation mask. The *lift-off* step follows, i.e., the resist is removed with the metal film on top. The final result is identical to that one of scheme (a). For selective etching of the substrate (c), steps 1 to 3 are identical to (b). Then, the patterned resist is used as an etch mask for the substrate. We now discuss these fabrication steps in further detail.

Defining patterns in resists

The two standard techniques for imposing a pattern into a resist are optical lithography and electron beam lithography.

1. Optical lithography
   By this we mean illumination of a photoresist by visible or ultraviolet light. The sample is coated with a thin and homogeneous photosensitive resist. This is done by dropping some resist solution onto the sample, which is then rotated for about one minute at high speed, typically a few thousand rpm. The spinning speed and the viscosity of the solution determine the thickness of the resist layer, which is of the order of 1 micron.

   After baking the resist, the sample is mounted into a mask aligner, a device designed for adjusting the sample with respect to a mask that contains the structure to be illuminated. The mask aligner is equipped with a strong light source that illuminates the resist film through the mask, see Fig. 4.9. The pattern sizes are Doppler limited, which means

4.1 Sample fabrication

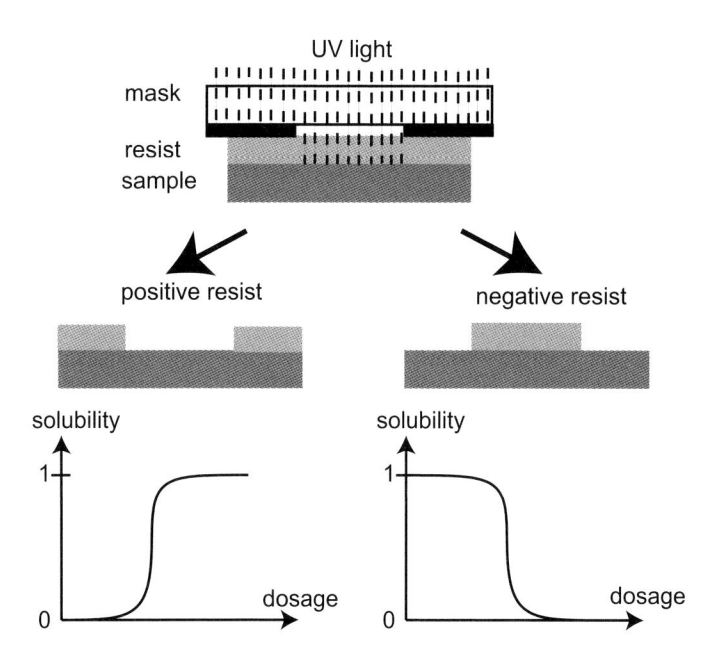

Figure 4.9: Top: contact illumination. The mask pattern is transferred into the resist via illumination and subsequent development. Center: resulting resist cross section for positive (left) and negative (right) resist. Bottom: Solubility characteristics for the two resist types.

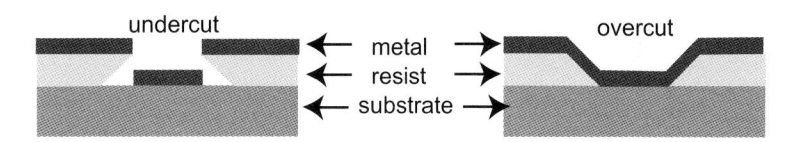

Figure 4.10: Overcut (left) and undercut (right) resist profile after illumination, development, and surface metallization.

that the smallest feature sizes are about half the wavelength (≈ 150 nm), divided by the index of refraction of the resist (≈ 1.5), which limits the resolution to roughly 100 nm. The mask can be a quartz plate coated with a thin chromium film, which contains the pattern to be illuminated. In the contact illumination scheme, the Cr film is in mechanical contact with the resist and blocks the light, such that the resist underneath the Cr remains unexposed.

During contact illumination, the mask suffers contaminations due to dust particles on top of the resist, as well as by resist adhesion. This can be avoided by projection illumination, where the mask pattern is transferred into the resist via lenses. This technique is widely used in industry, but somewhat unusual in research labs.

The photoresists can be classified as *positive* and *negative*. The solubility of the exposed areas increases for a positive resist, while it decreases in negative resist, see Fig. 4.9. Immersing the sample into a suitable developer removes the corresponding sections of the resist film. Both types of resists have in common that their solubility as a function of the illumination dosage is a step-like function. This ensures high resolution and sharp edge profiles. It may seem irrelevant at first what kind of resist is used in a particular process. There may, however, be some process specific requirements which favor one type or the other. Most importantly, negative resist predominantly produces an *undercut profile*, which means that after development, the resist area in contact with the sample is smaller than the area at the resist surface, Fig. 4.10. This is a consequence of the approximately exponentially decreasing intensity of the illuminating light as it penetrates into the resist. An undercut profile is highly desirable for subsequent metallization steps, in which the resist itself serves as mask. After the metallization, the resist including the metal film on top usually has to be removed in a *lift-off* step, which is bound to fail for resists with an overcut profile, since the metal on the sample and that one on top of the resist are connected. An undercut profile avoids this problem, provided the thicknesses of metal layer and resist are properly selected. [2]

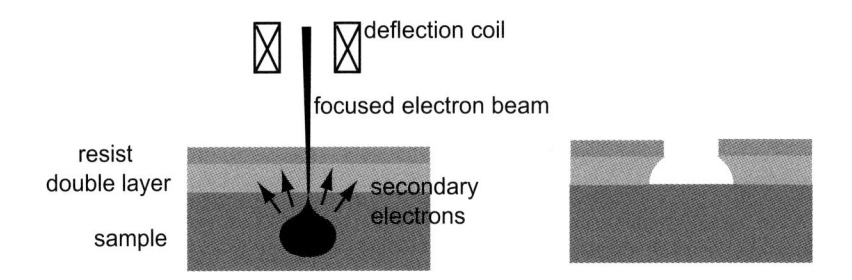

Figure 4.11: Left: a focused electron beam is scanned across the sample surface with a pattern generator that drives the deflection coils, which are part of the electron optics of the electron microscope. The electrons get scattered both elastically and inelastically in the substrate, and secondary electrons are generated, which have a large cross section for resist illumination. The resulting profile of a two-layer electron sensitive resist after illumination is shown to the right.

In principle, the resolution can be increased by using shorter wavelengths. In X-ray lithography, resists are illuminated with wavelengths in the 10 nm regime. While significant progress has been achieved over the past decade, severe technological obstacles have to be overcome before this version of optical lithography can be widely used. Photoelectrons limit the resolution to several 10 nm, and optical components as well as masks are difficult to fabricate, since metals get transparent in the UV. The ultimate limit of such lithographic techniques is set by the resolution of the resists, which contain organic

---

[2]It should be mentioned that techniques exist for achieving undercut profiles with positive resist as well.

4.1 Sample fabrication

polymers. The cross linking of the polymers is enhanced or reduced by the light, which modifies their solubility accordingly. Thus, the resolution cannot become better than the size of the corresponding monomers, which is of the order of 0.5 nm. For feature sizes

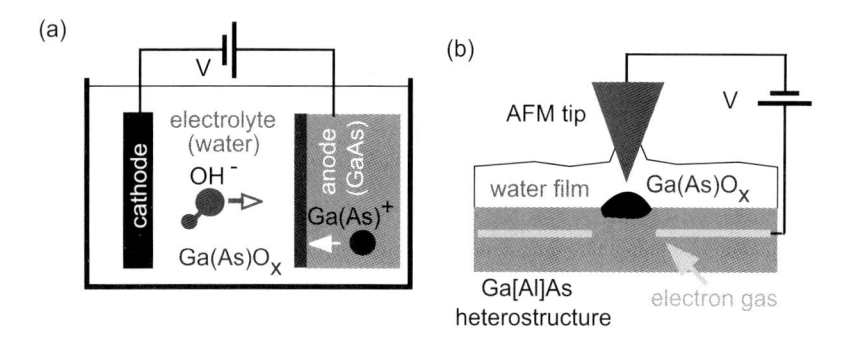

Figure 4.12: (a): scheme for conventional anodic oxidation of GaAs. (b): downscaled version of anodic oxidation, with the conductive tip of an atomic force microscope as the cathode, and the water film on top of the sample as electrolyte.

below $\approx$ 150 nm, electron beam lithography is the current technique of choice.

2. Electron beam lithography
   Instead of light, electrons may be used as well for illuminating resists, which are in this case polymers like PMMA (poly-methyl metacrylate) with a well-defined molecular weight. In a positive resist, the electron beam breaks the bonds between the monomers, and an increased solubility results. In negative resists, on the other hand, the electron beam generates inter-chain cross linking, which deceases the solubility. In that respect, electrons ave a very similar effect as UV light on the resist. A typical experimental setup is shown in Fig. 4.11. A focused electron beam is scanned in a predefined pattern across the sample using deflection coils in the electron optics. In contrast to optical lithography, this is a serial and therefore a slow process. However, structure sizes of 50 nm and even below can be fabricated. Many research groups use electron beam lithography in the lab for all feature sizes below 2 microns or so, because the technique gives very good and reproducible results. One type of electron beam lithography uses a high energy beam of electrons (about 30 keV or larger), which produces extremely small spot sizes of about 1 nm only. However, the illumination resolution is worse than this, since the spatial distribution of secondary electrons backscattered from the substrate actually illuminate the resist, Fig. 4.11. Since the intensity of those electrons drops from the substrate towards the surface of the resist, an undercut profile is intrinsic to this process. The undercut is often enhanced by a two- layer electron beam resist with different dosages.

Direct writing methods

Per definition, such methods do not require resists. Rather, the sample is patterned directly by the exposure. The number of process steps (see Fig. 4.8) is reduced from 5 or 6 to just one.

We briefly discuss two methods.

1. Focused ion beam writing
   The experimental setup resembles the electron writing system, with the electron source replaced by an ion source (e.g., gallium). The ions are implanted in the substrate and localize the electrons in the exposed areas. Highly resistive regions can be defined this way. However, the lateral depletion is rather large, typically above 100 nm. Suitable ion beams can also be used to dope the sample locally.

2. Scanning probe lithography
   As an example of current research activities, we briefly discuss lithography techniques based on scanning probe microscopes (SPMs) [Binnig1982]. Recently, tremendous progress has been made in this respect. Since SPMs achieve atomic resolution, they are highly promising tools for achieving a further, significant size reduction. Meanwhile, SPMs have been used in a wide variety of operational modes in order to modify surfaces [Marrian1993]. Moving single atoms with an SPM tip [Eigler1990], material deposition from the tip on the substrate [Mamin1990], has been demonstrated experimentally, for example. Amazing nanodevices can also be fabricated by scratching [Irmer1998]. Another widely investigated technique is local oxidation of the substrate. [Dagata1990] oxidized a variety of substrates locally by applying a negative voltage to the tip of an SPM with respect to the grounded substrate. Local oxidation with an atomic force microscope has also been used to pattern the electron gas in Ga[Al]As heterostructures directly [Held1998]. Anodic oxidation is a standard process to oxidize surfaces of metals and semiconductors. The setup for local oxidation with an AFM is essentially identical (Fig. 4.12). Here, the water film forming under ambient conditions on top of the substrate provides the electrolyte. A conductive AFM tip acts as cathode, while the chip to be nanostructured is grounded. As a result, the sample surface oxidizes in close vicinity to the AFM tip. The 2DEG is depleted underneath the oxide lines in shallow HEMT structures, provided the distance between 2DEG and sample surface is smaller than about 50 nm. The underlying mechanism can be understood in a simple picture: as the cap layer is oxidized, the semiconductor surface gets closer to the 2DEG, while the surface area, and thus the number of surface states, is slightly increased. In samples with the 2DEG so close to the surface, only $\approx 10\%$ of the donor electrons from the doping layer go into the 2DEG, while the remaining 90% fill the surface states. A small reduction of the distance between surface and the 2DEG changes the internal electric fields and can lead to depletion. We have seen several examples of Ga[Al]As-HEMTs patterned by this technique in the introduction.

Etching

An important technique of transferring the resist pattern into the sample is etching. Patterned resists can be used as etch masks, provided the etchant is sufficiently selective. We distinguish between dry etching and wet chemical etching.

1. Dry etching
   The setup for dry etching techniques consists of a vacuum chamber with two electrodes

4.1 Sample fabrication

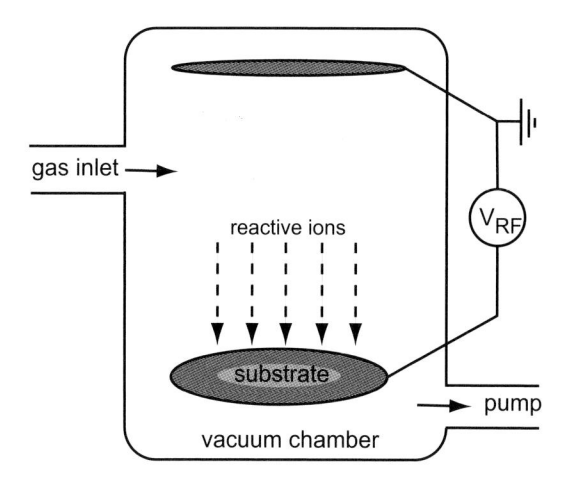

Figure 4.13: Scheme of a vacuum chamber for reactive ion etching.

at the top and the bottom. The sample is placed at the bottom, which may be the anode or the cathode, depending on the process. A gas discharge is ignited, and the ions of the etch gas hit the sample (Fig. 4.13). One speaks of plasma etching if the reaction is purely chemical. Oxygen plasma etching is often used to remove resist layers. The low energy ions avoid damage of the semiconductor and metal components of the sample. A purely physical technique, on the other hand, is ion etching. Here, suitably selected ions are generated and strongly accelerated towards the sample. The physical impact removes sample atoms. Here, resists may serve as masks for a limited time. Radiation damages in the sample, combined with the required high vacuum and the large rate of material deposition at the walls, make this a rather unusual technique. Widely used, however, is *reactive ion etching*. Here, both the physical and the chemical aspects of the ionic exposure are important. A very convenient side product in this kind of etching a polymer formation at the etched walls, which prevent lateral removal of material. As a consequence, very steep and deep grooves can be etched.

2. Wet chemical etching
Wet chemical etching means immersing the sample in a suited etchant solution. In contrast to metals, the majority of the common semiconductors are not attacked by pure acids. Therefore, the etch typically consists of a mixture of an oxidizer, such as H_2O_2, an acid, like HCl, and water. H_2O_2 oxidizes the semiconductor, while the acid removes the freshly formed oxide. The oxidation and etch rates depend on the etch composition as well as on the crystal direction. The resulting edge profile can thus be tuned accurately. For many purposes, an overcut edge profile is desirable, since often, thin metal layers have to be deposited on the surface later on. A metal layer thinner than the etched depth may get disconnected across an etched step with undercut profile.

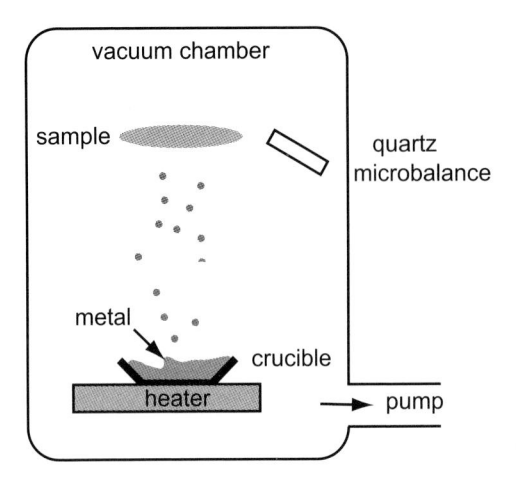

Figure 4.14: Scheme of an evaporation system for metallizations.

4.1.4 Metallization

By metallization, we mean the deposition of metal films on the semiconductor surface. This is usually done by evaporation of the metal in a vacuum chamber. The metals are molten (or sublimed, respectively) in a crucible made of tungsten of carbon, which can be done by heating the crucible with a current, or by focussing an electron beam onto the metal, see Fig. 4.14. At sufficiently large temperature, the metal vapor pressure is so high that a metal film grows at the exposed surfaces with a rate of the order of a nanometer per second. The film thickness is monitored by an oscillating quartz plate. As the metal gets deposited on the quartz, its resonance frequency gets smaller. This effect can be calibrated, and the film thickness can be measured with high accuracy. For lift-off processes, the film thickness should be smaller than the thickness of the resist, for obvious reasons. Typical metallization layers measure thicknesses between 20 nm and a few microns.

Of particular importance for the fabrication of nanostructures is the so-called *angle evaporation technique* [Dolan1977], because feature sizes below the lithographic resolution can be made this way. The trick is to evaporate successive layers of metals from different angles and use the resist as a shadow mask. The technique is illustrated in Fig. 4.15. Overlap areas as small 30 nm by 30 nm can be fabricated routinely by angle evaporation.

As pointed out in the previous chapter, metal-semiconductor interfaces form Schottky barriers for the vast majority of material combinations. In order to obtain an Ohmic contact, a suitable metal film is evaporated and afterwards alloyed into the semiconductor. "Suitable" means that the Schottky barrier should be small, the metal should have a low melting point and should act as a dopant in the semiconductor. For GaAs, the $Au_{0.88}Ge_{0.12}$[3] eutectic alloy is a standard Ohmic contact material. The Schottky barrier of the $Au_{0.88}Ge_{0.12} - GaAs$ system is only 0.3 eV, In addition, eutectic AuGe has a melting point of 360 °C, and already at

---

[3]The numbers here mean the weight fraction.

4.1 Sample fabrication

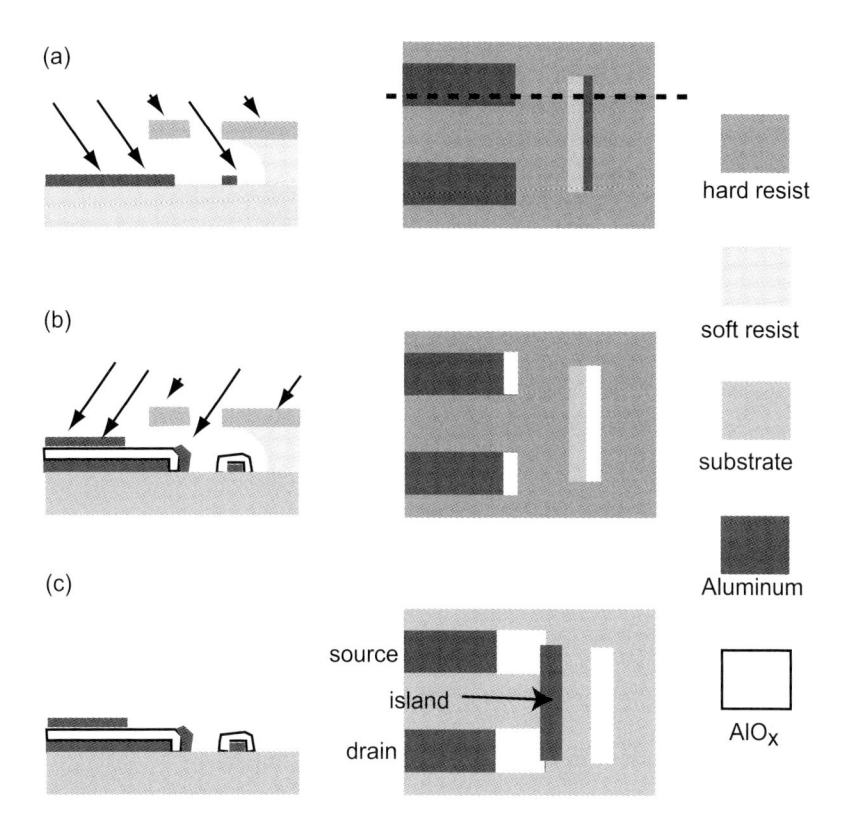

Figure 4.15: Angle evaporation. The right column shows a top view of a sample section after illumination by electron beam lithography, development, and subsequent evaporation of aluminum under a certain angle. To the left, a cross sections of the layers along the dashed line in the right figure is shown. A layer of electron beam resist with low dosage is covered by a resist with a higher dosage. This leads to cage formation as an extreme version of an undercut profile. The upper resist layer is free-standing over a certain area. In (b), the Al gets oxidized, and a second Al layer is evaporated on top at a different angle, as indicated by the arrows. A sandwich structure with small overlap areas results, which can be below the pattern sizes in the resist mask. In (c), the resulting structure, a small Al island coupled to two leads via small-area tunnel barriers, is shown after the resist layers have been removed. Such islands will be investigated further in chapter 9.

420 °C, it begins to alloy into GaAs. Ge atoms diffuse into the Ga and act as donors. This diffusion can be enhanced by adding a small fraction of Ni to the alloy. The resistivity of such a contact is of the order of 10^{-6} Ωm. The low process temperatures are important, since they ensure Ohmic contact formation well below critical temperatures for other processes, such as Si dopant migration in GaAs, which would damage the modulation doping profile.

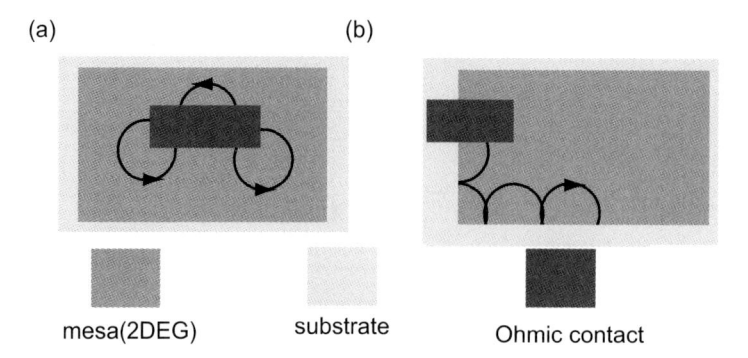

Figure 4.16: In strong magnetic fields, electrons move in cyclotron orbits and thus remain localized close to the Ohmic contact (a), unless the contact crosses the mesa edge (b).

Finally, one small technical note should be made here. Since in many cases, mesoscopic transport experiments involve application of strong magnetic fields, it is very important that the Ohmic contacts extend across the mesa edge. Otherwise, the contact resistance increases sharply in strong magnetic fields, since the electrons move in cyclotron orbits in the electron gas, and localize within a small area around the contact, see Fig. 4.16. This problem arises in particular in two-dimensional electron gases and at cyclotron radii below the mean free path.

4.1.5 Bonding

Once the sample is patterned and everything looks good, the last step in the fabrication process is to mount the sample into a chip carrier and connect wires to the Ohmic contacts and to the gate electrodes. Two versions of this so-called *bonding* are widely used. In ball bonding, the tip of a gold wire is molten locally by a discharge or a flame, and is pressed against a bond pad defined on the sample surface. The sample is heated to a moderate temperature, say 200 °C, and a connection forms via thermo - compression. The second scheme is known as wedge bonding, see Fig. 4.17. Here, the wire is pressed against the bond pad and rubbed across it with an ultrasonic frequency. The friction force is sufficient to locally melt the materials, and an alloy is formed which holds the wire in place. After the second bond, the wire is pulled and breaks at the weakest point, which is right after the position of the wedge.

4.2 Elements of cryogenics

Helium is the only element that remains liquid when cooled to the lowest possible temperatures (well below 1 mK) under atmospheric pressure. It is therefore the prime candidate as a refrigeration medium for temperatures below the condensation temperature of nitrogen (77 K). The vast majority of mesoscopic transport experiments are performed in this tempera-

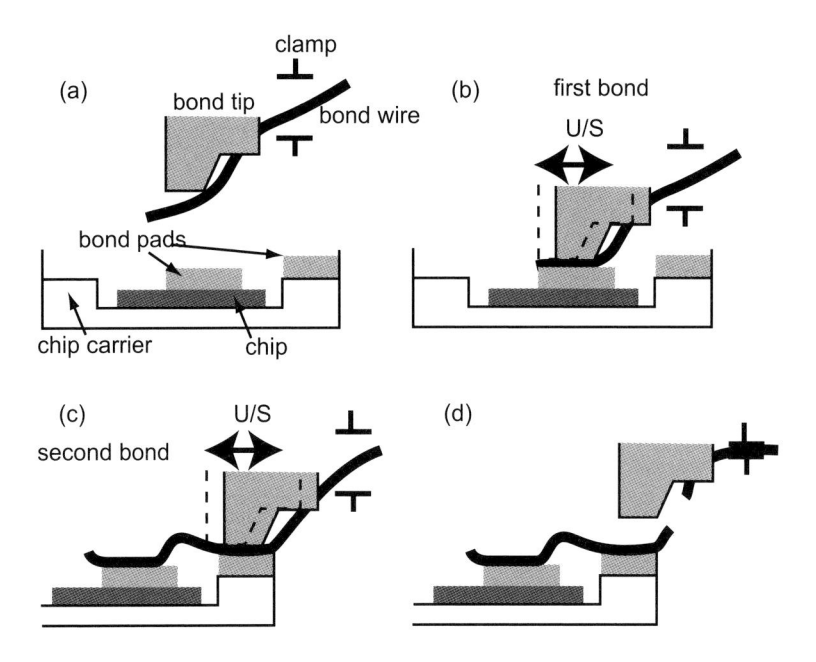

Figure 4.17: Wedge bonding. (a): the bond tip containing the wire is positioned on top of the bond pad on the sample. (b): the wire is wedged onto the bond pad, and the tip is retracted with the wire clamp open. (c): the second bond on the pad integrated in the chip carrier. Here, the tip retracts with the clamp closed, and the wire breaks behind the second wedge.

ture range. The latent heat that has to be paid when liquid helium is evaporated is the cooling power made use of in helium cryogenics. Continuous evaporation of liquid is possible by pumping off the vapor pressure. Therefore, we will look a the properties of liquid helium (LHe), as well as cryostats, the devices used to establish low temperatures.

4.2.1 Properties of liquid helium

The physics of LHe is extremely interesting and rich, and experimentalists working on transport in nanostructures will almost inevitably get in contact with its unusual properties.

Helium comes in two isotopes, the boson ^{4}He and the fermion ^{3}He. The mono-isotopic liquids have therefore very few properties in common. As a liquid is cooled, kinetic energy is taken away from the atoms. At the condensation temperature, the attractive interatomic van der Waals forces start to dominate in any liquid other than LHe, and the crystallization sets in. He is the only element for which the van der Waals force is smaller than the kinetic energy of the atoms due to zero-point fluctuations. The van der Waals forces in He are particularly weak since the atoms have no dipole moment. On the other hand, the zero-point fluctuation energy is particularly large, due to the small atomic mass. Only by applying a pressure above ≈ 30 bars, the atoms are pressed sufficiently close together such that crystallization sets in.

So much as for the common properties of ^{3}He and ^{4}He. We now look at some properties of the pure isotopic liquids, before we turn to the interesting issue of ^{3}He/^{4}He mixtures.

Some properties of pure ^{4}He.

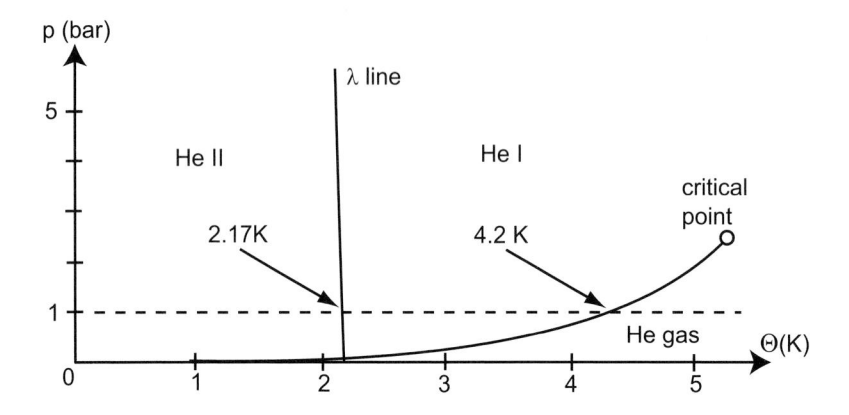

Figure 4.18: Phase diagram of the ^{4}He.

Fig. 4.18 shows the phase diagram of ^{4}He. Under atmospheric pressure, it liquifies at $\Theta = 4.2$ K. The density of the liquid $\rho(L^4He) = 125$ kg/m^3. The vapor pressure drops approximately exponentially as LHe gets colder, and reaches 1 mbar at $\Theta = 1.2$ K. As we cool the liquid, we cross the λ *line* at some temperature, which for atmospheric pressure happens at $\Theta_\lambda = 2.17$ K, also known as the λ point. The λ point got its name from the specific heat as a function of Θ around this transition, a function that looks like this omnipresent Greek letter. For $\Theta > 2.17$ K, ^{4}He behaves just like any ordinary liquid. As we lower Θ and cross the λ point, ^{4}He undergoes a phase transition and develops highly remarkable properties. L^{4}He in this phase is often referred to as HeII. In fact, the phase transition at the λ point can be modelled as a Bose-Einstein condensation, i.e., the condensation of a boson gas. Within such a model, the ^{4}He above the λ point is described as a gas, which is not a bad approximation, considering the weak interactions. At $\Theta = 0$, on the other hand, all atoms of HeII are in the ground state. At higher temperatures, the energy levels in a Bose-Einstein condensate (BEC) are occupied according to the Bose-Einstein distribution function

$$f_{BE}(E, \Theta) = \frac{1}{e^{(E-\mu)/k_B \Theta} - 1}$$

A pure BEC, however, cannot explain the observed behavior of HeII. Rather, [London1938] proposed a two-liquid model, which treats HeII as a mixture of a normal fluid and a superfluid, which interpenetrate on a microscopic length scale, similar to the electronic state in a type II superconductor. The normal fluid behaves just like ^{4}He above the λ point. In particular, it has a non-vanishing entropy and viscosity. The superfluid, on the other hand, has zero entropy and viscosity, which means, for example, that there is no flow resistivity. Furthermore the

4.2 Elements of cryogenics

thermal conductivity of the superfluid is infinitely large. How the composition of HeII changes with temperature has been measured by [Andronikashvili1946]. In this seminal experiment, a torsion pendulum made of a stack of thin disks was immersed in HeII, and the damping of the oscillation was measured as a function of temperature. Since the normal fluid is viscous, it adds to the moment of inertia of the system via the law of Hagen-Poiseuille, while the superfluid does not. The measured composition of HeII is shown schematically in Fig. 4.19. As the temperature is lowered, the normal fluid fraction rapidly vanishes and an almost pure superfluid remains for $\Theta \lesssim 0.7$ K. This two-component mixture has some unique properties

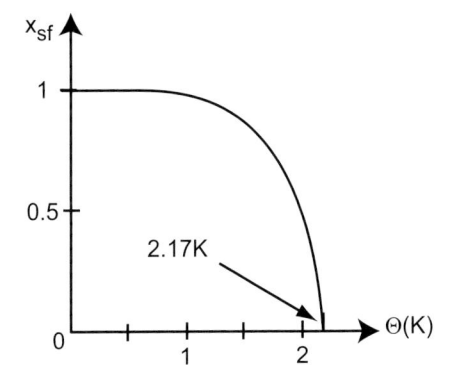

Figure 4.19: Superfluid fraction x_{sf} of HeII as a function of temperature. After [Andronikashvili1946].

we should know, in order to appreciate its behavior in cryogenic equipment.

1. Absence of bubbling.
   If we heat a conventional liquid, it starts bubbling, since the liquid evaporates at some random spot, and the gas bubble rises to the surface. In HeII, the thermal conductivity is very large, and evaporation takes place at the surface only. Hence, HeII is perfectly quiet, even if it boils off. In a simple picture, we can understand the extremely high thermal conductivity as follows. Imagine we connect heat reservoirs to both ends of a tube filled with HeII. At the end with higher temperature, superfluid is transformed into normal fluid, with a final ratio in accordance with Fig. 4.19. The heat is transferred to the low-temperature end by normal fluid convection. Here, the normal fluid is re-transformed into superfluid. Since the superfluid carries no heat (its entropy is zero), all the heat is thereby absorbed by the heat sink. The heat transfer is therefore very efficient. The heat conductivity is further increased by the extremely low viscosity of HeII, which means there is vanishingly small friction during the convection process.

2. HeII osmosis.
   Consider two chambers filled with HeII, connected to each other by a *superleak*, i.e., a connection only permeable for superfluid helium, see Fig. 4.20. Such connections can be made by extremely fine capillaries, or by tubes stuffed with powder. This setup immediately reminds us of an osmotic pressure cell, with the semipermeable membrane being

the superleak, the solvent being the superfluid, and the normal fluid component starring as the solute. Recall that in osmosis, the solute can be thought of a gas, and that the osmotic pressure evolves due to the tendency of the solvent to equalize the concentrations in both chambers. As we heat HeII in one chamber, the fraction of normal fluid increases, and superfluid will enter this chamber, in order to dilute it. Consequently, a pressure difference is built up. In equilibrium, the hydrostatic pressure will compensate the osmotic pressure, and the surface positions in the two chambers will differ by Δh.

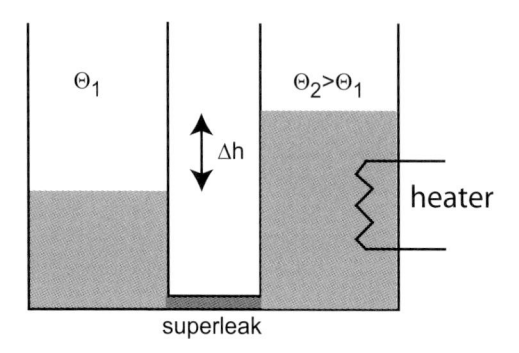

Figure 4.20: Sketch of a HeII osmotic cell.

3. Superfluid film creeping.
   HeII tends to creep over any wall of reasonable height, as long as its temperature stays below the λ point. Therefore, containers filled with HeII to different heights will equilibrate their surface levels, see Fig. 4.21. This effect has it origin in the extreme adhesion of HeII to surfaces. Within the framework of liquid-solid interfaces, this is known as "complete wetting". Since the shape of the liquid surface is determined by the condition that the tangential forces vanishes, this effect occurs for $-\sigma_{ls} > \sigma_{gl}$, where $\sigma_{ls}(\sigma_{gl})$ denote the liquid-solid (gas-liquid) interface tension.

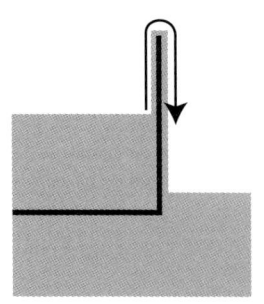

Figure 4.21: Superfluid film creeping across a wall.

4.2 Elements of cryogenics

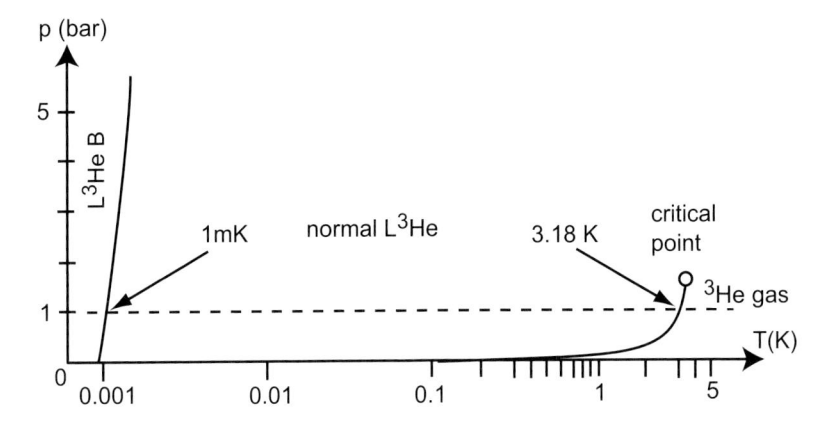

Figure 4.22: Phase diagram of ^{3}He.

Some properties of pure ^{3}He.

In Fig. 4.22, the phase diagram of ^{3}He is sketched. For our purposes, the additional phases occurring at extremely low temperatures below 2 mK are irrelevant.[4] L ^{3}He has a density of $\rho\,_{^3\mathrm{He}} = 59$ kg/m^3. Under atmospheric pressure, it liquifies at $\Theta = 3.19$ K. This boiling point is about 1 K below that one of ^{4}He, which can be easily understood, since its mass is smaller, and thus the atoms have a larger average velocity at the same temperature. Consequently, the vapor pressure is also higher at identical temperatures. It drops to 10^{-3} mbars at about $\Theta = 270$ mK. ^{3}He atoms are fermions, and the liquid can be approximated by a Fermi gas, with many analogies to electron gases.

Question 4.1: Calculate the Fermi energy of ^{3}He!

Within the Fermi liquid picture, we can imagine that each ^{3}He is surrounded by a screening cloud, which results in quasiparticles with an effective mass given by the interactions. At atmospheric pressure, $m^*(^3\mathrm{He}) \approx 3m(^3\mathrm{He})$. For practical cryogenic purposes, ^{3}He behaves as an ordinary liquid.

A further important point concerning ^{3}He is its near-to complete natural absence on earth. It can be generated by nuclear reactions, and is consequently extremely expensive. Therefore, all ^{3}He cryostats keep it in a closed cycle.

The ^{3}He/^{4}He mixture.

Let us first look at the phase diagram of this mixture, Fig. 4.23. For $\Theta > 860$ mK, nothing spectacular happens. The main effect of the ^{3}He is to reduce the λ point of the homogeneous mixture. Below the λ - line, ^{3}He dissolved in HeII can be just thought of an additional fraction

---

[4]For $\Theta < 2$ mK, the ^{3}He atoms form Cooper pars and undergo a Bose-Einstein condensation into superfluid ^3He. Further phases exist at high pressures.

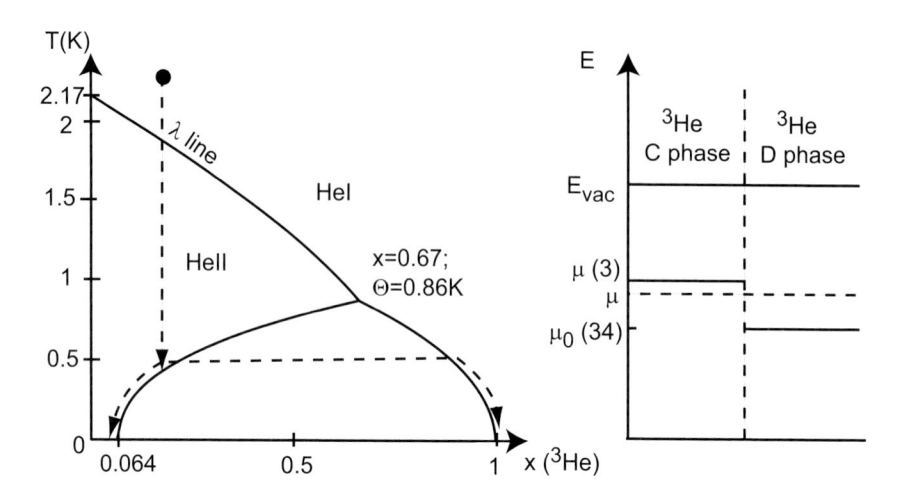

Figure 4.23: Left: phase diagram of the ^{3}He/^{4}He mixture vs. ^{3}He concentration x and temperature Θ. The tricritical point is at $x = 0.67$ and $\Theta = 860$ mK. At lower temperatures, the mixture segregates into a ^{3}He-concentrated phase (the C phase) and a ^{3}He dilute (D) phase. Right: sketch of the chemical potential of the two phases at $\Theta = 0$.

of the normal fluid component. For temperatures below 860 mK, a remarkable phase separation into a ^{3}He - poor phase (called the *dilute phase - D phase* in the following) and a ^{3}He - rich phase (the *concentrated - C phase*) takes place. At these temperatures, the pure HeII is almost completely superfluid, and the dissolved ^{3}He forms a normal fluid component.

A qualitative understanding of the phase separation can be obtained by recalling that ^{3}He is a Fermi liquid, while ^{4}He in this regime is a Bose-Einstein condensate. The ^{3}He dissolved in ^{4}He can be thought of a dilute Fermi gas with an effective mass given by the interaction between the ^{3}He atoms and the surrounding ^{4}He, which is m*(^{3}He in ^{4}He) ≈ 2.4m(^{3}He). Since superfluid ^{4}He has zero viscosity, the ^{3}He atoms can move around without friction, once the ^{3}He $-^4$He interaction is included in the effective mass. L^{3}He can be regarded as a Fermi gas as well. We just have to establish the conditions for which the chemical potentials of the C phase and the D phase are identical. Here, the superfluid ^{4}He plays no role, as all these atoms are in the ground state. The problem somewhat resembles the alignment of chemical potentials at interfaces discussed in the previous chapter. Here, the common energy level is again the vacuum level, i.e., the energy of a ^{3}He atom at rest in the vacuum. The chemical potential $\mu(3)$ of the C phase is somewhat higher than that one ($\mu_0(34)$) of a single ^{3}He atom in ^{4}He, which can be understood by the fact that the (attractive) van der Waals forces are slightly larger in ^{4}He, since the average separation of the atoms is smaller. Hence, ^{3}He atoms will go into ^{4}He until the chemical potentials have aligned. This is the reason why even at $\Theta = 0$, the D phase contains still 6.4% of ^{3}He atoms. Note that is is energetically unfavorable for ^{4}He atoms to reside in the C phase.

4.2 Elements of cryogenics

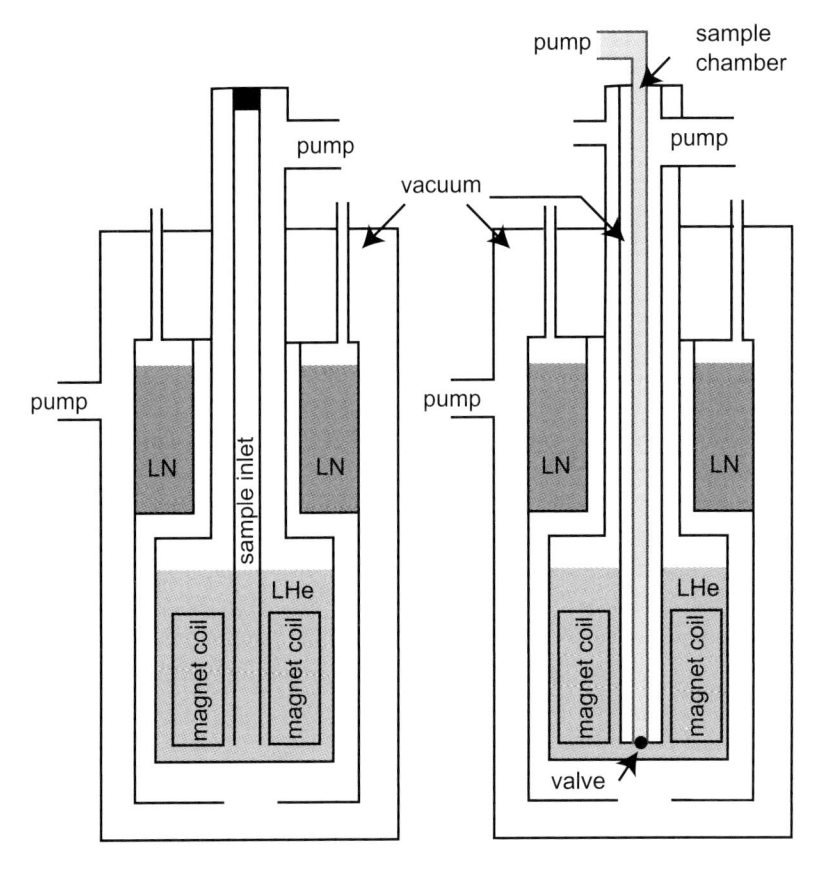

Figure 4.24: Sketch of a ^{4}He bath cryostat (left) and a ^{4}He gas flow cryostat (right).

4.2.2 Helium cryostats

Helium cryostats can be classified according to the kind of helium mixture for which they are designed. Occasionally, liquid nitrogen cryostats are used as well, for temperatures between 77 K and larger. However, from our discussion of the ^{4}He cryostat, their design should be pretty obvious. We begin with the "high-temperature" helium cryostats.

^{4}He cryostats

Helium has a small latent heat, which means it boils off easily. Therefore, the LHe cryostat has to be thermally decoupled from the environment. This is achieved by several means. Separating the He vessel from the outer world by a vacuum avoids heating via convection. Second, the LHe container is made of a material with a poor thermal conductivity, such as glass or stainless steel. Finally, the thermal radiation from the environment is shielded by surrounding the LHe vessel with liquid nitrogen, in order to reduce the temperature of the blackbody radiation that hits the He dewar. Alternatively, it is possible to wrap the dewar in

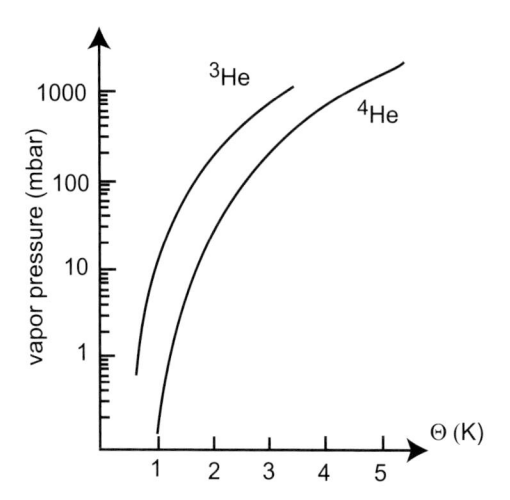

Figure 4.25: Vapor pressure of ^{4}He and ^{4}He.

"super-insulating" foil, which is a multilayer of insulating foil, where each layer is coated with a metal on one side. Examples of L^{4}He cryostats are shown in Fig. 4.24. In a *bath cryostat*, the sample is simply immersed in the LHe. The liquid, and with it the sample, can be cooled by pumping away the He vapor. This causes LHe to evaporate, which costs the latent heat and thus cools the liquid. The pumping speed and the incoming heat flux essentially determine the lowest possible temperature. To be somewhat more quantitative, recall the Clausius-Clapeyron equation, which gives the slope of the vaporization as a function of temperature as

$$\frac{dp}{d\Theta} = \frac{L}{\Theta \times (V_{gas} - V_{liquid})} \quad (4.1)$$

Here, the latent heat per atom is given by $L = \Theta \times (S_{gas} - S_{liquid})$, where S denotes the atomic entropy of the gas and the liquid, respectively. We have further assumed here that L does not depend on temperature which is a reasonable approximation for LHe. If we neglect the volume of the liquid (for LHe at 4.2 K, it is a factor of 750 smaller than the volume of the vapor), and model the gas as an ideal gas, $pV = nk_B\Theta$, it is found by integration of (4.1) that the vapor pressure p drops exponentially as Θ decreases, i.e.,

$$p(\Theta) = p_0 exp(-\frac{L}{k_B\Theta}) \quad (4.2)$$

The cooling power P is simply the latent heat taken from the liquid per evaporated atom, multiplied by the number of atoms evaporated per time.

$$P = \frac{dn}{dt}L \quad (4.3)$$

Since $\frac{dn}{dt}$ is determined by the pumping speed $\frac{dV}{dT}$ of the pump used via

$$\frac{dn}{dt} = \frac{1}{m_{He}}\frac{dM}{dt} = \frac{1}{m_{He}}\rho\frac{dV}{dt} = \frac{p(\Theta)}{k_B\Theta}\frac{dV}{dt}$$

4.2 Elements of cryogenics

the cooling power drops exponentially as Θ decreases.

Question 4.2: The latent heat of ^{4}He is 88 J/mol. What is the cooling power at $\Theta = 1.2$ K when a pump with a pumping speed of 200 m^3/h is used?

The steady state is reached when the cooling power equals the heat load of the LHe. With a conventional pump with a pumping speed of, say 10 m^3/h, a temperature of about 1.2 K can be reached. Lower temperatures somewhat below 1K are possible, but require very powerful pumps. Therefore, if this temperature range is needed, people usually prefer a ^{3}He cryostat or a dilution refrigerator. Sometimes, temperatures *above* 4.2 K are required. The device of choice is then a *gas flow cryostat*. Here, the sample sits in a flow of cold helium gas, which enters the sample chamber via a needle valve. The sample chamber itself is thermally decoupled from the LHe by an additional vacuum chamber. The sample temperature can be adjusted by controlling the power applied to a heater for the gas, in combination with the gas flow rate. Continuous variation of the temperature between 1.2 K and room temperature is possible in gas flow cryostats. Many cryostats are equipped with a superconducting magnet, which is cooled below the critical temperature by the LHe. Most of these magnets are made from Nb alloys, since they have very large critical magnetic fields. A typical magnet is able to generate magnetic fields of the order of 10 T, although 20 T are commercially available. Experiments at higher magnetic fields can be carried out at some national and international high magnetic field laboratories.

^{3}He cryostats

Below 1K, the vapor pressure of ^{3}He is much higher than that one of ^{4}He. Therefore, temperatures down to about 270 mK can be reached easily by pumping L ^{3}He. In a ^{3}He cryostat, the ^{3}He is isolated from the ^{4}He precooling stage by an inner vacuum chamber (Fig. 4.26). As mentioned already, the ^{3}He is kept in a closed cycle. The pumped ^{3}He gas is collected in a storage vessel. Measurements can be performed until all the L^{3}He has been pumped, which results in measurements intervals up to one day. The ^{3}He gas can be condensed by a small, pumped ^{4}He pot, which is connected to the ^{4}He bath via a needle valve, such that its temperature stays well below 3.2 K, the condensation temperature of ^{3}He. Some cryostats are equipped with a continuous flow option. Here, the pumped ^{3}He is immediately recondensed. For the price of a somewhat higher base temperature due to the additional heat load, the measurement period becomes unlimited this way.

^{3}He/^{4}He dilution refrigerators

This type of cryostat uses the special properties of ^{3}He/^{4}He mixtures in a clever way, and makes possible temperatures as low as 1 mK and even below. Since the D phase of the mixture is approximately a dilute Fermi gas, it can be thought of the ^{3}He vapor of the C phase, with a significant vapor pressure even at $\Theta = 0$. Since the C phase has a smaller density, the "liquid" will float on top of the "gas", though. Pumping the ^{3}He atoms out of the D phase will surely cause ^{3}He from the C phase to evaporate, which pulls the corresponding effective latent heat out of the mixture. This is the cooling mechanism in a dilution refrigerator as

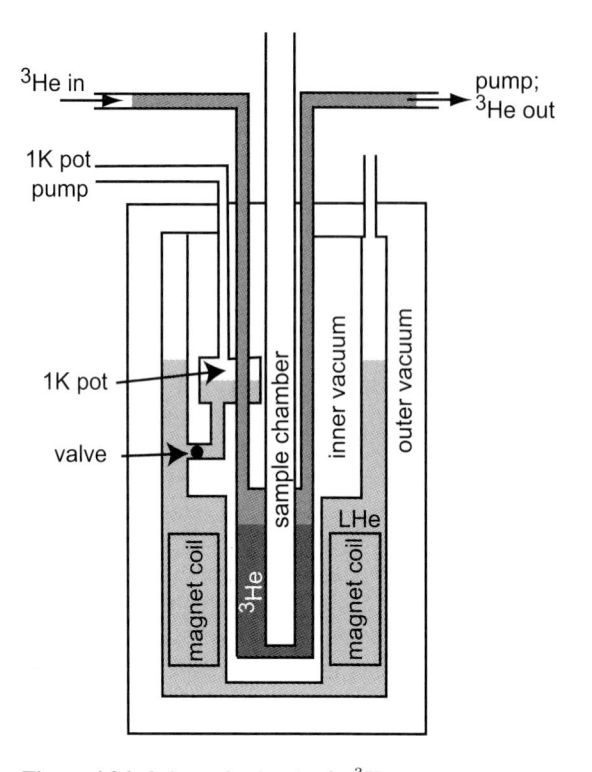

Figure 4.26: Schematic sketch of a ^{3}He cryostat.

sketched in Fig. 4.27. The mixture rests in the *mixing chamber*. The D phase is connected through a tube with the *still*, a pot that gets heated to about 600 mK. At this temperature, the vapor pressure of ^{3}He is significant, while that one of ^{4}He is negligible. The still therefore effectively distills ^{3}He from the D phase. The missing ^{3}He in the D phase gets delivered by "evaporation" across the C-D phase boundary, and the mixture in the mixing chamber gets colder. Usually, the evaporated ^{3}He is recondensed into the mixing chamber by a pot filled with ^{4}He, that gets pumped temperatures below the condensation temperature of ^{3}He. This is the "1K pot". The freshly condensed ^{3}He has, of course, still a much higher temperature than the mixture. The heat flow in the mixing chamber is therefore optimized by a flow impedance in the condenser line. In addition, the outgoing gas at the still temperature is used to further precool the condensed ^{3}He via heat exchangers. Virtually all mesoscopic transport experiments below 270 mK have been carried out by thermally coupling the sample to the mixing chamber, either by immersing it directly, or by mounting it in the vacuum at the outside wall of the mixing chamber.

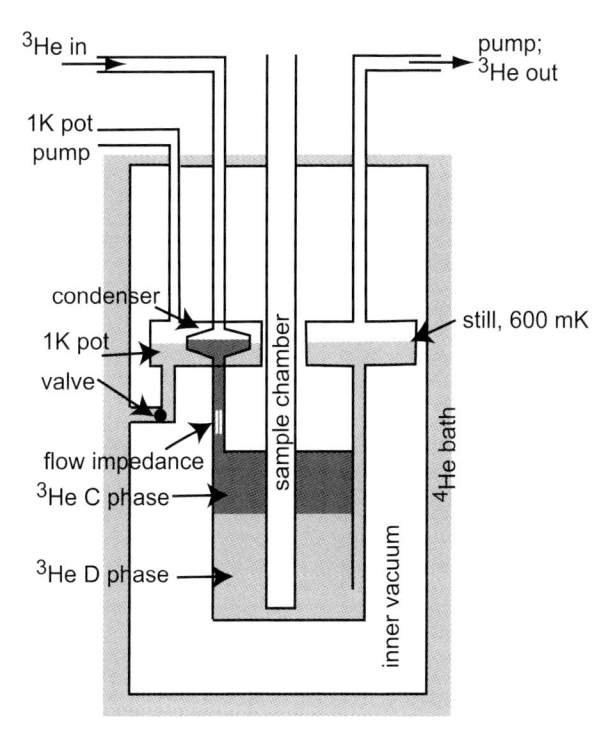

Figure 4.27: Essential components of a $^3He/^4He$ dilution refrigerator.

4.3 Electronic measurements on nanostructures

Measuring the resistances and conductances of a sample requires the application of currents and/or voltages, as well as the detection of voltage drops and/or currents, respectively. Conceptually, these measurements are very simple. The greatest efforts in practice are usually related to the reduction of the electronic noise level. This is done by avoiding ground loops, filtering, and choosing the right cables, among other important issues. As in the previous section, we are not that much interested in these technical details. This topic has been dealt with in great detail in excellent books (see the reference section at the end of this chapter). Our goal here is to present in brief some basic setups, just enough for the reader to know what type of setup has been used in the experiments to be discussed. In the previous section, we have seen that the cryostats available set some limitations to the temperature range. Likewise, the measurement setup limits the physical quantities, as well as their ranges, that can be measured. The present section, together with the previous one, should put us in a position to judge why a particular experimental setup has been used, and how it affects the parameter ranges. We begin by showing how the samples are actually mounted in the low-temperature environment, before we discuss the most important electronic measurement setups.

4.3.1 Sample holders

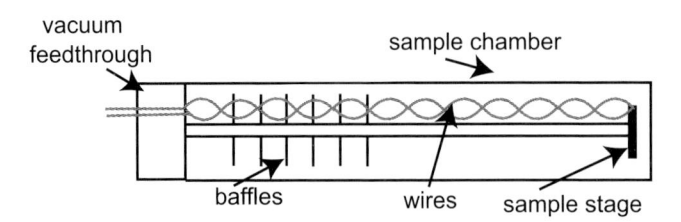

Figure 4.28: Sketch of a sample holder used to stick a specimen in the sample space of a cryostat.

A sample holder contains the sample in an appropriate way, and is mounted in the sample space of the cryostat. Its basic components are sketched in Fig. 4.28. The sample is mounted in some kind of carrier, which is placed inside the cryostat, in the center of the magnetic field. Cables are brought into the sample space via a vacuum feedthrough at the top end. Typically, the wires run in twisted pairs, which reduces the currents induced by the magnetic field due to vibrations, since the magnetic flux through adjacent loops points in opposite directions. Furthermore, the sample holder contains baffles, i.e., polished metal plates which reflect the thermal radiation from the top. Some sample holder are equipped with a rotator, which permits the sample to be tilted with respect to the magnetic field (which points in the vertical direction in most cryostats).

4.3.2 Application and detection of electronic signals

General considerations

For many experiments, measuring in a low temperature environment only makes sense when the electric signals are kept sufficiently small. Suppose, for example, we plan to investigate the transmission properties of a tunnel barrier. The low temperature reduces the thermal smearing of the Fermi function, which corresponds to an energy scale of $\delta E = 3.52 k_B \Theta \approx 300\,\mu\text{eV/K} \cdot \Theta$. Therefore, the voltage drop across the barrier at, e.g., $\Theta = 1$ K, should be small compared to 300 μV. For larger voltage drops, the temperature does no longer determine the energy resolution.

Measurements can be performed AC or DC. AC measurements have the advantage that a lock-in amplifier can be used, a device that selectively detects signals with the source frequency, within a narrow band width. In addition, phase-sensitive measurements are possible, such that, e.g., capacitance measurements can be performed by measuring the voltage drop with a phase shift of $\pi/2$ with respect to the source signal. Although the frequency selection greatly reduces the noise, it is not always best to use an AC signal. For example, imagine the sample has a very large resistance, such that the capacitances, which are always present in the leads, cause significant phase shifts. This makes it hard to determine the resistive part of the impedance. Also, theoretical results are often obtained for DC transport.

Furthermore, it has to be clearly distinguished between the resistance and the resistivity (conductance and conductivity, respectively). The plain result of, say, applying a current I and

4.3 Electronic measurements on nanostructures

measuring the voltage drop ΔV is the resistance $R = \Delta V/I$. If the sample is macroscopic, we can assume that the voltage drops homogeneously in between the voltage probes, and we can translate the resistance into a resistivity, an intrinsic property of the sample, by taking the sample geometry into account. This is no longer true in mesoscopic samples. Here, the measurement does not average over a large volume of randomly distributed scatterers, and the sample simply does not have a resistivity.

Voltage and current sources

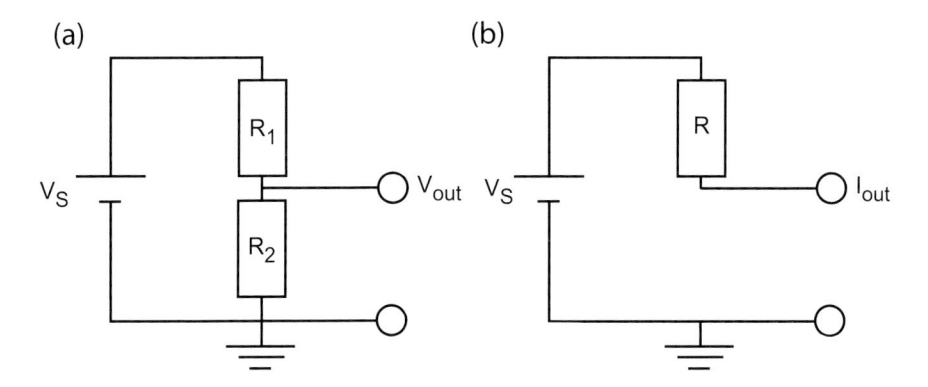

Figure 4.29: A voltage divider (left) and a voltage to current conversion (right).

High quality commercial voltage sources typically provide voltages in the regime of Volts, with an accuracy of, say 10^{-6}. Hence, some conversion to smaller voltages, or to a small current, is often necessary. This is done by a voltage divider, or a voltage-to-current conversion, respectively, see Fig. 4.29. The voltage divider simply consists of two resistors in series connected to the output voltage V_S of the commercial voltage source. The potential in between the two resistors is applied to the sample with respect to ground. This voltage is given by

$$V_{out} = V_S \frac{R_2}{R_1 + R_2}$$

In order to divide V_S by a few orders of magnitude, R_1 must be much larger than R_2. This immediately implies an experimental limitation, since R_1 adds to the effective internal resistance of the voltage source. Connecting a sample with a resistance of R_s, causes the applied voltage to drop to

$$V_{out} = \frac{R_2}{R_1 + R_2 + \frac{R_1 R_2}{R_s}}$$

The circuit to the left in Fig. 4.29 is only a good voltage source for $R_s \gg R_1$. R_1 should be chosen as small as possible. The required output voltage then determines R_2. These resistors cannot be arbitrarily small, however, since a minimum current of $I = V_S/(R_1 + R_2)$ must be provided by the voltage source. Hence, only samples of high resistances should be voltage-biased with such a setup.

An analog consideration holds for current biasing the sample, see the circuit of Fig. 4.29 (b). The current I_in is simply given by $I_in = V_S/R$. The setup is a good current source only if the sample resistance is small compared to the conversion resistance. On the other hand, the available voltage sources, the noise, and the minimum currents needed limits R to $R \lesssim 100$ mΩ. Consequently, samples with low resistances should be current biased.

Question 4.3: Calculate how the sample resistance modifies the current in the setup of Fig. 4.29 (b)!

Signal detectors

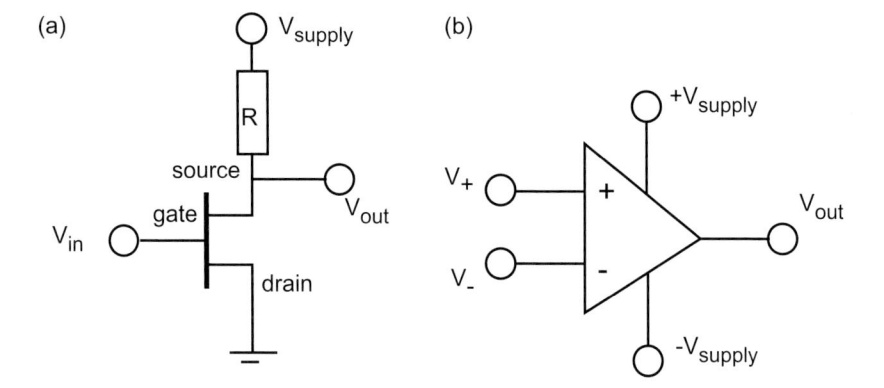

Figure 4.30: (a): scheme of a field effect transistor circuit used to amplify the input voltage V_{in}. (b): symbol of an operational amplifier. The supply voltage is typically $\pm 15V$.

A signal detector should not modify the measurement, which implies that the input resistance of a voltage detector should large compared to the sample resistance, while that one of a current detector should be small. The simplified setup shown in Fig. 4.30(a) shows the principle of voltage amplification with a transistor, which we suppose to be a Si MOSFET or a Ga[Al]As HEMT. Properly designed, they will operate at low temperatures as well, and can be integrated into the chip that hosts the experiment, which is useful in some cases. The advantage is an enhanced sensitivity and reduced thermal noise, The voltage to be amplified is superimposed to the gate voltage which defines the operating point of the transistor. The output voltage is the voltage drop between source and drain, which is highly sensitive to the gate voltage. A supply voltage is applied between source and drain, in series with a resistor R at the source side. Hence,

$$V_{out} = V_{supply} - RI_{SD}$$

Here, I_{SD} is the current that flows from the supply to drain. We have assumed that the current that flows between the gate and source or drain is small compared to the current provided by

4.3 Electronic measurements on nanostructures

the supply, which is reasonable in field effect transistors. A small[5] input voltage V_{in} changes the current by

$$\Delta I_{SD} = tV_{in}$$

where $t = \partial I_{SD}/\partial V_{in}$ denotes the *transconductance* of the transistor at the operating point. Consequently, the output voltage changes according to

$$\Delta V_{out} = -RtV_{in}$$

The amplification is thus $a = -Rt$. Since t is typically of the order of 10^{-3} A/V, an amplification by $\approx 10^3$ can be obtained this way.

More common, however, are detector circuits that rely on operational amplifiers. A scheme is shown in Fig. 4.30(b). Operational amplifiers are three-terminal devices with two inputs and one output. In addition, they require a bipolar supply voltage of typically ± 15 V. Their internal structure is of no further interest to us here. We consider them as a back box with the following features:

- The input resistance is very high, e.g. 10^{12} Ω.

- The output resistance is small, typically of the order of 100 Ω.

- For the circuit of Fig. 4.30(b), the output voltage is proportional to the difference of the two input voltages: $V_{out} = a_0(V_{in,+} - V_{in,-})$. The amplification $a_0 \approx 10^6$ to 10^8 is the *open loop gain*. Note, however, that the output voltage cannot exceed the supply voltages.

- If a fraction of the output is, via some circuit elements, coupled back into one of the inputs, the operational amplifier adjusts its V_{out} such that $(V_{in,+} - V_{in,-}) = 0$.

These properties make operational amplifiers extremely useful. We study how operational amplifiers can be used conceptually to measure voltages and currents. The circuit of Fig.4.31(a) is a voltage amplifier. Why? Suppose $R_1 \gg R_2$. The voltage to be amplified is connected to the + input, and the output assumes the value $V_{out} = a_0(V_{in} - V_-)$. The voltage divider connected to the output determines $V_- = V_{out}R_2/(R_1 + R_2)$. Hence,

$$V_{out} = (V_{in} - V_{out}\frac{R_2}{R_1 + R_2})a_0 \Rightarrow$$

$$V_{out}(\frac{1}{a_0} + \frac{R_2}{R_1 + R_2}) \approx V_{out}\frac{R_2}{R_1} = V_{in}$$

The approximation is valid for $a_0 \gg R_1/R_2$. Hence, we see that the feedback reduces the open loop gain to the amplification R_1/R_2, which can be chosen within wide ranges. Note that in this circuit, the input resistance of the volt meter is very high, namely that one of the + - input. Note further that this example implies $a_0 = \infty$ for an ideal operational amplifier, which leads to the condition $(V_{in,+} - V_{in,-}) = 0$.

---

[5] By "small", we mean voltages in the microvolt regime

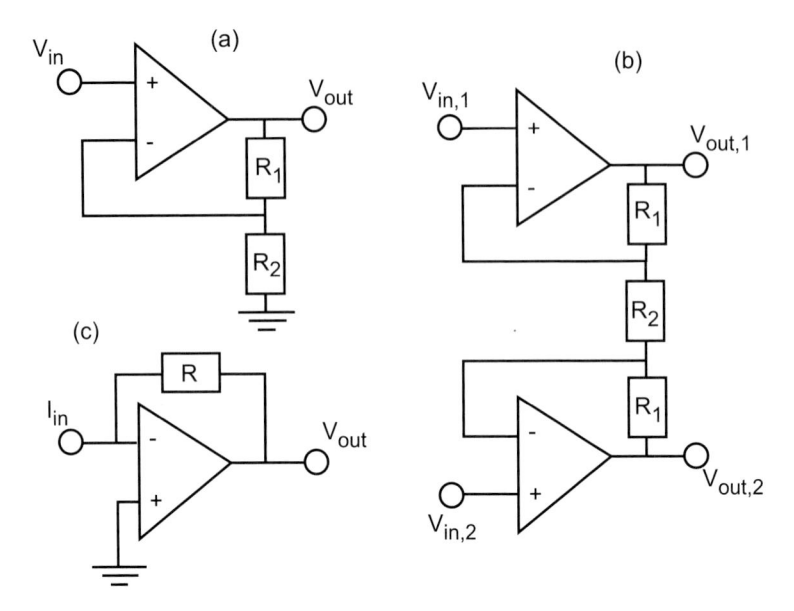

Figure 4.31: A voltage amplifier (a), a differential amplifier (b), and a current-to-voltage converter (c).

In many experiments, voltage differences are to be measured, either to exclude contact resistances (see below) from the measurements, or to measure without a direct reference to ground. Also, differential measurements limit the pickup noise, since a large fraction of this noise will be identical in both measurement wires. Differential measurements can be made with a differential amplifier as shown in Fig. 4.31(b). It should now be easy for you to work out the amplification of this circuit:

Question 4.4: Verify that the amplification of circuit 4.31(b) is

$$\frac{V_{out,1} - V_{out,2}}{V_{in,1} - V_{in,2}} = 1 + 2\frac{R_1}{R_2}$$

To conclude this brief section on operational amplifiers, let's have a look at the current meter depicted in Fig. 4.31(c). Here, the "+" - input is grounded, and an input *current* is applied to the "-" - input. There is no place for the current to go, and it thus charges up the input capacitance C_-, which is inevitably present. The total current that arrives at the "-" - input is

$$I_-(t) = I_{in}(t) + \frac{1}{R}V_{out}(t)$$

4.3 Electronic measurements on nanostructures

On the other hand,

$$V_{out}(t) = a_0(V_{in}^+ - V_{in}^-) = -\frac{a_0}{C_-}\int_{t'=0}^{t} I_-(t')dt'$$

We differentiate this expression with respect to t and substitute $I_-(t)$ with the previous equation. This gives

$$-\frac{C_-}{a_0}\frac{dV_{out}}{dt} = I_{in}(t) + \frac{1}{R}V_{out}(t)$$

The left hand side is approximately zero, due to the large open loop gain. Consequently,

$$V_{out}(t) = -RI_{in}(t)$$

The input current is converted into a voltage with a conversion ratio determined by the resistor. Its resistance can be very high, like $R \approx 1$ GΩ, since the condition is that R must be small compared to the input impedance. Thus, the output voltage adjusts in such a way that there is no charge buildup at the input. The current is effectively drained at the "-" - input. For this reason, the "-" input is sometimes referred to as *virtual ground*. In cryogenic experiments, the conversion resistor R is sometimes mounted inside the cryostat, in order to reduce the thermal noise.

Some important measurement setups

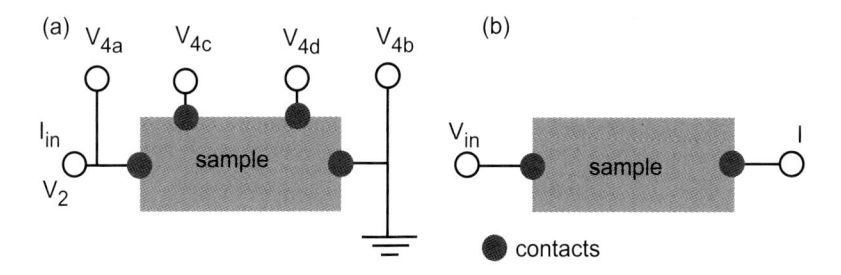

Figure 4.32: (a): two-point and four point resistance measurements. (b): setup for a conductance measurement.

As we have just seen, low-resistance samples should be investigated by applying a current and detecting the voltage drop (Fig. 4.32). This can be done in a two-probe configuration, where the voltage drop is measured at the connections used to apply the current and ground the sample. This has the disadvantage that the not only the sample is measured, but also leads and the contacts, which, in case of a 2DEG, are the Ohmic contacts between the sample surface and the electron gas. In a quasi four-probe setup, two wires are connected to both contacts used. Applying a current I_{in} in Fig. 4.32(a) and measuring the voltage between

4a and 4b eliminates the wire resistance, but not the contact resistances. Therefore, a true four-probe configuration is preferable, where the contacts used for measuring the voltage are different from those used to pass the current through the sample. In Fig. 4.32(a), this setup corresponds to measuring $V_{4c} - V_{4d}$. True four-probe setups are not always possible, though. It is then difficult or even impossible to discriminate between the contact resistances, which may be quite high, and the resistance of the sample. Note that in a two-probe measurement, we measure $R_{xx} + R_{xy}$. The individual components of the resistivity tensor can be measured only with the corresponding four-probe configurations.

A conductance measurement, on the other hand, is usually a 2-terminal experiment. Here, the current meter is in series with the sample. This setup is preferable for samples with a high resistance above ≈ 1 MΩ.

Sometimes, it is convenient to be able to measure a differential quantity, such as the differential conductance dI/dV, or the transconductance $t = dI/dV_G$, of a transistor. Here, I is the source-drain current, V the source drain voltage, and V_G denotes the gate voltage. This can be done by superposing a small AC voltage to the DC voltage that is tuned, and detect only those signals with the superimposed frequency with a lock-in amplifier. Schematic setups are shown in Fig. 4.33. Such differential measurements often give a higher resolution and a lower noise level, since the lock-in technique can be used where absolute measurements must be performed DC. Of course, we obtain the current as a function of the gate voltage, or of the source-drain voltage, by simple numerical integration of the measured differential trace.

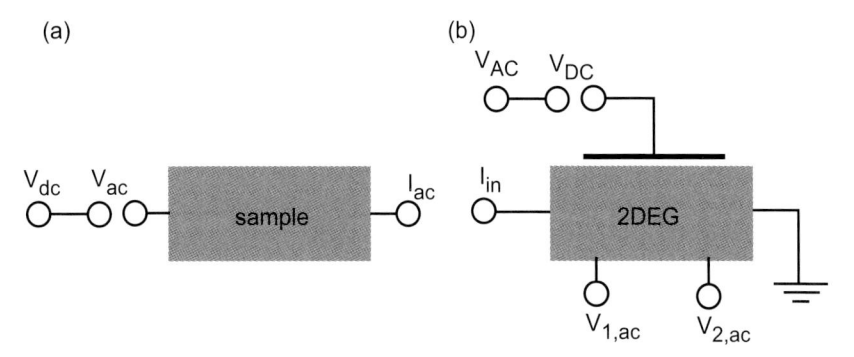

Figure 4.33: Setup for measuring the differential conductance (a), and the transconductance of a field effect transistor (b).

Papers and Exercises

P 4.1: In [Wegscheider1997], nanostructures of dimension 2,1 and zero are fabricated by cleaved edge overgrowth. Work out how the authors managed to do this, and discuss their way of detecting the dimensionality!

P 4.2: A 3-dimensional fcc lattice of self-assembled quantum dots has been grown by [Springholz1998]. Focus on the mechanism behind the ordering. How is the lattice constant tuned?

P 4.3: A clever way of illuminating resists by an electron beam is presented by [Fulton1983]. Describe how it works and what the advantages are!

P 4.4: The *van der Pauw* technique, named after the author of the original proposal ([Pauw1958]), is an important concept for measuring semiconducting samples. Use [Vries1995] to discuss this technique.

E 4.1: Pressure considerations for molecular beam epitaxy
Use the kinetic gas theory to show that gas molecules hit a unit area with a rate F, given by

$$F = \frac{p}{\sqrt{2\pi m k_B \Theta}},$$

p is the partial pressure of molecules with mass m. How long does it take until a monolayer of oxygen has been formed on the surface? Assume an O_2 partial pressure of $p = 10^{-10}$ mbars. Assume further that all molecules that hit the surface remain adsorbed (i.e., the sticking coefficient is 1).

E 4.2: Some considerations concerning clean rooms
Dust particles with a size above > 500 nm frequently cause trouble in the microchip fabrication. They rest on the resist and generate defects, like interrupted connections, or short circuits. They are therefore also known as "killer defects". The yield Y is the fraction of working microchips. Y can be written by

$$Y = \frac{1}{(1 + A \cdot D)^n}$$

where D denotes the density of killer defects, A is the chip area, and n the number of relevant process steps (steps which involve resist illuminations).

  (i) Suppose that $n = 12$ for fabricating a certain microchip. What is the maximum D when a yield of at least 0.5 is needed for a chip of size $A = 2$ cm^2?

  (ii) How many defects are, under these conditions, acceptable on an 8 inch wafer?

  (iii) For a rough estimate, assume that during a process step, 1/6 of all dust particle inside a volume of (8 in.)^3 get deposited on the wafer. What clean room class is needed in order to obtain a yield above 90% ?
  The class of a clean room is the number of particles with sizes above 500 nm, in a volume of one cubic foot.

E 4.3: An electron beam resist has a sensitivity of $S = 2\,\mathrm{C/m^2}$. The pattern generator places the focus of the electron beam at positions in a grid of $2^{13} \times 2^{13}$ points (a "13-bit resolution"). The writing field A is chosen via the magnification of the electron microscope, and selected to an area of $100\,\mu\mathrm{m} \times 100\,\mu\mathrm{m}$. Calculate the *dwell time*, the time the electron beam has to rest at each position. What is the minimum size of a single illuminated that guarantees homogeneous illumination?

E 4.4: Analyze the operational amplifier circuit of Fig. 4.34! How is the output voltage related to a *time-dependent* input voltage? Discuss the response of the output voltage to a step-like input voltage $V_{in}(t) = V_0 \theta(t - t_0)$! Can you imagine a possible application of this circuit?

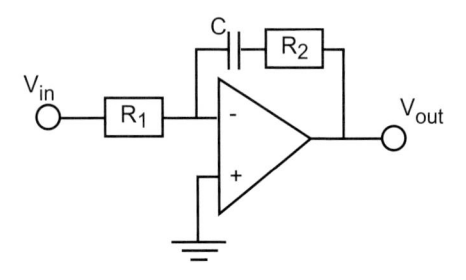

Figure 4.34: Circuit of [E 4.4].

Further Reading

An overview over both the technology and the applications of semiconductor devices is given in [Sze1985]. A review of silicon processing technology has been given in [Hattori1998]. For details of GaAs processing, see [Williams1990]. For both Si and GaAs processing, see e.g., [Ghandhi1994]. A very nice review of many aspects of matter at low temperatures can be found in [McClintock1984]. The book [Wilks1970] treats the amazing properties of helium in a more rigorous way. In [Lounasmaa1974], you find an extensive discussion of dilution refrigerators. If you need recipes and practical tips for measurements in a cryogenic environment, you will find almost certainly what you need to know in [Richardson1988], or in [Rose-Innes1973]. You are encouraged to read through an introductory textbook on electronics, like, e.g., [Horowitz1989], or [Franco1997]!

5 Important Quantities in Mesoscopic Transport

It has been pointed out in the introduction that the mesoscopic regime is characterized by certain scales in space, time, and energy. They will be introduced in this short chapter. We will frequently refer to these definitions later on.

- **The Fermi wavelength** λ_F

  The *Fermi wavelength* $\lambda_F = 2\pi/k_F$ is the de Broglie wavelength of the electrons at the Fermi edge. Size quantization thus takes place at length scales comparable to λ_F, although we will see systems with characteristic sizes of $10\lambda_F$ that still show size quantization. The fermi wavelength decreases as the electron density n_d (d denotes the dimensionality of the electron gas) increases, while the exact relation depends on d. For a spin degeneracy of 2 and within the effective mass approximation, one finds

  $$d = 3: \quad \lambda_F = 2^{3/2}\left(\frac{\pi}{3n_3}\right)^{1/3}$$

  $$d = 2: \quad \lambda_F = \sqrt{\frac{2\pi}{n_2}}$$

  $$d = 1: \quad \lambda_F = \frac{4}{n_1} \tag{5.1}$$

  Thus, the Fermi wavelength is directly obtained from the electron density, which can be determined via Hall measurements. Note that λ_F does not depend on the effective mass.

- **The elastic scattering times and lengths**

  The *quantum scattering time* τ_q is the average time between subsequent elastic scattering events of arbitrary strength. It is related to the *quantum scattering length* ℓ_q via $\ell_q = v_F \tau_q$. Here, v_F denotes the Fermi velocity of the electrons at the Fermi edge, i.e. $v_F = \sqrt{2E_F/m^*}$. Hence, the ℓ_q is just the average distance the electrons at the Fermi energy travel without being elastically scattered. The quantum scattering length does not determine the resistivity, though. For momentum relaxation (see chapter 2), the scattering *angle* is important as well. In fact, weighing each scattering event with the scattering angle ϕ by the factor $(1 - \cos\phi)$ leads to the momentum relaxation time (which we will frequently denote as the Drude scattering time) τ as introduced in Section 2.6.2.

  The *elastic mean free path* ℓ_e is defined as $\ell_e = v_F \tau$ and represents the average distance an electron moves in between two subsequent, strong scattering events, also referred to as large-angle scattering events.

In the relevant case of a two dimensional electron gas, $\ell_e = \frac{\hbar}{e}\mu\sqrt{2\pi n_2}$. Inserting typical numbers for a 2DEG in a GaAs-HEMT at low temperatures, one finds $\ell_e \approx 8\ \mu m$. The ratio τ_q/τ is determined by the relevance of various scattering mechanisms. Of course, $\tau_q/\tau \leq 1$ must hold. If this ratio is small compared to 1, the scattering potential is weak on the scale of the Fermi energy. For a GaAs-HEMT at low temperatures, for example, one typically finds $\tau_q/\tau \approx 0.1$.

The Drude scattering time follows directly from the resistivity, once the electron density is known. The quantum scattering time can be extracted from $\rho_{xx}(B)$, provided that magnetoresistivity oscillations can be observed. This, by the way, is also a method to determine the effective mass. This analysis is discussed in chapter 6.

- **The diffusion constant** D

The *diffusion constant* D originates from the diffusion equation

$$\frac{dn}{dt} = D\frac{d^2 n}{dx^2} \qquad (5.2)$$

which tells us that gradients in the in the electron density cause diffusion, see e.g. [Reif1985]. We discuss the diffusion constant in a one-dimensional model; the extension to higher dimensions is straightforward. Due to the Brownian motion, the electrons experience a fluctuating, Brownian force $b(t)$, which averages to zero in large time intervals. This can be included in the equation of motion for the electrons (see chapter 2) by simply adding $b(t)$ to the forces $F(t)$ exerted by the external fields. The result is the Langevin equation for the electrons

$$m^*\left(\frac{dv}{dt} + \frac{v}{\tau}\right) = F(t) + b(t) \qquad (5.3)$$

Suppose the external forces are zero. From statistical physics [Reif1985], it is well known that the diffusion constant is obtained from the autocorrelation function[1] $C_v(t)$ of the electron velocities

$$D = \int_{t=0}^{\infty} C_v(t)dt; \qquad C_v(t) = \int_{t'=0}^{\infty} v(t')v(t'+t)dt' \qquad (5.4)$$

We take the derivative of $C_v(t)$ with respect to time and replace $\frac{dv}{dt}$ using the Langevin equation, which gives

$$\frac{dC_v(t)}{dt} = \frac{1}{m*}C_{vb}(t) - \frac{1}{\tau}C_v(t) \qquad (5.5)$$

Here, the first term on the right hand side vanishes[2], as there is no correlation between the velocity at time t' and the Brownian force at time $t' + t$. One therefore obtains a differential equation for the velocity autocorrelation function with the solution

$$C_v(t) = C_v(0)e^{-t/\tau} \qquad (5.6)$$

---

[1]Correlation functions are introduced in Appendix B.
[2]The terminology for correlation functions of the type $C_{fg}(x)$ is introduced in Appendix B as well.

Consequently, within the approximations made, the diffusion constant equals

$$D = C_v(0)\tau \tag{5.7}$$

Since $C_v(0)$ represents the averaged square of the electron velocity, we can make use of the equipartition theorem of statistical mechanics and write

$$\frac{m}{2}C_v(0) = \frac{1}{2}k_B\Theta \Rightarrow$$

$$D = \frac{k_B\Theta}{m^*}\tau = \frac{k_B\Theta}{e}\mu \tag{5.8}$$

This is the *Einstein relation*. A calculation of $C_v(0)$ [Reif1985] gives the result

$$D = \frac{1}{2}v_F^2\tau \tag{5.9}$$

A typical value for electron gases in Ga[Al]As-HEMTs at low temperature is $D = 0.1 \text{ m}^2/\text{s}$.

Diffusion constant and mobility are thus intimately related. A gradient in the electrostatic potential, as well as a gradient in the chemical potential, lead to a spatially varying electrochemical potential and cause drift or diffusion, respectively, see Fig. 5.1. The Einstein relation tells us how the drift is related to the diffusion.

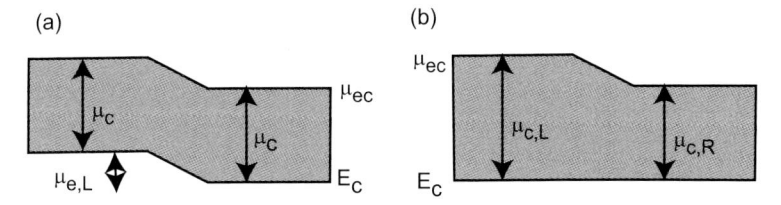

Figure 5.1: An electrostatic potential gradient (a) and a chemical potential gradient (b) can generate identical gradients in the electrochemical potential.

It is worth mentioning that due to the Einstein relation, we can calculate the conductivity σ from the velocity autocorrelation function. We recall that $\sigma = ne\mu$ and write the conductivity as

$$\sigma = \frac{ne^2}{k_B\Theta}\int_{t=0}^{\infty} C_v(t)dt \tag{5.10}$$

This is the simplest version of the *Kubo formula* [Kubo1957]. It is frequently used in numerical simulations of the conductivity, where electrons are injected in random directions in the potential landscape and $C_v(t)$ is calculated. For a review of this technique, see [Jacoboni1985].

The diffusion constant enters in some of the scales to be introduced below. The Einstein relation makes clear that D is obtained experimentally from the mobility.

- **The dephasing time τ_ϕ and the phase coherence length ℓ_ϕ**
  Elastic scattering events, which determine the electron mobility at low temperatures, do not cause dephasing, since the phase shift experienced by the scattered electrons is reproducible. If the scattering event can, in any way, be regarded as a measurement of the electron's location, dephasing takes place. This is the case in spin-flip scattering events at magnetic impurities, or for electron-phonon scattering. Electron-electron scattering does cause dephasing, since energy is transferred between the scattering partners. However, the latter kind of scattering does not, in general, cause resistance, since the total momentum of the electron system remains unchanged. Therefore, electron-electron scattering gives to a first approximation no contribution to the resistivity. Transport effects which rely on interference of electronic wave functions, however, can be used to determine the *dephasing time* τ_ϕ. The theory for the magnitudes and the parametric dependence of τ_ϕ is developed in [Altshuler1982]. For two dimensions and in the diffusive regime, one finds

$$\frac{1}{\tau_\phi} = \begin{cases} \frac{\pi}{2} \frac{(k_B\Theta)^2}{\hbar E_F} \ln(\frac{E_F}{k_B\Theta}); & k_B\Theta > \hbar/\tau_D \\ \frac{k_B\Theta}{2m^*D} \ln(\frac{m^*D}{\hbar}); & k_B\Theta < \hbar/\tau_D \end{cases} \quad (5.11)$$

This linear relation between $1/\tau_\phi$ and T at low temperatures, which changes to a quadratic dependence at higher temperatures, is found experimentally in reasonable agreement with the theoretical expressions, see e.g. [Choi1987].

The *phase coherence length* ℓ_ϕ is the distance the electrons travel before their phase is randomized. For $\tau_\phi < \tau$, $\ell_\phi = v_F\tau_\phi$. For $\tau_\phi > \tau$, however, the electrons get scattered elastically within the phase coherence time, and the distance they travel within τ_ϕ gets reduced. For $\tau_\phi \gg \tau_D$, which is often the case at low temperatures, $\ell_\phi = \sqrt{D\tau_\phi}$. Typical phase coherence times in mesoscopic samples are of the order on 1 ps.

We will meet the dephasing time in chapter 8, where we will also see how it can be determined experimentally.

- **The electron-electron scattering time τ_{ee}**
  The *electron-electron scattering time* is the average time of flight for the electrons between successive electron-electron scattering events. In a simple picture [Ashcroft1985], we expect that $\tau_{ee} \propto 1/\Theta^2$. Let's assume a single electron (1) has an energy Δ above the Fermi energy of an electron gas at zero temperature: $E_1 = E_F + \Delta$. Consider how this electron can scatter with electrons in the Fermi sea. For the energy of the second electron 2, $E_2 \leq E_F$ holds. In addition, both final states (with energies E_3 and E_4) the electrons occupy after the scattering must be empty: $E_3, E_4 > E_F$. Since $E_1 + E_2 = E_3 + E_4 > 2E_F$, it follows $(E_1 - E_F) + (E_2 - E_F) = \Delta + (E_2 - E_F) > 0$. Therefore, only electrons within a shell of thickness Δ below the Fermi edge can scatter at 1. This is a fraction Δ/E_F of all electrons. Furthermore, E_3 and E_4 must be inside $[E_F, E_F + \Delta]$. This means that the electron-electron scattering probability is $\propto (\frac{\Delta}{E_F})^2$. For a thermally smeared Fermi gas, we can identify $\Delta \approx k_B\Theta$, and $\tau_{ee} \propto 1/\Theta^2$ results. This argument has been verified to be approximately true by more sophisticated

calculations of [Guilani1982], which derived for the ballistic regime

$$\frac{1}{\tau_{ee}} = \frac{E_F}{4\pi\hbar}(\frac{\Delta}{E_F})^2 \left[ln(\frac{E_F}{\Delta}) + ln(\frac{2k_{TF}}{k_F}) + \frac{1}{2} \right]$$

where k_{TF} is the Thomas-Fermi screening vector, see chapters 2 and 3.

Although interactions are not treated explicitly, τ_{ee} will occasionally pop up, in particular in chapter 8. It can be determined experimentally by magnetoresistivity measurements, as carried out, e.g., by [Choi1986].

- **The thermal length ℓ_Θ**
  The *thermal length* specifies the length scale over which thermal smearing takes place. From the uncertainty relation $\hbar \leq k_B\Theta \cdot \tau_\Theta$ one obtains a "thermal time", below which it is not possible to determine the energy of the electron better than $k_B\Theta$. This time scale corresponds to a length scale $\ell_\Theta = \sqrt{D\tau_\Theta} = \sqrt{\frac{\hbar D}{k_B\Theta}}$.

  We will meet the thermal length in particular when phase coherence effects are discussed, namely in chapter 8.

- **The localization length ℓ_ξ**
  This is the average length over which an electronic state extends in a sample. The disorder potential localizes the states on this length scale. As we shall see in the following chapter, magnetic fields can tune the localization length over wide ranges.

- **The interaction parameter (or gas parameter) r_s**
  The *interaction parameter* is the ratio of the (unscreened) Coulomb energy between two electrons at their average distance, and their kinetic energy at the Fermi edge. In two dimensions,

  $$r_s = \frac{e^2/4\pi\epsilon\epsilon_0 r}{m^* v_F^2/2} = \frac{e^2 m^*}{\epsilon\epsilon_0 h^2}\frac{1}{\sqrt{n_e}}$$

  For 2DEGs in GaAs, $r_s \approx 1$. In Si-MOSFETs as well as in hole gases residing in GaAs, r_s can be much larger, and values up to $r_s \approx 20$ have been reported.

  Although we essentially treat the electron gases as non-interacting or weakly interacting, it is worth keeping in mind that this is not really true in low-dimensional electron gases. Strictly speaking, the validity of Fermi liquid theory in systems with $r_s > 1$ is questionable.

- **The magnetic length ℓ_B and the magnetic time τ_B**
  A magnetic field sets a length scale as well, namely the spatial extension of wave functions in the magnetic field. It is given by the *magnetic length* $\ell_B = \sqrt{\hbar/eB}$, which corresponds to the width of the ground state of a quantizing magnetic field, as will be discussed in more detail in the next chapter. Also of importance is the *cyclotron radius* r_c, i.e. the radius of the circle the electrons follow in a magnetic field: $r_c = k_F \ell_B^2$, as can be easily checked. The magnetic time is the time an electrons needs to diffuse across the area $\frac{1}{2}\ell_B^2$. It is given by $\tau_B = \frac{\ell_B^2}{2D}$.

  The magnetic length will be of particular importance in chapters 6 and 7.

Exercises

E 5.1: The quantities listed in Table 5.1 have been determined experimentally in Si-MOSFETs and in GaAs - HEMTs. Calculate the scattering time τ, the diffusion constant D, the

Table 5.1: Characteristic transport parameters for Si and GaAs.

	GaAs (Θ = 4.2 K)	Si (Θ = 4.2 K)
electron density [10^{15} m^{-2}]	4	0.7
electron mobility [m^2/Vs]	100	4
effective mass [m_e]	0.067	0.19
phase coherence time [10^{-12} s]	30	10

Fermi wavelength λ_F, the phase coherence length ℓ_ϕ, the inelastic scattering length ℓ_{in}, the thermal length ℓ_T, and the interaction parameter r_s.

Which material would you prefer for ballistic electron experiments, for investigations of phase coherent electrons, and for studying electron-electron interactions?

Further Reading

More about the relevant quantities in mesoscopic transport can be found in Datta's book [Datta1997], as well as in the review article by Beenakker and van Houten [Beenakker1991]. For an introduction to the Kubo formula, see [Ziman1992].

6 Magnetotransport Properties of Quantum Films

Transport experiments in external magnetic fields are very common in mesoscopic physics. With the magnetic field, we can reversibly and - if we perform our experiment carefully enough - non-destructively tune various scales, such as the magnetic length, the effective mass, or the cyclotron energy. We have seen already that measuring the Hall resistivity in small magnetic fields allows us to determine the carrier density. In this chapter, we are mainly interested in strong magnetic fields, such that $\omega_c \tau \gtrsim 1$. This condition simply means that the electrons can complete at least one cyclotron orbit before they get scattered. It is immediately clear that new effects can be expected in this regime. Recall Bohr's atomic model: discrete states are obtained from interferences of electronic waves circulating around the nucleus. For an interference to be constructive, the circumference of the trajectory must be an integer multiple of the electronic wavelength. A similar thing happens in strong magnetic fields. Here, the electrons are forced to circulate in cyclotron orbits. The result, known as Landau quantization, is discussed in section 1. In particular, it is a very important ingredient to the quantum Hall effect, which is the topic of section 2. In section 3, we return to intermediate magnetic fields and show how the magneto-oscillations observed in longitudinal direction (Shubnikov - de Haas oscillations) can be analyzed to obtain the quantum scattering time and the effective mass. In the subsequent section 4, we give a small selection of further magnetotransport experiments. Up to that point, the magnetic field has been perpendicular to the plane of the electron gas. The basic effects in parallel magnetic fields are presented in section 5, which concludes this chapter.

If not stated otherwise, the magnetic field is homogeneous and points in the z-direction. In this case, we call it B. Magnetic fields in the plane of the electron gas are referred to as parallel magnetic fields, and are denoted by $B_\parallel$.

6.1 Landau quantization

6.1.1 2DEGs in perpendicular magnetic fields

The Schrödinger equation of a free 2DEG in a magnetic field B reads [1]

$$\left[\frac{(\vec{p}+e\vec{A})^2}{2m^*} + V(z)\right]\Phi(\vec{r}) = E\Phi(\vec{r}) \qquad (6.1)$$

where $\vec{A}$ denotes the vector potential. The z-direction is of no further interest for us, since B does not influence the electronic motion in that direction. We therefore assume that the

---
[1] We define e is positive, the electronic charge is thus $-e$.

z-direction can be treated separately, leading to a quantized energy E_z, which is the conduction band bottom of the two-dimensional electron gas. We choose the Landau gauge $\vec{A} = (-By, 0, 0)$ for mathematical simplicity.[2] The xy-Hamiltonian emerging from eq. (6.1)

$$H_{xy} = \frac{1}{2m^*}\left[(p_x - eBy)^2 + p_y^2\right] \tag{6.2}$$

With the ansatz

$$\Phi(x, y) = \Psi(y)e^{ik_x x} \tag{6.3}$$

a one-dimensional Schrödinger equation in y-direction is obtained

$$\left[-\frac{\hbar^2}{2m^*}\frac{\partial^2}{\partial y^2} + \frac{\hbar^2 k_x^2}{2m^*} - \frac{\hbar k_x eBy}{m^*} + \frac{e^2 B^2 y^2}{2m^*}\right]\Psi(y) = (E - E_z)\Psi(y) \tag{6.4}$$

while plane waves are the eigenfunctions in x-direction of the separated Hamiltonian. The Schrödinger equation for the y-direction can be mapped onto the harmonic oscillator equation

$$\left[-\frac{\hbar^2}{2m^*}\frac{\partial^2}{\partial v^2} + \frac{1}{2}m^*\omega^2 v^2\right]u(v) = Eu(v) \tag{6.5}$$

by

$$\omega \to \omega_c = \frac{eB}{m^*}$$

$$v \to y - \frac{\hbar k_x}{m^*\omega_c} \tag{6.6}$$

The cyclotron frequency ω_c is the angular frequency of the electron in the magnetic field. For simplicity, let us consider a rectangular sample in the x-y-plane of area $L_x \cdot L_y$. Since k_x quantizes as a consequence of the boundary conditions according to in x-direction as

$$k_{x,n} = \frac{n\pi}{L_x} \tag{6.7}$$

The harmonic oscillators in y-direction are centered at the positions

$$y_n = \frac{n\pi\hbar}{m^*\omega_c L_x} \tag{6.8}$$

The eigenfunctions of the xy - Hamiltonian are thus plane waves in x-direction, multiplied with Hermite polynomials in y-direction, as shown schematically in Fig. 6.1.

Question 6.1: Check that the full width at half maximum (FWHM) of the ground state in y-direction equals the magnetic length!

---
[2]It is instructive to solve this problem in the symmetric gauge $\vec{A} = 0.5(-By, Bx, 0)$, and use polar coordinates. See, e.g., [Ezawa2000] for details.

6.1 Landau quantization

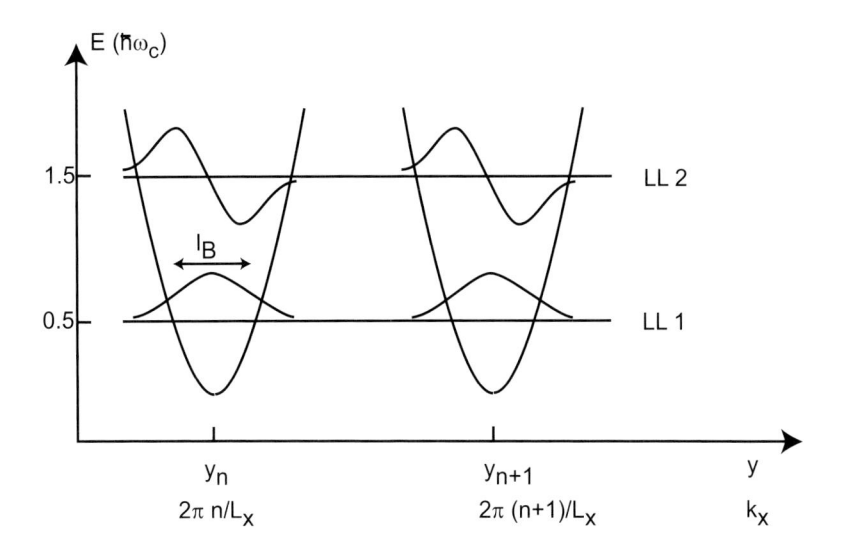

Figure 6.1: Electronic states in a Landau level. The positions of the harmonic oscillator potentials y_n in y-direction are given by the wave numbers k_x that satisfy the boundary condition.

The corresponding energy eigenvalues are

$$E_j = \hbar\omega_c\left(j - \frac{1}{2}\right) \tag{6.9}$$

with j being a positive integer. Besides spin and valley degeneracies, the degeneracy of each energy level is given by the number of allowed wave numbers in x-direction. The states of energy E_j form the jth Landau level. In order to determine the degeneracy of a Landau level, we can use the fact that the integrated density of states is independent of the magnetic field. Hence, the number of states per unit area in a Landau level must be $\frac{g_s m^*}{2\pi\hbar^2}$, multiplied by $\hbar\omega_c$, see Fig. 6.2. A degeneracy of $g_s/(2\pi\ell_B^2)$ per unit area in each energy level is obtained (g_s counts both the spin and valley degeneracies). Hence, the density of states of an ideal 2DEG in a perpendicular magnetic field reads

$$D(E) = \frac{g_s}{2\pi\ell_B^2}\delta(E - E_j) \tag{6.10}$$

with E_j given by eq. (6.9).

Question 6.2: Calculate the degeneracy of a Landau level by counting the states with (6.8). Use the condition that all the harmonic oscillators must have their center y_n inside the sample! Assume further that the magnetic length is small compared to the sample size.

In real samples, the Landau levels are broadened to an approximately Gaussian shape by potential fluctuations, and split via the two possible alignments of the electron spin with respect to the magnetic field (Fig. 6.2). The spin splitting is described by the effective g-factor g^*. For bulk GaAs, $g^* = -0.44$. [3]

In general, the Landau level at the Fermi energy is only partly occupied, and it is thus useful to introduce a quantity that measures the degree of filling of Landau levels. This is the task of the *filling factor* ν, defined as

$$\nu = \frac{g_s E_F}{\hbar \omega_c} \tag{6.11}$$

For the frequent case of $g_s = 2$ (spin degeneracy), $\nu = 2j$ means that j Landau levels are completely filled. Furthermore, in sufficiently strong magnetic fields, the spin degeneracy

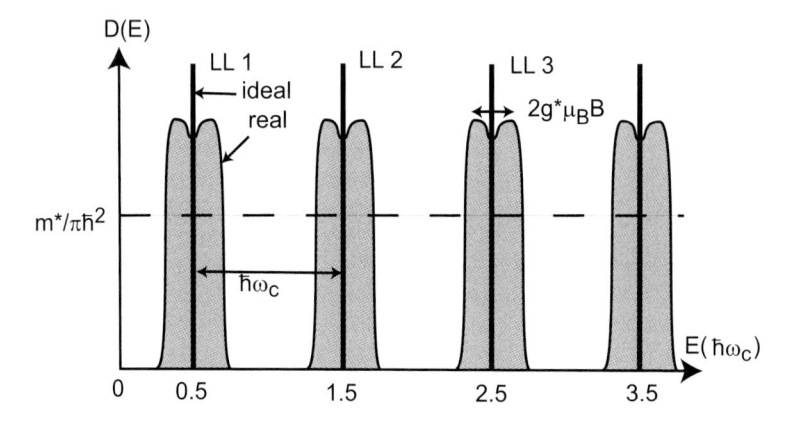

Figure 6.2: Ideal and real density of states of a Landau-quantized 2DEG which is spin-degenerate at $B = 0$. The δ-functions broaden due to fluctuations of the conduction band bottom, while the spin degeneracy is lifted, and a Zeeman doublet results for non-zero effective g-factors g^*.

may be lifted due to the Zeeman effect. In that case, an odd integer value of ν means that one spin direction of Landau level $j = \nu/2$ is full, while the other is empty.

6.1.2 The chemical potential in strong magnetic fields

In experiments, it is common to vary the Landau level occupation by tuning either the electron density or the magnetic field. Suppose B is fixed, the temperature is zero, and we tune the electron density. Let us for simplicity assume that there is no valley degeneracy, i.e. $g_s = 2$, and there is no spin splitting. The integrated density of states in each Landau level D_{LL} is then given by

$$D_{LL} = \frac{1}{\pi \ell_B^2} \tag{6.12}$$

---

[3] A magnetic field can strongly modify g^*.

6.1 Landau quantization

The Fermi level $\mu = E_F$ is independent of the electron density n as long as there are empty states in the highest occupied Landau level. In this case, $E_F = (j - 1/2)\hbar\omega_c$. At electron densities $n_j = jD_{LL}$, all Landau levels are either full or empty, and the Fermi energy equals $E_{F,j} = j\hbar\omega_c$ (Fig. 6.3(a)). In the latter case, we would classify the system as an insulator, since the density of states at the chemical potential is zero. Remarkably, a 2DEG in magnetic fields experiences a sequence of metal-insulator transitions as a function of n! In our ideal system with the density of states composed of δ-functions (eq. (6.10)), the insulating phases are just points along the n-axis. As we will see shortly, the insulating behavior extends over non-zero intervals in real samples, since the electronic states in the wings of a peak in $D(E)$ tend to be localized for a real sample. Similar parametric metal-insulator transitions are, of

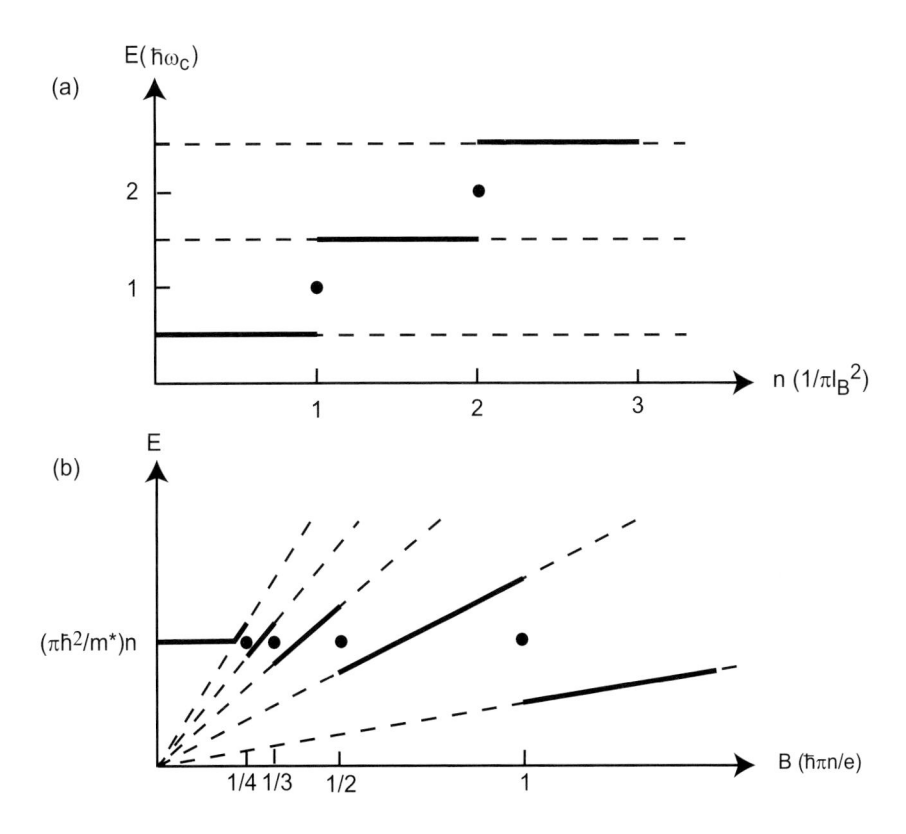

Figure 6.3: Evolution of the Fermi level as a function of electron density in a fixed magnetic field (a), and as a function of the magnetic field with n fixed (b). An ideal density of states is assumed. In both scenarios, metal-insulator transitions exist, with insulating phases of zero width in the parameter coordinate.

course, also found as a function of B with the electron density fixed. Here, the insulating points correspond to magnetic fields $B_j = \frac{\pi \hbar n}{e} \frac{1}{j}$. It is easily verified that at these magnetic fields, E_F changes from

$$E_F(B_j - \delta B) = \frac{\hbar e B_j}{m^*}(j - \frac{1}{2})$$

via

$$E_F(B_j) = \frac{\hbar e B_j}{m^*} j = E_F(B = 0)$$

to

$$E_F(B_j + \delta B) = \frac{\hbar e B_j}{m^*}(j + \frac{1}{2})$$

where δB denotes an arbitrarily small magnetic field. This behavior is depicted in Fig. 6.3(b).

The density of states and its evolution in magnetic fields can be nicely detected by a powerful tool known as *capacitance spectroscopy*. The experimental setup resembles somewhat that one used to measure differential conductances (sketched in Fig. 4.33(a)), see Fig. 6.4(a). The sample used for the experiments in 6.4 was a GaAs - $Al_xGa_{1-x}As$ interface without modulation doping. Instead, a highly doped layer was defined 100 nm away from the heterointerface, towards the substrate. In this structure, a 2DEG can be generated at the heterointerface by applying positive voltages between the top gate and the doping layer. The electrons reach the interface by tunneling across the undoped GaAs spacer layer. In the experiment, a dc voltage is applied between the gate and the doping layer. In addition, a small ac signal is superimposed on the dc voltage, and the current at a phase difference of $\pi/2$ is measured with a lock-in amplifier. In order to deposit charge on the capacitor formed by the 2DEG and the gate, the voltage source has to do both electrostatic and chemical work on the system. The density of states in the metal electrode is very large. Hence, its chemical potential will remain constant. The density of states of the 2DEG, however, is much lower. As charge is added, the Fermi level in the electron gas will therefore change significantly, namely by an amount that depends on the density of states, as well as on the total charge added. Hence, the voltage can be split in two parts V_{chem} and V_{elstat} that perform the chemical and the electrostatic work, respectively. Changing the charge on the capacitor by dq thus requires

$$dV_{elstat} = \frac{1}{C} dq$$

plus

$$dV_{chem} = \frac{1}{e} d\mu = \frac{1}{e} \frac{d\mu}{dn} dn = \frac{1}{e^2} \frac{d\mu}{dn} dq = \frac{1}{e^2 D(E)} dq$$

Here, dn is the electron density change in the 2DEG, and μ denotes its chemical potential. Thus, the total voltage change equals

$$dV = dV_{elstat} + dV_{chem} = \left[\frac{1}{C} + \frac{1}{e^2 D(E)}\right] dq$$

6.2 The quantum Hall effect

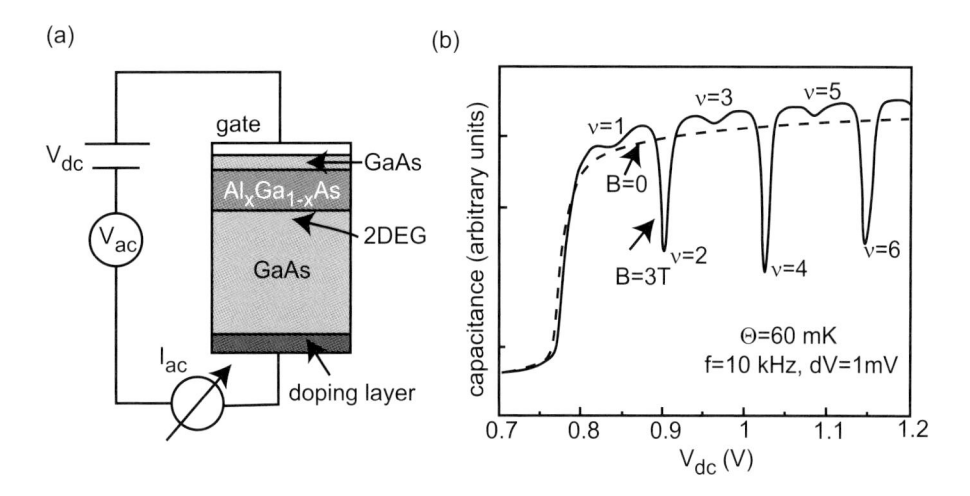

Figure 6.4: Schematic setup for measuring the capacitance of an electron gas (a). The quantum well is empty at $V_g = 0$ and can be filled with electrons by applying a positive voltage between the gate and the doping layer. (b): the measured capacitance shows the filling of the 2DEG at $V_g = 0.77$ V, as well as the modulated density of states in perpendicular magnetic fields. Adapted from [Drexler1994]).

It becomes apparent that the effective capacitance is the geometric capacitance in series with a "chemical capacitance", i.e.,

$$\frac{1}{C_{eff}} = \frac{1}{C} + \frac{1}{e^2 D(E)} \tag{6.13}$$

and we can determine the density of states by capacitance measurements. This explains the observations in Fig. 6.4(b). In the absence of magnetic fields, the capacitance essentially experiences a jump as the gate voltage fills the potential well at the heterointerface. Since $D(E)$ of a 2DEG is constant, no further structure is observed. This changes as a magnetic field is applied, due to the formation of Landau levels. Once the geometrical capacitance is known, it is straightforward to extract the density of states.

Capacitance spectroscopy is also possible by applying the voltage directly between the top gate and a 2DEG. This method is inferior in high magnetic fields, though. For a discussion of this issue, see [P 6.1].

6.2 The quantum Hall effect

6.2.1 Phenomenology

Back in 1980, K. von Klitzing and coworkers [Klitzing1980] discovered a quantization of the Hall resistance in 2DEGs residing in Si - MOSFETs. Examples are shown in Fig. 1.3, as well as in 6.5. Plateaux were observed at integer filling factors, i.e., for

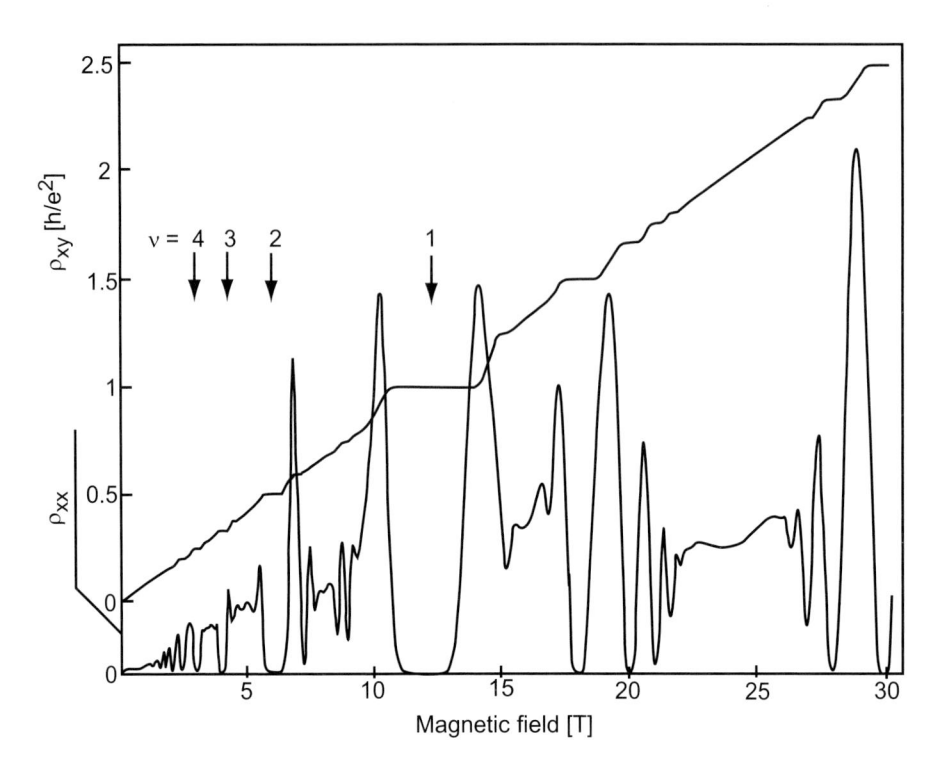

Figure 6.5: The quantum Hall effect and Shubnikov-de Haas oscillations in a Ga[Al]As-HEMT, measured in a dilution refrigerator at a temperature of 100 mK. A filling factor of $\nu = 1$ is reached at $B \approx 12$ T. Besides the integer quantum Hall effect, pronounced structures at fractional filling factors are observed (After [Willett1987]).

$$\rho_{xy}(B) = \frac{1}{\nu}\frac{h}{e^2}$$

Subsequent experiments showed that this quantization is *universal*, in the sense that it is observed in all kinds of materials, provided the electron gas is two-dimensional. In fact, the accuracy $\delta\rho_{xy}/\rho_{xy}$ can be of the order of 3×10^{-10} (Fig. 6.6), such that the $\nu=1$ - quantum Hall plateau has been chosen as the resistance standard, with a resistance of $R_Q = 25812.807 \, \Omega$ *per definition* [Quinn1989]. Furthermore, another quantization of ρ_{xy} has been discovered by [Tsui1982]. In extremely high-quality samples, additional resistance plateaux are observed at $\rho_{xy} = h/ke^2$, with k being a rational number. This type of quantization is very pronounced in Fig. 6.5, but its discussion is beyond our scope. For further information on this *fractional quantum Hall effect*, the reader is referred to specialized literature listed at the end of this chapter. Interestingly, the accurate determination of e^2/h is highly relevant for quantum electrodynamics, since this ratio is contained in the fine structure constant, $\alpha = \frac{\mu_0 c}{2}\frac{e^2}{h}$, which

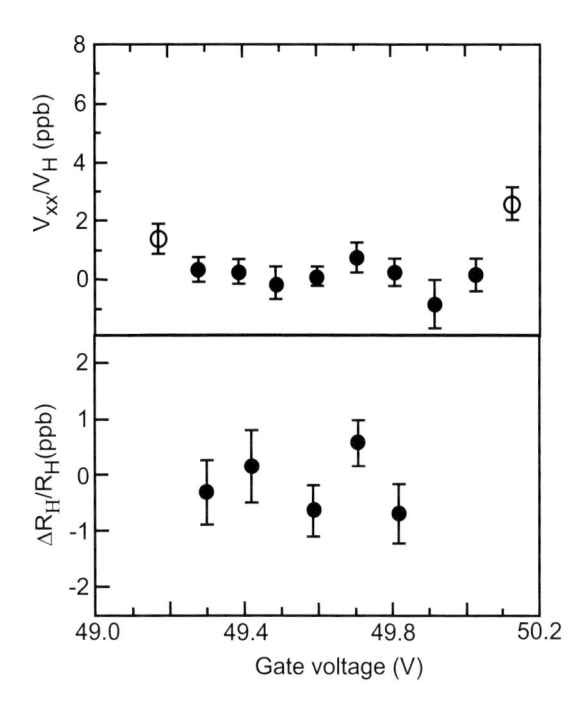

Figure 6.6: The relative accuracy of the ρ_{xy} in the $\nu = 1$ quantum Hall plateau, and the corresponding longitudinal voltage drop, measured as a function of the electron density, which is changed via a gate voltage (after [Jeckelmann1997]). For the accuracy $\Delta R_H/R_H$, a value of $3 \cdot 10^{-10}$ is obtained.

describes the coupling of elementary particles to electromagnetic fields, and as such represents the expansion parameter in this theory.

The QHE is closely connected to another highly remarkable effect, namely the magneto-oscillation of the longitudinal resistivity ρ_{xx}, see Fig. 6.5. These oscillations are known as *Shubnikov - de Haas oscillations*. In fact, in the regions of quantized Hall resistance, ρ_{xx} becomes zero, with an accuracy comparable to that one of ρ_{xy} (Fig. 6.6). The striking correlation between ρ_{xx} and ρ_{xy} suggests a common explanation. The relation between these two quantities will become clear in chapter 7.

6.2.2 Origin of the integer quantum Hall effect

Besides the Landau quantization, disorder is an essential ingredient in understanding the origin of the quantum Hall effect. Suppose we have adjusted the 2DEG to one of the insulating points, characterized by $B_j = \frac{hn}{2e}\frac{1}{j}$. The classical Hall resistivity at such a point is given by

$$\rho_{xy,j} = -\frac{B_j}{en} = -\frac{h}{2e^2}\frac{1}{j} \tag{6.14}$$

which are the resistances of the observed plateaux.[4] Clearly, in an insulating regime, charge

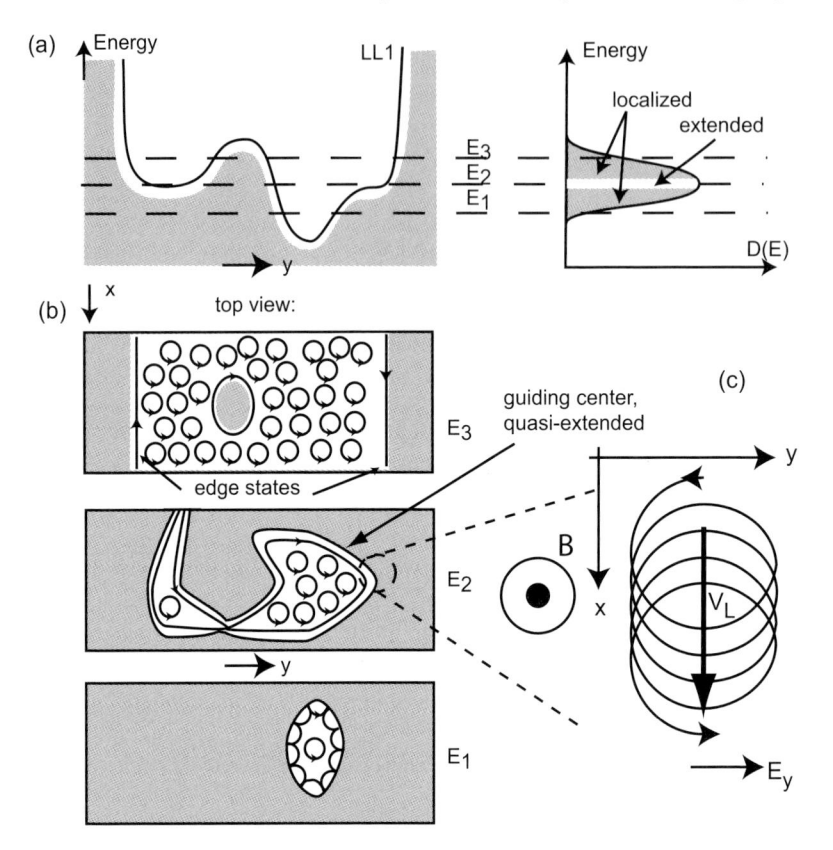

Figure 6.7: Cross section of the potential landscape in y-direction across a Hall bar (a). The first Landau level (LL 1) follows the energy of the conduction band bottom. The resulting peak in the density of states is shown to the right. In (b), top views of the samples are sketched at energies E_i, as indicated in (a). Gray areas denote regions where the energy of the first Landau level is larger than E_i. At energy E_3, the states are localized at potential maxima. At low energy E_1, the electrons are caught in potential minima. At the intermediate energy E_2, however, the electron puddles merge, and electrons may travel across the whole sample via skipping orbits at the edges of the Fermi sea. This leads to extended states around the center of the peak in D(E). In (c), a close-up of the electronic motion in the presence of an electric field in y-direction is shown. The cyclotron motion is superimposed on the motion of the guiding center, which moves perpendicular to both fields at constant velocity. Note for the orientation of the electronic motion that we are looking at the sample in z-direction, opposite to the direction of the magnetic field.

can not flow inside the sample, and the differential Hall conductance should be zero. There-

---

[4]Recall that in our discussion of the Landau quantization, we have assumed spin degeneracy. For spin-split Landau levels, we obtain $\rho_{H,\nu} = -\frac{h}{e^2}\frac{1}{\nu}$.

6.2 The quantum Hall effect

fore, we expect the Hall conductance to remain constant. But in our model system, the insulating interval is only a point! It is here where the "real" properties of the sample enter the explanation: there is disorder in the system, which broadens the δ - functions in the density of states. With the help of Fig. 6.7, we argue that the states in the wings of the peaks of $D(E)$ tend to be localized. Consider a cross section of the potential landscape in y-direction. As long as the magnetic length is small compared to the length scale of the potential fluctuations, the energies of the Landau levels will just follow the potential fluctuations. We study the first Landau level as an example. The same line of arguing holds for higher Landau levels. Clearly, the states inside one Landau level now have different energies, and the shape and width of the peak in the density of states reflects the energy distribution of the disorder potential. Consider states in the low-energy wing of the peak in D(E), i.e., at the energy E_1 in Fig. 6.7. Apparently, the electrons reside in minima of the potential landscape. Close to the center of such puddle, where the local electric field is weak or zero, the electrons move in cyclotron orbits. At the puddle edge, though, they skip along the potential wall during their cyclotron motion.

In order to see this, it pays off to revisit the $\vec{E} \times \vec{B}$ drift, i.e. the motion of charged particles under crossed electric and magnetic fields. In the equation of motion for electrons in the diffusive regime (eq. 2.46), it was assumed that $\omega_c \tau \ll 1$, such that the magnetic field deflects the electrons that move through the sample with the drift velocity, which is determined by the electric field on the one hand, and by the friction term on the other hand. Now, however, the magnetic field is much stronger, and the electrons are able to complete cyclotron circles without being scattered, which means $\omega_c \tau > 1$. Consider the motion on a time scale between $1/\omega_c$ and τ. There is no diffusive scattering, and the classical equation of motion now reads

$$m^* \frac{d^2 \vec{r}}{dt^2} = -e(\vec{E} + \frac{d\vec{r}}{dt} \times \vec{B}) \tag{6.15}$$

where $\vec{r}$ is the position of the electron in the xy-plane. Suppose the electric field points in y-direction, $\vec{E} = (0, E_y, 0)$, and the magnetic field, as usual, points in the z- direction. Equation (6.15) is then solved by

$$x(t) = \frac{E_y}{\omega_c B} \sin(\omega_c t) - \frac{E_y}{B} t + X_0$$

$$y(t) = -\frac{E_y}{\omega_c B} \cos(\omega_c t) + Y_0 \tag{6.16}$$

It is common to separate the motion in the motion of the *guiding center*

$$\bigl(X(t), Y(t)\bigr) = \bigl(X_0 - v_L t, Y_0\bigr) \tag{6.17}$$

and the motion relative to the guiding center

$$\bigl(x_r(t), y_r(t)\bigr) = \bigl(r_c \sin(\omega_c t), -r_c \cos(\omega_c t)\bigr) \tag{6.18}$$

Here, v_L denotes the drift velocity due to the Lorentz force, and $r_c = E_y/\omega_c B$ is the cyclotron radius. We thus see that the electrons perform a cyclotron motion around the guiding center, which moves at constant velocity in the direction perpendicular to both fields, as sketched in

Fig. 6.7 (c). This is nothing but the $\vec{E} \times \vec{B}$ drift. It is straightforward to write down the components of the conductivity tensor for such a situation. Here, we average over the (fast) cyclotron motion, such that only the motion of the guiding center gives a contribution. Since $j_x = -ne(dx/dt) = \sigma_{xx}E_x + \sigma_{xy}E_y$ and $j_y = \sigma_{yx}E_x + \sigma_{yy}E_y$, it follows that

$$\sigma_{xy} = \frac{-ne}{B}$$

$$\sigma_{yy} = 0 \qquad (6.19)$$

Likewise, we obtain the remaining two component by considering an electric field in x-direction. The corresponding components of the resistivity tensor are $\rho_{xx} = \rho_{yy} = 0$ and $\rho_{xy} = -\rho_{yx} = -B/(ne)$. It may seem strange that both the conductivity and the resistivity can be zero at the same time. Remember, however, that we are dealing with a resistivity and conductivity *tensor*. The observation $\sigma_{xx} = 0$ implies that no current flows in x-direction when a voltage is applied in x-direction. $\rho_{xx} = 0$, on the other hand, means that applying a current in x-direction causes no voltage drop in x-direction. This is no contradiction, provided the equipotential lines are parallel to the x-direction if a voltage or current is applied along this axis, Fig. 6.8. It has been verified experimentally that this is in fact the case inside a quantum Hall plateau [Knott1994].

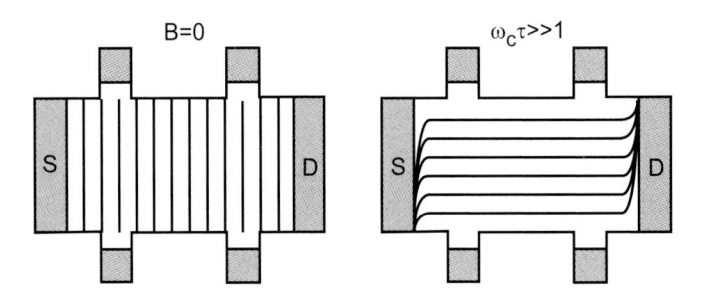

Figure 6.8: Equipotential lines in a 2DEG at B=0 (left) and in strong magnetic fields (right).

Suppose now we change the electron density of a perfectly clean system and apply a negligibly small voltage in x-direction. The only electric field of relevance is the Hall field in y-direction. The electrons move in parallel to the x direction, as long as we are in a metallic state. As we fill one initially empty Landau level, i.e., $\delta n = D_L L$, the Hall conductance changes by $\delta \sigma_{xy} = -2e^2/h$. Nevertheless, $\sigma_{xy}(n)$ is a straight line, since the insulating regions are just points on the n axis. Also, σ_{xx} is zero for all electron densities.

The electrons in the low-energy tail of a peak in $D(E)$, however, are localized. They either perform cyclotron orbits, or they circle around at the edge of the puddle, in the direction of rotation opposite to that one of the cyclotron motion. Such orbits are called *skipping orbits*, which we will revisit in chapter 7. In any case, all the electrons in such a puddle, cannot carry a current across the sample. Thus, we should modify our definition of a metal accordingly and speak of a metal only if the density of *extended* states at the Fermi energy is non-vanishing. In such a situation, we can fill the Landau level while the system is in an insulating state, which means that $\delta\sigma_{xy}(\delta n) = 0$.

A somewhat different kind of localization occurs in the high-energy wing of the peak in $D(E)$. Consider the situation at energy E_3. The states at this energy correspond to skipping orbits that circle around maxima in the potential landscape. Hence, we conclude that also for Fermi energies in the high-energy tail of the density of states peak, the system behaves insulating.

At an intermediate energy E_2, the electron puddles and the potential hills have about equal sizes on average, and the localized skipping orbits of adjacent structures will merge. At this energy, the electron may *percolate* across the whole sample, and consequently, the sample behaves metallic. Under these conditions, σ_{xy} can change and σ_{xx} becomes larger than zero, since the electrons may traverse the sample from source to drain. That such a state does in fact exist is shown more rigorously in the specialized literature (see the references at the end of the chapter). Here, we just state that energy E_2 represents a percolation threshold [Stauffer1995], and that the electron gas undergoes a percolation transition as the Fermi energy is swept across energy E_2.

Thus, the disorder does something truly remarkable: it increases the insulating regions of the parameter range (the parameter is the electron density or the magnetic field) from points (Fig. 6.3) to extended intervals in real samples, while the extended metallic regions of the ideal sample are reduced to very small intervals. This allows the observation of the conductance steps.

Question 6.3: Sketch $\sigma_{xy}(n)$ and $\sigma_{xx}(n)$ for an ideal sample and for a real sample with disorder! Here, n is the two-dimensional electron density.

A more fundamental explanation of the quantum Hall effect has been provided by [Laughlin1981]. This explanation derives the exact quantization from (i) the gauge invariance and (ii) the existence of a mobility gap around the Fermi energy. A different approach by [Thouless1981] derives the exact quantization of the Hall resistance by treating the random potential within perturbation theory. These models are beyond our scope, however. We will instead discuss another point of view in the following chapter, which is based on transport in quantum wires that effectively form at the sample edge [Halperin1982] as indicated in Fig.6.7(b).

6.2.3 The quantum Hall effect and three dimensions

By now, you may wonder whether the quantum Hall effect occurs only in quantum films. In fact, it vanishes in three-dimensional free electron gases. The reason is illustrated in Fig. 6.9. A magnetic field pointing in z-direction quantizes the motion in the xy-plane, where the electrons perform cyclotron orbits. The motion in z direction, however, remains unaffected. The Fermi sphere "condenses" into cylinders of radii

$$k_{xy} = \sqrt{k_x^2 + k_y^2} = \sqrt{\frac{2m^*\omega_c(j + 1/2)}{\hbar}}$$

which extend along the z-direction. The Landau levels have evolved into Landau bands with a one-dimensional density of states. Thus, no matter what magnetic field we apply or how large the Fermi energy is, there are always states at the Fermi energy, there are no insulating regions in the parameter space, and consequently, there are no metal-insulator transitions. Confining the electron gas in z-direction corresponds to a Fermi circle parallel to the xy-plane, the cylinders are projected onto circles, and parametric metal-insulator transitions are again possible. Note that this line of arguing is based on a *free* electron gas in z-direction. The density of states in this direction has been modified by [Stormer1986], who grew a periodic sequence of GaAs quantum wells in z-direction, separated by $Al_{0.18}Ga_{0.82}As$ barriers. This periodic superlattice generated bands of width b, with band gaps in the meV regime, i.e., comparable to $\hbar\omega_c$ for moderate magnetic fields. The corresponding density of states, shown in Fig. 6.9(b), thus develops gaps in sufficiently large magnetic fields, such that the quantum Hall effect should be visible as soon as b gets smaller than $\hbar\omega_c$. This has been experimentally verified by [Stormer1986]. Hence, although the electron gas does not need to be strictly two-dimensional, a sufficiently large anisotropy is a prerequisite for the quantum Hall effect to occur.

6.3 Elementary analysis of Shubnikov-de Haas oscillations

We now turn our attention to the Shubnikov-de Haas (SdH) oscillations at small and intermediate magnetic fields, i.e., for $\omega_C \tau < 1$. Here, the quantum Hall effect is weak, and $\rho_{xx}(B)$ oscillates, but does not vanish, see Fig. 1.3. In this regime, the magnetic field causes just a weak modulation of the density of states. As a consequence, the density of states at the Fermi level, as well as the screening properties of the electron gas, oscillate as the electron density or the magnetic field is tuned. This is reflected in the longitudinal resistivity. The scattering theory of such a system is hampered by several difficulties, which are discussed in the papers by [Ando1974]. For short-range scattering potentials that are weak compared to the Fermi energy, however, Ando has derived an analytic expression for $\rho_{xx}(B)$, which in the limit $\omega_c \tau \ll 1$ reads

$$\rho_{xx}(B) = \rho_{xx}(0)\left[1 - 4\cos\frac{E_F}{\hbar\omega_C} \cdot D(m^*, T) \cdot E(m^*, \tau_q)\right] \tag{6.20}$$

which we call the Ando formula. The resistivity according to this formula is illustrated in Fig. 6.10. The expression $D(m^*, T)$ is known as Dingle term. It contains the temperature dependence and depends on the effective mass:

$$D(m^*, T) = \frac{x}{\sinh x}; \qquad x = \frac{2\pi^2 k_B}{\hbar e B} m^* \Theta \tag{6.21}$$

The exponential term $E(m^*, \tau_q)$ in eq. (6.20) equals

$$E(m^*, \tau_q) = e^{-\frac{\pi}{\omega_c \tau_q}} \tag{6.22}$$

and depends on the quantum scattering time. Therefore, both the effective mass as well as τ_q can be extracted from measuring the temperature dependence of the SdH oscillations. In order

6.3 Elementary analysis of Shubnikov-de Haas oscillations

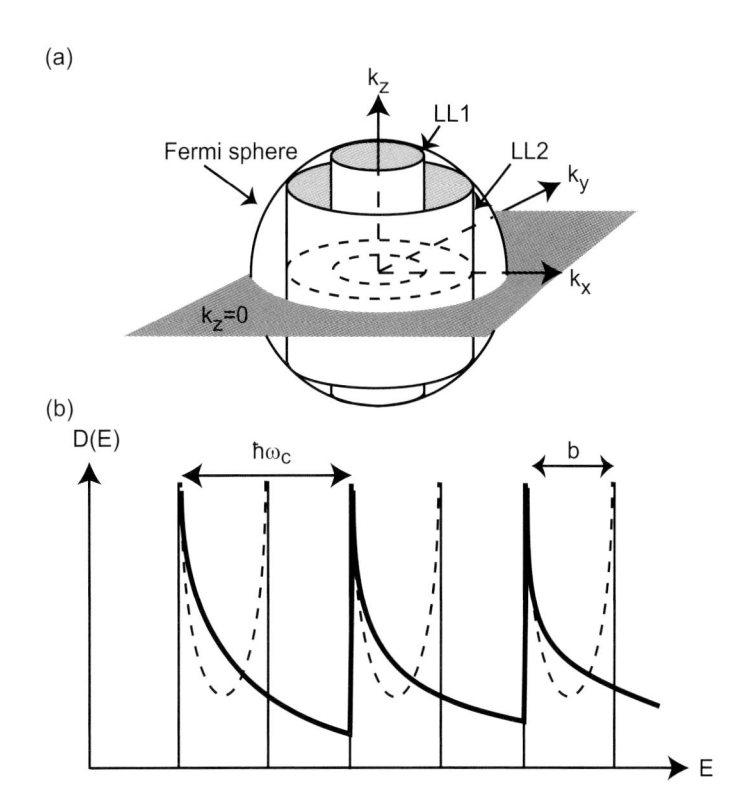

Figure 6.9: Landau quantization in three-dimensional systems, drawn for two occupied Landau bands. (a): under sufficiently large magnetic fields, the Fermi sphere of a free electron gas at B=0 condenses into Landau levels in the xy-plane, while k_z remains continuous. (b): the density of states for a free electron gas (bold line), and of a periodic superlattice in z-direction (dashed lines).

to measure m^*, we pick a single, suitable SdH resonance. Its amplitude is given by

$$A = 8\rho_{xx}(0)D(m^*,T)E(m^*,\tau_q) \tag{6.23}$$

This can be rewritten as

$$ln(\frac{A}{\Theta}) = C_1 - ln(\sinh\frac{2\pi^2 k_B}{\hbar eB}m^*\Theta) \tag{6.24}$$

Here, C_1 denotes a constant, which is no further interest for our purposes. For sufficiently small magnetic fields and sufficiently high temperatures (this is what we meant by "a suitable SdH resonance"), $ln(\sinh x) \approx x$. Thus, by plotting $ln(A/\Theta)$ vs. Θ, a straight line with a slope of

$$\frac{2\pi^2 k_B}{eB\hbar}m^* \tag{6.25}$$

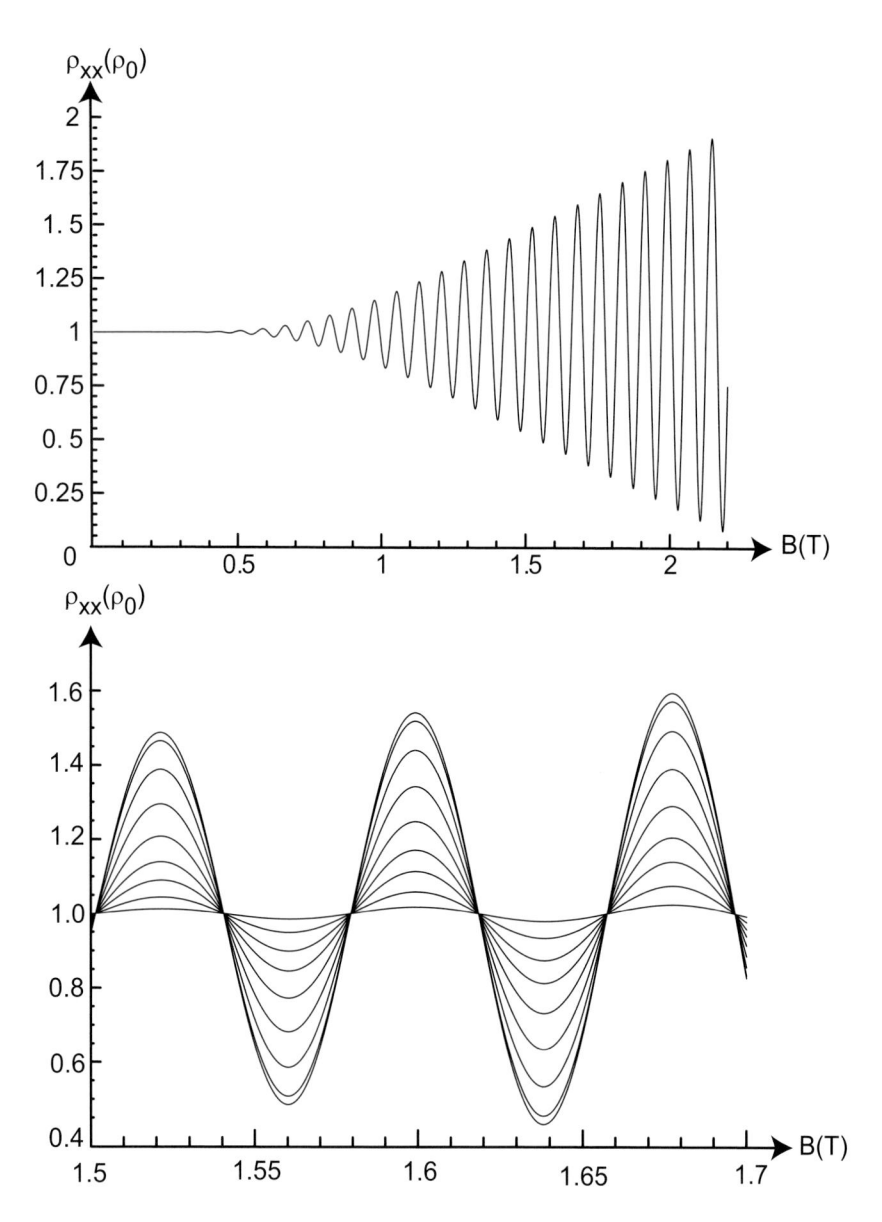

Figure 6.10: Top: SdH oscillations of a 2DEG as a function of B, as described by the Ando formula (eq. (6.20)). Bottom: temperature dependence of some oscillations. The temperatures are Θ = 1K, 2K, 4K, 6K, 8K, 10K, 12K, 15K, 20K. See also E 6.2.

is obtained. Besides analyzing cyclotron resonances, this is a common way to determine effective masses. Once we know m^*, we can exploit the exponential term and determine τ_q along similar lines. This time, a measurement of $\rho_{xx}(B)$ at fixed temperature, which extends over many SdH oscillations, is analyzed. We rewrite the expression for the oscillation amplitude as

$$AB \sinh \frac{2\pi^2 k_B \Theta m^*}{eB\hbar} = 8\rho_{xx}(0) \frac{2\pi^2 k_B \Theta m^*}{e\hbar} e^{-\frac{\pi}{\omega_c \tau_q}} \Rightarrow$$

$$Y = ln(AB \sinh \frac{2\pi^2 k_B \Theta m^*}{eB\hbar}) = C_2 - \frac{\pi m^*}{e} \frac{1}{\tau_q} \frac{1}{B} \quad (6.26)$$

and plot Y as a function of $1/B$ (known as *Dingle plot*). The slope of this straight line equals

$$\frac{\pi m^*}{e \tau_q} \quad (6.27)$$

Fig. 6.11 shows the results of such an analysis performed on a GaN $-$ Al$_x$Ga$_{1-x}$N-HEMT - structure.

6.4 Some examples of magnetotransport experiments

There are many further interesting magnetotransport experiments on quantum films, and we provide a few examples.[5]

6.4.1 Quasi-two-dimensional electron gases

For Fermi energies above the second quantized energy level of the confining potential in z-direction, the electron gas is no longer strictly two-dimensional, and we speak of a *quasi-2DEG*. While the Hall slope measures the total electron density, each two-dimensional subband causes a Shubnikov - de Haas oscillation, provided the scattering times are sufficiently large. This results in a modulation of SdH oscillations, Fig. 6.12. The scattering times in the upper subband, however, can be small, such that the corresponding SdH oscillation may not be observable. In this case, their occupation can be detected as a difference of the total electron density (as obtained from the Hall slope) and the density of the lowest subband (determined from the SdH oscillation period of the lower subband).

A further signature of multi-subband occupation is a parabolic and positive magnetoresistivity around B=0. The subbands can be regarded as resistors in parallel, such that the total conductivity tensor is obtained by simple addition of the individual subband conductivity tensors. The two subbands are characterized by different scattering times τ_1 and τ_2. We further assume that the effective masses in both subbands are identical, and that intersubband scattering can be neglected, i.e. the scattered electrons remain in their original subband during scattering events.[6] The total magnetoconductivity tensor is then given by

$$(\sigma) = \frac{n_1 e^2 \tau_1}{1+(\omega_c \tau_1)^2} \begin{pmatrix} 1 & -\omega_c \tau_1 \\ \omega_c \tau_1 & 1 \end{pmatrix} + \frac{n_2 e^2 \tau_2}{1+(\omega_c \tau_2)^2} \begin{pmatrix} 1 & -\omega_c \tau_2 \\ \omega_c \tau_2 & 1 \end{pmatrix}$$

---

[5] We will see another very important example in chapter 8, namely how electronic phase coherence manifests itself in the magnetoresistivity.

[6] Intersubband scattering can be included in the analysis, see [Zaremba1992]. It is important, as we have seen in Fig. 3.29.

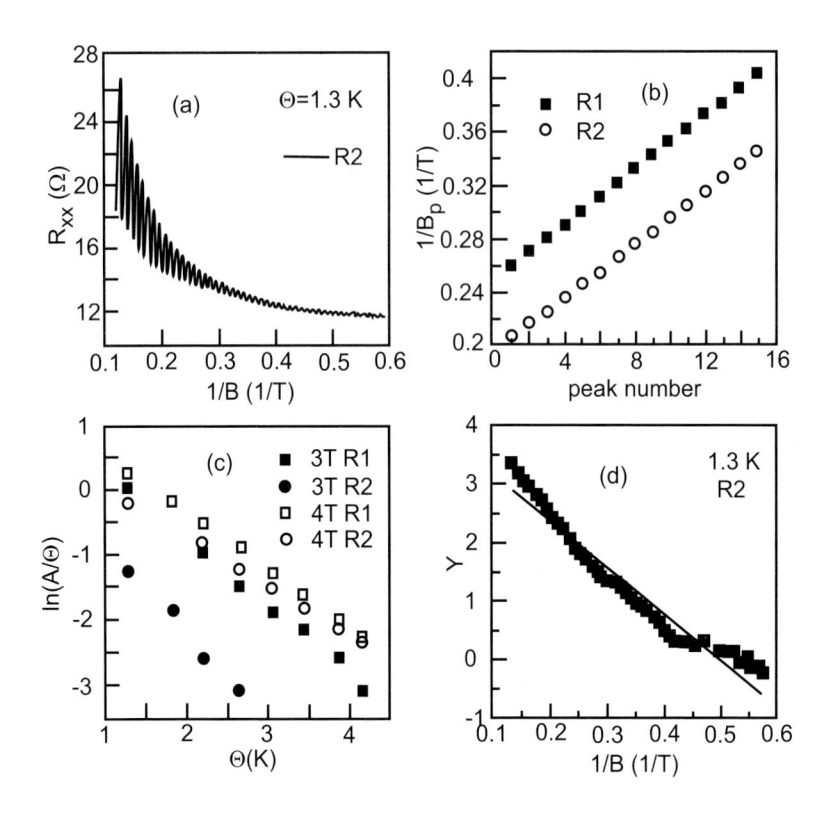

Figure 6.11: Standard analysis of SdH oscillations (a), performed on two different samples (R1 and R2) of a 2DEG at a modulation doped $GaN - Al_xGa_{1-x}N$ interface. The electron density was $n = 4.8 \cdot 10^{16}$ m^{-2} (b). A quantum scattering time of $\tau_q = 0.5$ ps (c) and an effective mass of $m^* = (0.215 \pm 0.006)m_e$ (Dingle plot, d) is found. After [Saxler2000].

The longitudinal resistivity ρ_{xx} is obtained by matrix inversion as discussed in chapter 2. For small magnetic fields, a Taylor expansion gives the expression

$$\rho_{xx}(B) = \rho_{xx}(0)\left(\left[1 + \frac{n_1\tau_1 n_2\tau_2(\tau_1 - \tau_2)^2 e^2}{m^{*2}(n_1\tau_1 + n_2\tau_2)^2}B^2\right]\right) \tag{6.28}$$

Typical experimental results show such a behavior, as illustrated in Fig. 6.12.

Question 6.4: Derive eq. (6.28), and determine the conditions for which the Hall slope measures the total electron density $n = n_1 + n_2$ of both subbands!

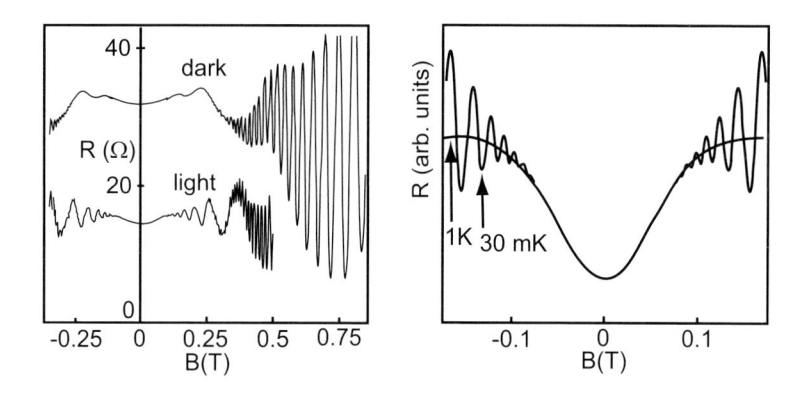

6.4 Some examples of magnetotransport experiments 155

Figure 6.12: Longitudinal magnetoresistivity of a 2DEG in a GaAs $-$ Al$_x$Ga$_{1-x}$As-HEMT with two occupied subbands (left). Here, "light" indicates that the sample has been illuminated by light of a frequency below the GaAs band gap. This ionizes residual neutral donors and thus increases the electron density in the 2DEG. The magneto-oscillations are modulated in both cases. The two SdH frequencies correspond to the partial electron densities in the two subbands. In addition, a positive magnetoresistivity around $B = 0$ is observed (right). After [Houten1988a].

6.4.2 Mapping of the probability density

Perturbation theory tells us that inserting a δ-function $U_0\delta(z-z_0)$ in a potential generates, to first order, energy shifts ΔE_i of the energy eigenvalues E_i, where ΔE_i is proportional to the probability density of the corresponding eigenstate at z_0:

$$\Delta E_i = U_0 \mid \psi(z) \mid^2 \tag{6.29}$$

With U_0 known, $|\psi(z)|^2$ can therefore be determined by measuring ΔE_i [Marzin1989]. If we could scan the δ-function across the potential, the probability density could be mapped this way. This idea can in fact be realized experimentally in parabolic quantum wells, where the conduction band bottom varies parabolically in growth direction. It is a unique property of a parabolic potential that superposition of a constant electric field displaces the potential without changing its shape, see [E 6.3]. In the sample depicted in Fig. 6.13, the quantum well contains a Al$_{0.3}$Ga$_{0.7}$As spike at its as-grown center, while the parabola can be shifted by applying a voltage applied to the two electrodes haloing the parabola [Salis1997]. In the experiment shown in Fig. 6.13, two subbands were occupied, and by measuring the two subband densities via SdH oscillations, the energy shifts have been detected. Hence, $|\psi_2(z)|^2 - |\psi_1(z)|^2$ has been measured, see Fig. 6.13(c,d).

6.4.3 Displacement of the quantum Hall plateaux

According to our model in 6.2, the quantum Hall plateaux are centered around integer filling factors. In other words, if we extrapolate the classical Hall slope into the quantum Hall regime,

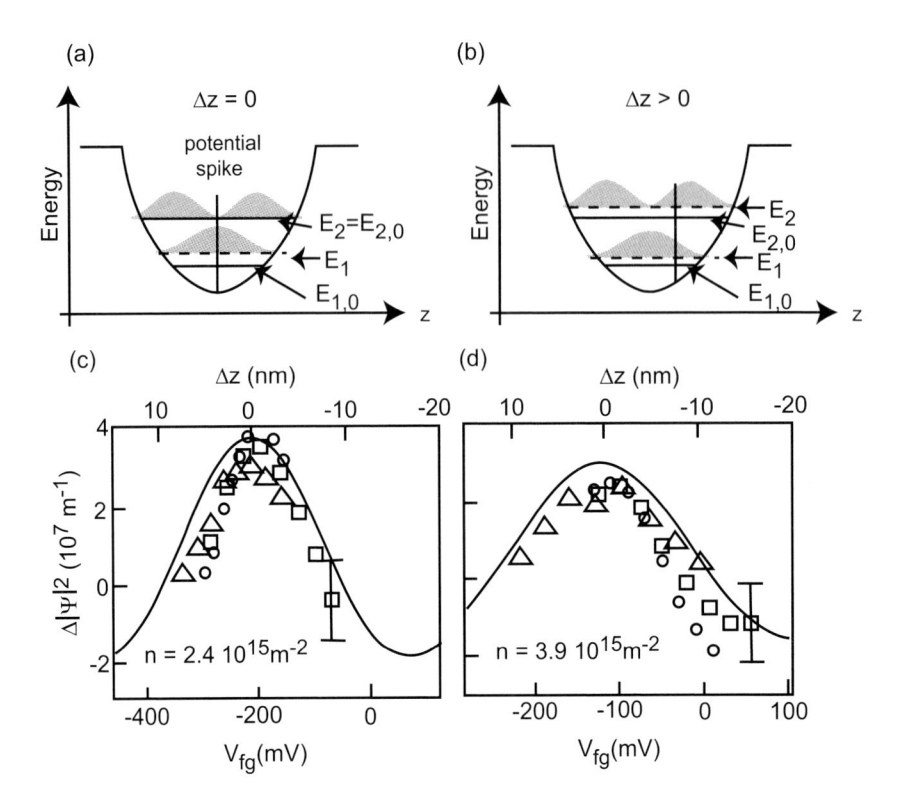

Figure 6.13: (a): conduction band of a parabolic Ga[Al]As quantum well. A potential spike has been grown at the center of the parabola, which shifts the energy levels as compared to the same potential without the spike (dashed lines). The shift is proportional to $|\Psi^2|$ at the position of the spike. By applying a constant electric field in z-direction, the confining parabola is displaced with respect to the spike, and the energy levels shift accordingly. Measured differences of the probability density between subband 1 and 2 as a function of z for two different electron densities. The different symbols denote different spike heights, i.e. a Al concentration of $x = 0.05, 0.1$ and 0.15, respectively. Adapted from [Salis1997].

it should intersect the plateaux at their center. Here, we have implicitly assumed that the peaks of the density of states are symmetric, which is not necessarily the case. Their shape depends of the character of the scatters. For example, predominantly repulsive scatterers shift the center of gravity of a Landau level towards higher energies, as shown schematically in Fig. 6.14(a). This asymmetry affects the position of the quantum Hall plateaux. Suppose the scatterers are predominantly repulsive, we start out from LL j completely filled, and we increase the magnetic field. As long as we deplete localized states, no changes in the resistivities are observed. As can be seen from Fig. 6.14(c), this means that the delocalized states reach the Fermi level at larger magnetic fields as compared to the symmetric situation, and the jump in ρ_{xy}, as well as the peak in ρ_{xx}, are shifted correspondingly with respect to the classical Hall slope. This effect has been studied systematically by [Haug1987] (Fig. 6.15) and is

6.5 Parallel magnetic fields

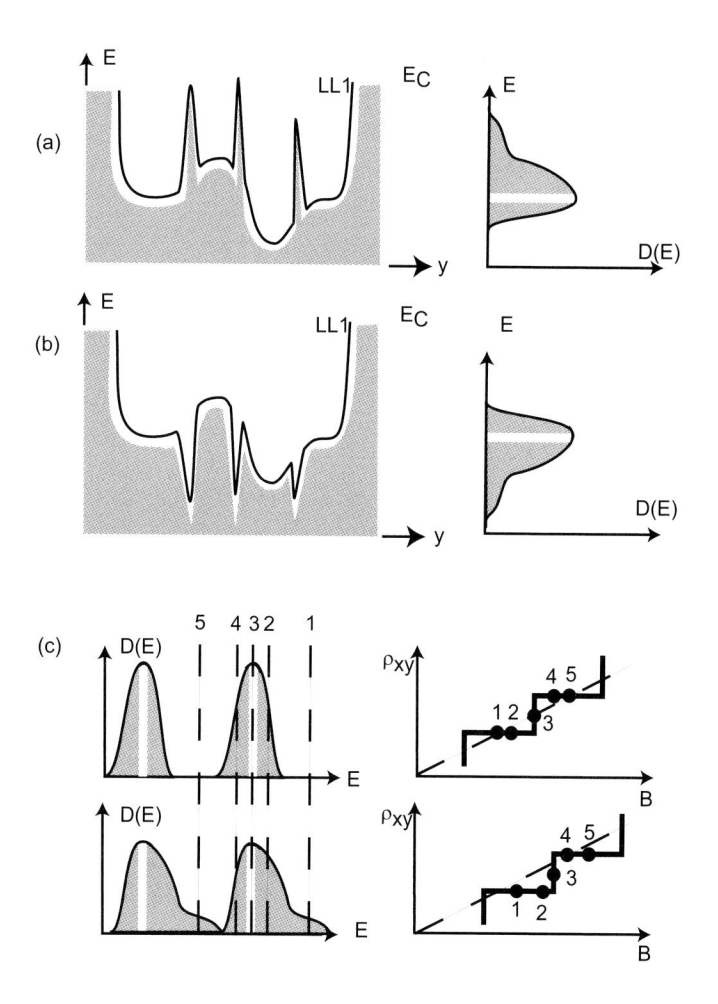

Figure 6.14: Repulsive scatterers (a) shift a fraction of the states within a peak of the density of states to higher energies, which results in a shift of the quantum Hall plateaus to larger magnetic fields (c). Likewise, predominantly attractive scatterers (b) shift the quantum Hall plateaus to smaller magnetic fields.

occasionally used to obtain further information on the scatterers.

6.5 Parallel magnetic fields

In comparison to the quantum Hall effect, a magnetic field in the plane of the 2DEG produces much less spectacular results. Nevertheless, investigating the transport properties as a function of a parallel magnetic field $B_\parallel$ is a useful tool. The density of states can be tuned, and spin effects can be investigated without being buried in the dominating orbital effects. Here, we

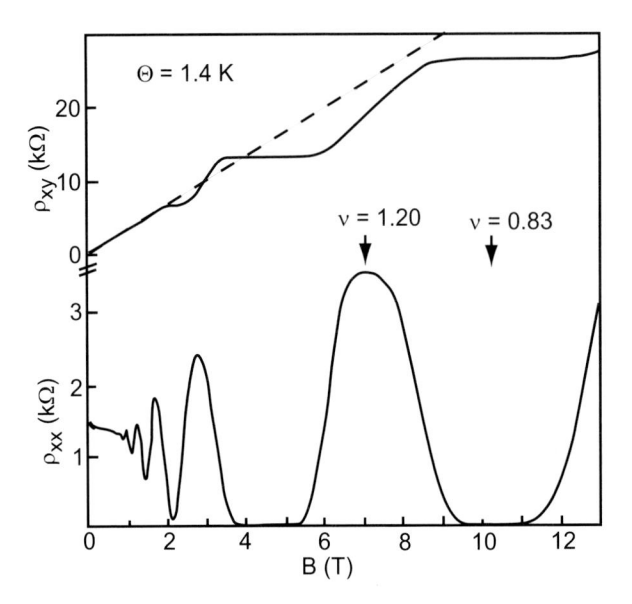

Figure 6.15: Quantum Hall plateaux can be shifted with respect to the extrapolated classical Hall trace (dashed line). Adapted from [Haug1987].

study how $B_\parallel$ affects the density of states and discuss the consequences.

A 2DEG in a homogeneous magnetic field of arbitrary orientation is a very complicated problem in its most general form. For example, the classical dynamics of a square well in tilted magnetic fields is chaotic [Fromhold1995], which leads to the corresponding quantum mechanical signatures. The evolution of the energy levels as a function of a tilted magnetic field can be studied using perturbation theory [Bhattacharya1982],[Salis1998]. An analytical solution is possible for a parabolic confinement [Maan1984]. Besides tuning the spin splitting, $B_\parallel$ has two effects. First of all, the parabolic confinement generated by $B_\parallel$ adds to the electrostatic confinement, such that the subbands shift to higher energies. This is sometimes referred to as *diamagnetic shift*. Second, the effective mass in the direction perpendicular to $B_\parallel$, but in the plane of the electron gas, increases. We have seen this behavior already in its extreme version in the QHE, where the effective mass goes to infinity.

Here, we restrict ourselves to a simple case, which reveals these properties analytically. Other potential shapes show a similar qualitative behavior. We consider a parabolic quantum well with an electrostatic potential in growth direction, given by

$$V(z) = \frac{1}{2} m^* \omega_0^2 z^2 \qquad (6.30)$$

The magnetic field is applied in x-direction, and we choose the gauge $\vec{A} = (0, -zB_\parallel, 0)$, which gives $\vec{B} = \vec{\nabla} \times \vec{A} = (B_\parallel, 0, 0)$. The Schrödinger equation now reads

6.5 Parallel magnetic fields

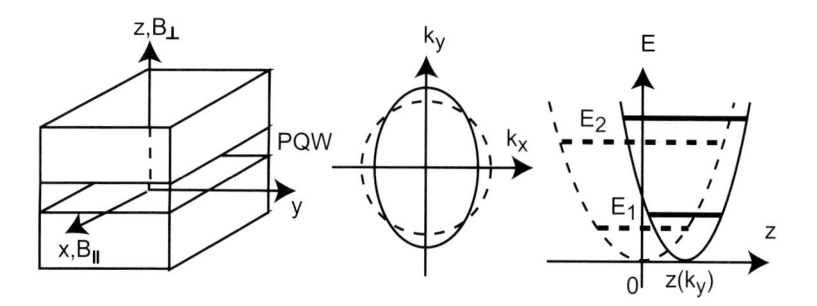

Figure 6.16: Left: definition of the coordinate system and the magnetic field directions for a parabolic quantum well in perpendicular and parallel magnetic fields. Center: corresponding shape of the two-dimensional Fermi sphere in a magnetic field applied in x-direction (full line), in comparison to the fermi sphere for $B = 0$. Right: the confining potential in z-direction for $B = 0$ (dashed line) shifts and narrows for a state with wave number k_y when a parallel magnetic field is applied (full line). This leads to a diamagnetic shift of the energy levels, and an increase in the subband separation, as indicated for the first two subbands.

$$\frac{1}{2m^*}(p_x^2 + (p_y - eB_\parallel z)^2 + p_z^2)\Psi(\vec{r}) + \frac{1}{2}m^*\omega_0^2 z^2 \Psi(\vec{r}) = E\Psi(\vec{r}) \tag{6.31}$$

by applying the momentum operators to the wave function ansatz

$$\Psi(\vec{r}) = exp(ik_x x)exp(ik_y y)\Phi(z)$$

this can be written as

$$\left[\frac{\hbar^2 k_x^2}{2m^*} + \frac{\hbar^2 k_y^2}{2m^*} + \frac{1}{2}m^*\omega^2(B)[z^2 + \frac{2\hbar k_y \omega_C}{m^*\omega^2(B)}z]\right]\Psi(\vec{r}) = E\Psi(\vec{r}) \tag{6.32}$$

with

$$\omega(B_\parallel) = \sqrt{\omega_0^2 + \omega_c^2} \tag{6.33}$$

To complete the square, we add and subtract

$$z^2(k_y) = \left[\frac{\hbar k_y \omega_C}{m^*\omega^2(B_\parallel)}\right] \tag{6.34}$$

and obtain

$$\left[\frac{\hbar^2 k_x^2}{2m^*} + \frac{\hbar^2 k_y^2}{2m^*}\frac{\omega_0^2(B)}{\omega^2(B)} + \frac{1}{2}m^*\omega^2(B)[z + z(k_y)]^2\right]\Psi(\vec{r}) = E\Psi(\vec{r}) \tag{6.35}$$

where

$$m_y^*(B) = m^*\frac{\omega^2(B_\parallel)}{\omega_0^2} \tag{6.36}$$

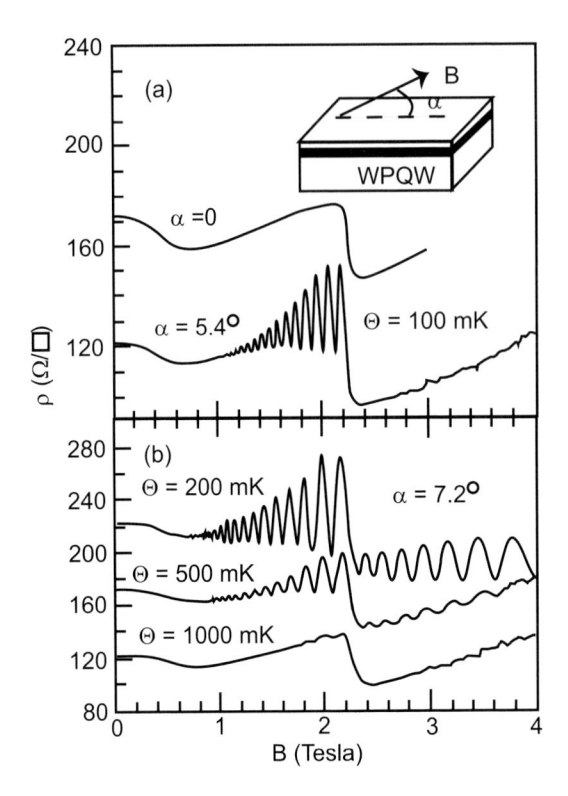

Figure 6.17: Magnetoresistivity measurements in a parabolic quantum well in tilted magnetic fields. In (a), $\rho_{xx}(B)$ is shown for a tilt angle of zero ($\theta = 0$). The minima can be attributed to the depletion of the subbands 3 and 2, respectively. As the sample is tilted, a perpendicular magnetic field component generates SdH oscillations. Temperature-dependent measurements (b) allow to determine the effective electron mass. After [Facer1997].

The solution of eq. (6.32) consists of free electrons in x-and y-direction, with an effective mass in y-direction which increases as $B_\parallel$ is increased. Intuitively, we can think of the electron trajectories being bent by $B_\parallel$ in z-direction, such that the electron has more difficulties moving in y-direction. Hence, the Fermi sphere is deformed into an ellipse, Fig. 6.16. As we have seen already during the discussion of the Fermi surface of Si in chapter 3, the average effective mass is given by

$$m^*(B_\parallel) = \sqrt{m_x^* m_y^*} = m^* \frac{\omega(B_\parallel)}{\omega_0} \tag{6.37}$$

Therefore, the density of states increases for each subband as $B_\parallel$ increases. The total density of states is given by

$$D_{PQW}(E, B_\parallel) = \frac{m^*(B_\parallel)}{\pi \hbar^2} \sum_{j=0}^{\infty} \theta(E - E_j(B_\parallel)) \tag{6.38}$$

6.5 Parallel magnetic fields

with the subband energies

$$E_j(B_\|) = (j + \frac{1}{2})\hbar\omega(B_\|) \tag{6.39}$$

Note further that the states in each subband are centered at positions given by $z(k_y)$. Electrons that move in $+k_y$-direction ($-k_y$-direction) are predominantly located at more negative (positive) z, see Fig. 6.16. This effect just describes the deflection of the moving electrons in a magnetic field. In order to check this model experimentally, we can apply a parallel magnetic field to a parabolic quantum well and probe the effective mass by temperature-dependent Shubnikov-de Haas measurements in an additional perpendicular magnetic field. True, the Hamiltonian has to be modified, but as long as the perpendicular magnetic field $B_\perp$ is sufficiently small, we can treat it as a perturbation, which leaves the modifications imposed by $B_\|$ unchanged. If more than one subband is occupied at $B = 0$, we should be able to see the magnetic depopulation of the upper subbands as $B_\|$ increases. In a standard experimental setup, however, there is only one magnetic field direction available. This problem can be solved by measuring the magnetoresistivity with the sample tilted with respect to the magnetic field direction. After performing this experiment for a sequence of different tilt angles, the magnetic field can be disentangled in its parallel and its perpendicular component and we are able to analyze $\rho_{xx}(B_\perp, B_\|)$. In Fig. 6.17, some raw data of such an experiment are shown. Here, three subbands were occupied in the parabolic quantum well at $B_\| = 0$. At zero tilt angle, the effect of a purely parallel magnetic field on the resistivity can be studied. One observes a minimum in $\rho_{xx}(B_\|)$ at $B_\| \approx 0.7$ T, and a sharp decrease at $B_\| = 2.2$ T, which are attributed to the depopulation of the third and second subband, respectively. While the origin of the minimum is not well understood, the resistivity drop at the depletion of the second subband has two reasons. First of all, the electrons in the second subband suffer a lot of scattering at low subband densities N_2, which increases the resistivity. Second, intersubband scattering is no longer possible for $B_\| > 2.2$ T. As the sample is tilted, the perpendicular magnetic field induces Shubnikov-de Haas oscillations, which can be used to determine the subband densities, once the oscillations have been attributed to a particular subband. In the example shown in Fig. 6.17, the oscillation in $1\,T < B < 2\,T$ is attributed to the second subband, while oscillations at higher magnetic fields stem from the first subband. In addition, Hall mesurements can be performed in order to determine the total electron density. Hence, the electron densities N_1 and N_2 can be determined, as shown in Fig. 6.18(a). As expected, the upper subband is depleted by $B_\|$, while N_1 approaches the total electron density at strong parallel magnetic fields. Furthermore, temperature dependent measurements can be used to determine the effective mass and the quantum scattering time, as shown in Fig. 6.18(b) and (c), respectively. While the average effective mass agrees reasonably well with the model described above, it is found that the quantum scattering time τ_q increases strongly as the second subband gets close to depletion, an effect which is poorly understood. Apparently, the electrons screen the small angle scattering potential much better when they are in the first subband.

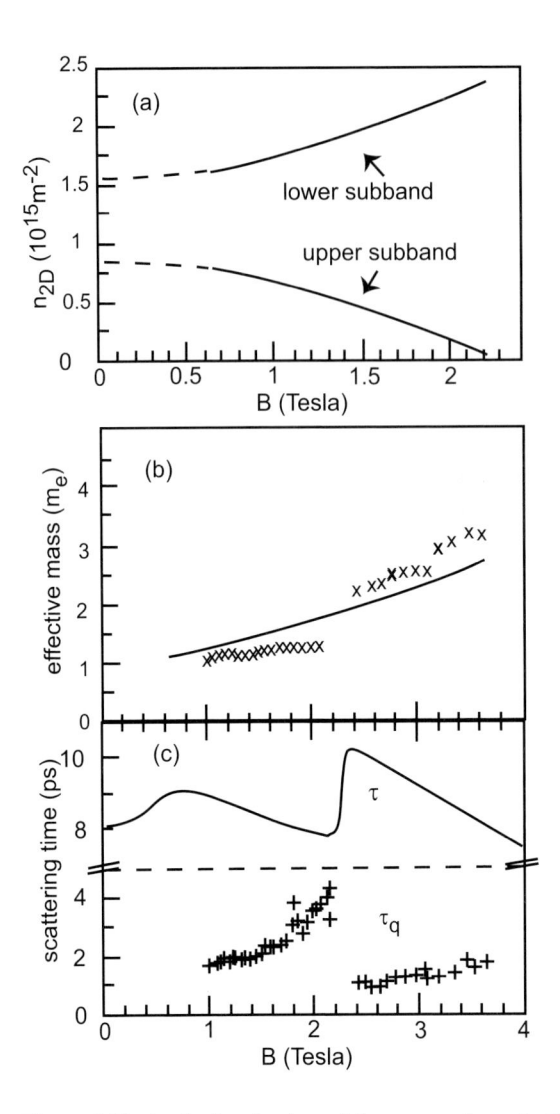

Figure 6.18: Analyzing the data of the type as shown in Fig.6.17 gives not only electron densities of the lowest two subbands (a), but also the effective mass (b) and the scattering times (c) as a function of $B_\parallel$. While $m^*(B_\parallel)$ is in reasonable agreement with the model described in the text (full line in (b)) and the behavior of the Drude scattering time τ is as expected, it is somewhat surprising that the quantum scattering time τ_q increases strongly in the second subband as this band gets depleted. After [Facer1997].

Papers and Exercises

P 6.1: Go through [Smith1986] and discuss the difficulties encountered when capacitances in high magnetic fields are measured by applying a voltage directly between a top gate and the electron gas.

P 6.2: The helium fountain effect (see section 4.2) has been used by [Klass1991] to detect the spatial distribution of dissipation in the quantum Hall effect. Discuss this entertaining experiment in relation to the equipotential lines sketched in Fig. 6.8!

P 6.3: An interesting relation between Shubnikov-de Haas oscillations and the quantum Hall effect has been reported in [Stormer1992]. What is the explanation suggested by [Simon1994]?

E 6.1: Analyze the data of Fig.1.3, measured on a GaAs-HEMT! Enumerate the quantum Hall plateaux. Determine the electron density from both the Hall slope and from the Shubnikov-de Haas oscillations? Extract the mobility and the scattering time. Estimate the effective g-factor.

E 6.2: Figure 6.10 shows measurements of the resistivity of a two-dimensional electron gas as a function of magnetic field and temperature. The electron density has been determined from the Hall slope to $n = 6.2 \cdot 10^{15}$ m^{-2}, while from measuring $\rho_{xx}(B = 0)$, the mobility is found to be $\mu = 77$ m^2/Vs

   (i) Determine τ, m^*, and τ_q!
   (ii) What material could it be? What does the ratio τ/τ_q tell you?

E 6.3: A one-dimensional harmonic oscillator (characterized by ω_0) is placed in a constant electric field F.

Solve the Schrödinger equation. Show that the parabola gets displaced in both space and energy, but maintains its shape. Determine the location and the energy of the potential minimum as a function of F.

Further Reading

The full story of the quantum Hall effect is much more complicated than the simple picture developed here. The fractional quantum Hall effect has been left aside. For a full theoretical discussion, the reader is referred to [Ezawa2000]. More elementary introductions are given by [Prange1990],[Hajdu1994], and by [Yoshioka2002]. More on quantum films in general can be found in [Ferry1997].

7 Quantum Wires and Quantum Point Contacts

Quantum wires (QWRs) are quasi-one-dimensional systems, i.e., their width w is comparable to the Fermi wavelength. In analogy to our previous notation in two dimensions, the wire is strictly one-dimensional if only the mode with the lowest energy is occupied. The wire is called diffusive if its length L is much larger than the elastic mean free path ℓ_e, Fig. 7.1(a). In this case, the electrons will suffer many elastic scattering events during their trip along the wire. Note that the trajectories indicated by an arrow in the figure are only meaningful for $\lambda_F \ll w$, which means that the number of occupied modes is sufficiently large, and a localized wave packet can be constructed. If only a few modes are occupied, the semiclassical picture

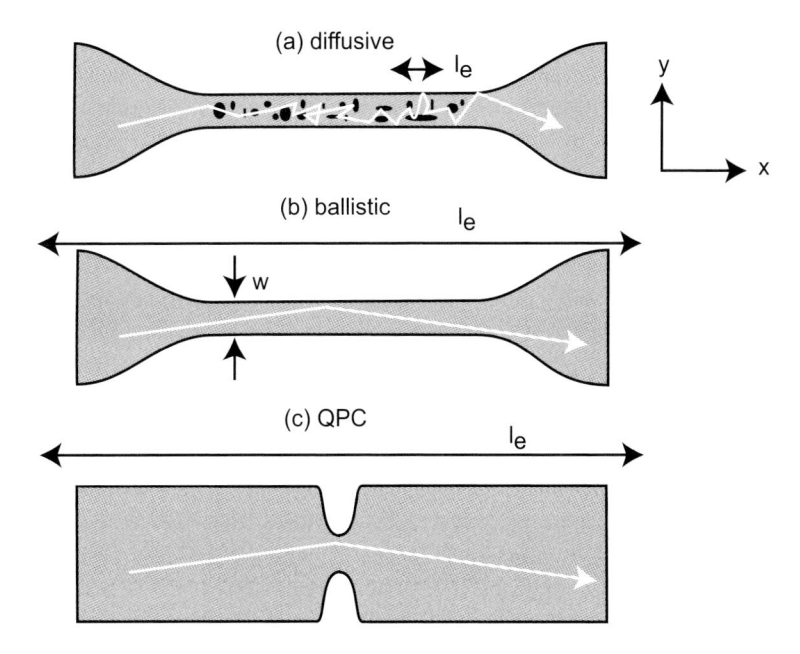

Figure 7.1: A diffusive wire (a) contains many scatterers. Its transport properties can be described by the Boltzmann equation. This clearly becomes questionable for wires of length $L \approx \ell_e$. A wire with $L > \ell_e$ is called ballistic (b): the electrons are scattered at the confining walls only. (c): a ballistic wire with $w \approx L \gg \ell_e$ is called a "quantum point contact" (QPC). Adapted from [Beenakker1991].

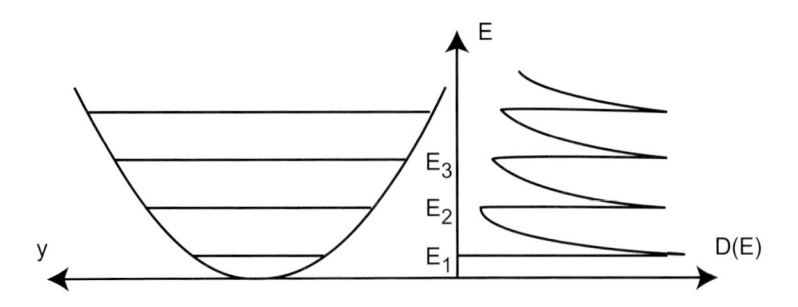

Figure 7.2: Schematic energy spectrum of a parabolic quantum wire, and the corresponding density of states.

breaks down, and we should think of the electrons as plane waves inside the QWR. We will study the basic magnetoresistance properties of diffusive quantum wires in section 7.1. In the opposite limit, $L \ll \ell_e$, there is no elastic scattering in the wire, except for boundary scattering at the walls, Fig. 7.1(b). Such wires are often very short and form a point-like contact between the left and the right reservoir. Such short ballistic quantum wires are usually called *quantum point contacts* (QPCs), Fig. 7.1(c). One of the central questions in this chapter is the resistance of ballistic quantum wires. "Well, there should be no resistance in a ballistic wire", you might say. Whether this is true or not depends on what exactly we mean by "the resistance of the wire". It turns out that a two-terminal measurement gives quantized resistances, which resemble very much the quantum Hall effect, and there is in fact a surprising relation. If, however, the resistance is measured in a four-terminal geometry using suitable voltage probes, the resistance vanishes. The formalism used to describe transport in ballistic wires has been developed by R. Landauer and M. Büttiker. An introduction to ballistic quantum wires is given in section 7.2. In section 7.3, we will discuss the quantum Hall effect and the Shubnikov-de Haas oscillations can be discussed in terms of transport through ballistic quantum wires. This includes introducing the Landauer-Büttiker formalism. It will be established that quasi-one-dimensional *edge states* carry the current in the quantum Hall regime. All the QWRs we will have discussed up to this point reside in semiconductor hosts, but this is not the only way of realizing a QWR experimentally. Some alternative approaches will be presented in section 7.4, which completes our discussion of QWRs.

Throughout this chapter, the wires extend in x-direction, while the motion is y-and z-direction is quantized. Furthermore, we use the effective mass approximation, except when carbon nanotubes are discussed in section 7.5. Hence, the density of states of a quantum wire is given by

$$D_{QWR}(E) = \sum_{j=1}^{\infty} D_1(E - E_j) \qquad (7.1)$$

where $D_1(E - E_j)$ denotes the one-dimensional density of states with a band bottom at energy E_j, see Fig. 7.2. In reality, the singularities at the mode bottoms are broadened by both disorder and temperature and thus do not cause any difficulties. The electron density in the ideal QWR is given by

$$n_{QWR} = \frac{2}{\pi\hbar}\sum_{j=1}^{\infty}\sqrt{2m^*(E-E_j)}\theta(E-E_j) \qquad (7.2)$$

Here, we have assumed a spin degeneracy of 2, and $\theta(E-E_j)$ denotes the Heaviside step function.

7.1 Diffusive quantum wires

7.1.1 Basic properties

An experimental realization of a diffusive quantum wire in a Ga[Al]As HEMT is shown in Fig. 7.3. For such wires, a parabolic confinement in y-direction is an excellent approximation (z is, as usual, the growth direction):

$$V(y) = \frac{1}{2}m^*\omega_0^2 y^2 \qquad (7.3)$$

In fact, it is not easy to detect experimentally a non-parabolic confinement. The characteristic quantities of the wire can be determined by - you guessed right - magnetotransport experiments. How does a magnetic field influence the energy levels in a quantum wire? The problem

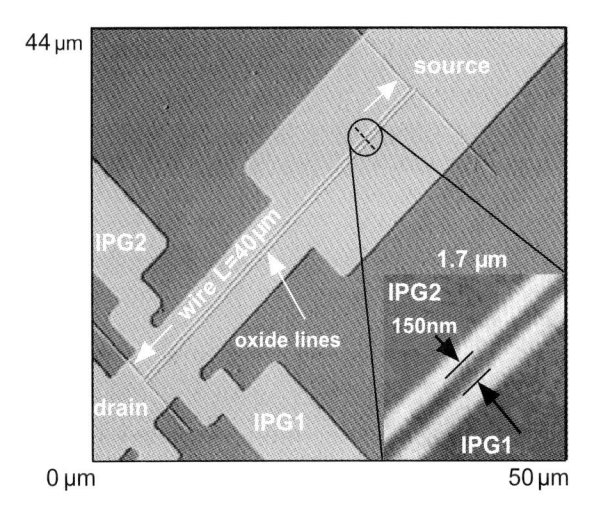

Figure 7.3: Top view of a diffusive quantum wire ($L = 40$ μm, geometric width $w_g = 150$ nm), patterned into a Ga[Al]As-HEMT by local oxidation. The bright lines define the walls (a close-up is shown in the inset). The electrostatic wire width w can be tuned by applying voltages to the two in-plane gates IPG1 and IPG2.

is similar to that one of a parabolic quantum well in a parallel magnetic field, as studied in section 7.5. We have to add the potential (7.3) to the Hamiltonian

$$H = \frac{1}{2m^*}p_y^2 + \frac{m^*\omega_c^2}{2}(y-y_n)^2 \tag{7.4}$$

known from the Landau quantization. We use the ansatz for the wave function

$$\Phi(x,y) = e^{ik_x x}\psi(y) \tag{7.5}$$

and obtain (see exercise E 7.1)

$$H_{xy} = \frac{1}{2m^*}p_y^2 + \frac{m^*\omega(B)^2}{2}(y-\bar{y}_n)^2 + \frac{\hbar^2 k_x^2 \omega_0^2}{2m^*\omega(B)^2}, \tag{7.6}$$

with

$$\omega(B) = \sqrt{\omega_0^2 + \omega_c^2}$$

and

$$\bar{y}_n = y_n \frac{\omega_c^2}{\omega^2(B)} \tag{7.7}$$

The last term in eq. (7.6) describes plane waves in x-direction. They have the usual free electron dispersion, with the magnetic mass

$$m^*(B) = m^*\left[\frac{\omega(B)}{\omega_0}\right]^2 \tag{7.8}$$

which is *larger* than the effective mass at $B=0$. The remaining terms in eq. (7.6) give an effective confinement in y-direction, characterized by $\omega(B)$. The energy eigenvalues therefore depend on B and are given by

$$E_j(B) = \hbar\omega(B)(j-\frac{1}{2}) \tag{7.9}$$

with $j = 1, 2, ...$ enumerating the one-dimensional modes. The resulting density of states is qualitatively sketched in Fig. 7.2.

Question 7.1: How do the wire modes evolve into Landau levels in the limit $\omega_c \gg \omega_0$? What happens to the magnetic mass and to the electron velocity in x-direction?

The magnetic field thus increases the confinement strength and may depopulate the magneto-electric modes by squeezing them above the Fermi level, very similar to the diamagnetic shift discussed in chapter 6. The density of states at the Fermi level therefore oscillates as function of B, which is reflected in the resistivity. In contrast to the SdH oscillations, the corresponding magneto-oscillations are not periodic in $1/B$. At large magnetic field ($\omega_0 \ll \omega_C$), the electrons feel an effectively two-dimensional potential governed by the magnetic confinement. In

7.1 Diffusive quantum wires

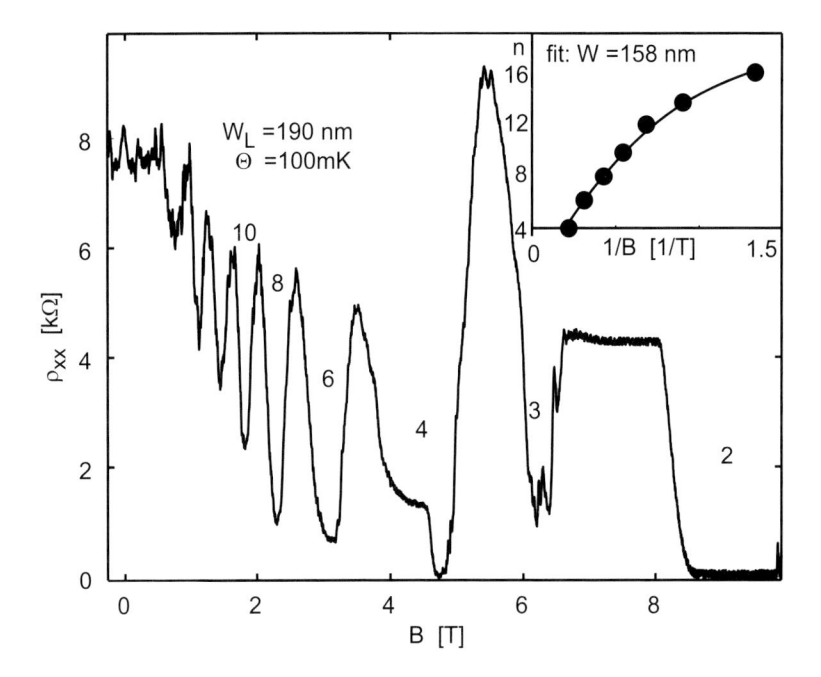

Figure 7.4: The magnetoresistivity of the wire shown in Fig. 7.3. The most prominent feature is the oscillation as a function of B, which detects the magnetic depopulation of the wire modes. Further features are a pronounced maximum at B=0, and fluctuations at small magnetic fields. These are due to phase coherence effects and will be discussed in chapter 8. The positions of the oscillation minima are plotted vs. $1/B$ as full circles in the inset. Here, the line is a least squares fit to eq. (7.10), which gives an electronic wire width of 158 nm.

this case, the magnetotransport properties should approach those of a 2DEG in the quantum Hall regime. At small magnetic fields, on the other hand, the mode spacing approaches $\hbar\omega_0$ as B is decreased, and the level spacing becomes approximately independent of B. The number j of occupied modes as a function of B, ω_0 and n_{QWR} has been calculated by [Berggren1988]. The authors obtain (see exercise E7.1)

$$j(n_{QWR}, \omega_0, \frac{1}{B}) = (\frac{3\pi}{4} n_{QWR}\omega_0)^{2/3} (\frac{\hbar}{2m^*})^{1/3} \frac{1}{\omega(B)} \quad (7.10)$$

Eq. (7.10) can be used to determine the characteristic wire parameters by fitting $j(n_{QWR}, \omega_0, \frac{1}{B})$, using ω_0 and n_{QWR} as fit parameters. A measurement of $\rho_{xx}(B)$, performed on the quantum wire of Fig. 7.3, including the result of such a fitting procedure, is shown in Fig. 7.4.

7.1.2 Boundary scattering

In quantum wires, the electrons hit the confining wall much more frequently than in a 2DEG. Whether this boundary scattering contributes to the resistivity is a legitimate question. This

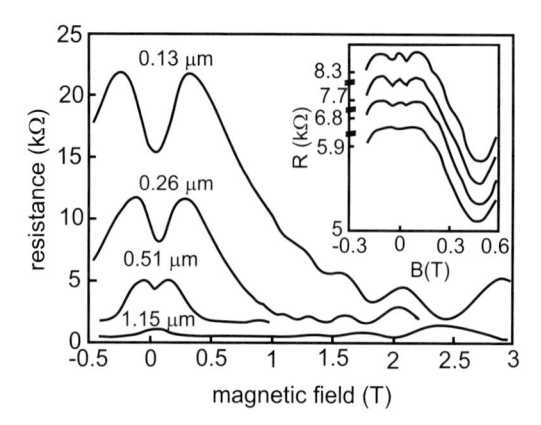

Figure 7.5: Diffusive scattering at the wire boundaries causes a magnetoresistivity peak at $w = 0.55 r_c$. The main figure shows these peaks for wires of different widths (stated in the figure), studied at QWRs made by ion beam implantation, which generates particularly rough, diffusive walls. Wire peaks in QWRs defined by top gates of similar widths (shown in the inset) are much weaker. The length of the wires was $L = 12\ \mu$m. After[Thornton1989].

would be the case if boundary scattering changed the electron momentum in x-direction. Smooth walls, i.e., walls which show spatial variations only on length scales much larger than the Fermi wavelength, scatter the electrons specularly and thus do not cause additional resistivity. The smoothness of the wall is hence a critical quantity. It has been experimentally demonstrated that usually, boundary scattering is almost completely specular, unless the walls are intentionally made rough. Such wires with highly diffusive walls have been fabricated by ion implantation [Thornton1989]. Here, a maximum in $\rho_{xx}(B)$ at $w \approx \frac{1}{2} r_c$, known as "wire peak", can be observed, see Fig. 7.5. It can be shown that the magnetic field determines how sensitive the electrons are to the specularity of the confinement. The highest sensitivity is reached for cyclotron radii of roughly twice the electronic wire width w, in detail $w = 0.55 r_c$. In a simple picture, we can imagine that at this magnetic field, the fraction of the electron trajectories close to the wire edge is maximized. For the details of this effect, which is beyond our scope here, see [Beenakker1991] and [Thornton1989], as well as references therein.

The measurements in Fig. 7.5 show a specularity of only 70 % (which means that 70% of the boundary scattering events take place in a specular fashion), while in top gate defined wires, the specularity is so high that observing a wire peak is quite hard (inset in Fig. 7.5). In fact, a wire peak is not visible in Fig. 7.4, although this particular wire is 40 μm long. Therefore, for QWRs defined by top gates, by cleaved edge overgrowth, or by local oxidation, we can safely neglect boundary scattering.

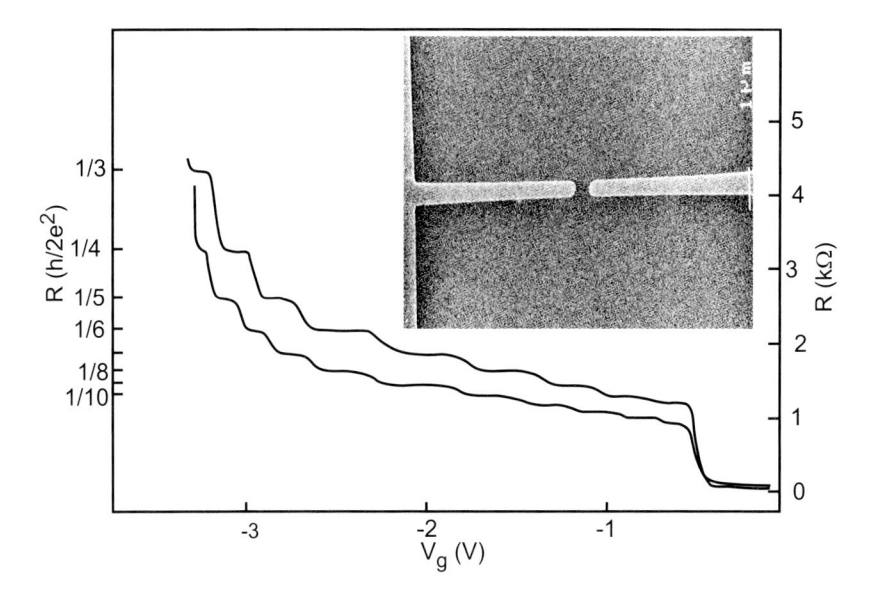

Figure 7.6: Inset: scanning electron micrograph of a split gate structure used to define a quantum point contact. The bright areas are gold electrodes on top of a Ga[Al]As - HEMT. Main figure: resistance as a function of the voltage V_G applied to the split gates with respect to the 2DEG. At a threshold voltage $V_{G,th} = -500$ mV, the resistance sharply increases by about 1 kΩ. Here, the 2DEG underneath the gates is depleted and the QPC is defined. As V_G is further reduced, quantized steps in the resistance are at $R_j = (1/j) \cdot (h/2e^2)$ are observed. The two traces are taken for two different carrier concentrations in the 2DEG, which has been changed by illumination with an infrared light-emitting diode. The temperature was 60 mK. The measurements are adapted from [Wharam1988].

7.2 Ballistic quantum wires

7.2.1 Phenomenology

In 1988, [Wees1988] and [Wharam1988] investigated the transport properties of quantum point contacts, of the shape sketched in Fig. 7.1(c). The QPCs were created by applying suitable voltages to a split gate, see the inset in Fig. 7.6, on top of a Ga[Al]As-HEMT structure. With such geometries, the QPC is imposed on the 2DEG by tuning the gate voltage to negative values, such that the electron gas underneath gets depleted. By further decreasing the gate voltage, the lateral electric stray field, and with it the lateral depletion zone around the gates, increases. This can be used to tune the electronic width, and with it the number j of occupied modes of the QPC, ideally all the way down to zero.

Once a small background resistance, which stems from the 2DEG between the QPC and the voltage probes, has been subtracted, the conductance of such QPCs turned out to be quantized in units of $j \cdot 2e^2/h$, *in the absence of magnetic fields*, see Fig. 7.6 (and Fig. 1.2, by the way). This quantization resembles very much the quantum Hall effect as a function of the electron density. In QPCs, however,

- there is no Landau quantization, and there are no scatterers in the region of interest;

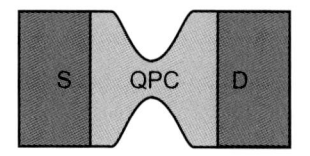

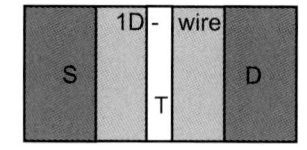

Figure 7.7: A QPC attached to source and drain (left) and its idealized model (right), where the transition regions are one-dimensional leads, and the constriction is a barrier with transmission probability T.

- the accuracy is typically of the order of $\delta R/(2e^2/h) \approx 10^{-2}$, much smaller than the accuracy of ρ_{xy} inside a quantum Hall plateau;

- the conductance quantization vanishes for longer quantum wires, typically for $L > 2\,\mu$m, as subsequent studies have revealed. However, signatures of the quantization have been observed wires up to 20 microns long [Worschech1999]. This is also in contrast to the QHE, which can be observed in samples of mm sizes and above.

As we shall see, there is in fact a close relation. Before we study this connection, some more obvious questions need to be discussed: why is there a resistance at all in QPCs, although there are no scatterers around? Shouldn't we just measure the resistance in series with the QPC? And why does the conductance of each mode in ballistic wires quantize in units of $2e^2/h$?

7.2.2 Conductance quantization in QPCs

Essentially, the previous question can be answered in two steps:

1. Neither inside the QPC nor at its exit, there is backscattering. Here, the exit is defined by the direction of the k-vector under consideration.

2. The occupation of the states in close proximity to the QPC is not described by a Fermi-Dirac distribution.

To begin with, we look in more detail at the geometric shape of a QPC and its environment, Fig. 7.7. Clearly, the QPC is connected to source S and drain D via a transition region which is quasi-1-dimensional. In a rather crude, but nevertheless very insightful approximation, we replace the transition region by ballistic, strictly one-dimensional QWRs and the QPC itself by a barrier with transmission probability T. We do so since we plan to calculate the conductance of the QPC from a scattering approach, where incoming electronic plane waves are scattered into outgoing plane waves, which are eigenfunctions of the one-dimensional wires. For simplicity, we assume an energy-independent transmission probability. The QPC is open for $T = 1$.

We now calculate the conductance for our model QPC. For this purpose, recall that the current density in its simplest form is given by $\vec{j} = ne\vec{v}$, where n is the three-dimensional carrier density and $\vec{v}$ the (energy-independent) velocity of the electrons. The corresponding

7.2 Ballistic quantum wires

one-dimensional expression is obtained by integrating over the cross section of the current; it reads $I = n_1 e \bar{v}$. Here, I is the current n_1 the one-dimensional electron density. This simple relation is generalized to our model system as follows: suppose a voltage $V = (\mu_S - \mu_D)/e$ drops between source and drain in our model system. The reservoirs fill the connected states of the wire with k-vectors pointing away from the reservoir (outgoing states), up to their respective electrochemical potentials. Now, I, n_1 and $\vec{v}$ depend on energy. Furthermore, the density of electrons at energy E, i.e., the spectral electron density, is given by the density of states on the side $j = S, D$ of the barrier, multiplied by the corresponding Fermi function:

$$n_j(E) = D_j(E) f(E - \mu_j)$$

In addition, an electron can only tunnel across the barrier if an empty state is available on the other side. The probability for this is $(1 - f(E - E_{j'}))$ The spectral current $I(E)$ is now given by

$$I(E) = eT \left[\vec{n}_S(E) v_S(E) - \overleftarrow{n}_D(E) v_D(E) \right] =$$

$$eT \, [\vec{D}_S(E) f(E - \mu_S)(1 - f(E - E_D) v_S(E) -$$

$$\overleftarrow{D}_D(E) f(E - \mu_D)(1 - f(E - E_S)) v_D(E)] \quad (7.11)$$

Here, $\vec{D}_S(E)$ ($\overleftarrow{D}_S(E)$) is the density of states for right-moving (left-moving) electrons in the part of the wire connected to source (drain), and T is taken to be independent of energy. Assuming just a spin degeneracy of 2, we have

$$\vec{D}_j(E) = \overleftarrow{D}_j(E) = \frac{1}{2} D_1(E - E_{1,j}) = \frac{1}{2\pi\hbar} \sqrt{\frac{2m^*}{E - E_{1,m}}} \quad (7.12)$$

where $E_{1,j}$ denotes the bottom of the mode on side j. Since the electron velocity in the mode is given by

$$v_j(E) = \sqrt{\frac{2(E - E_{1,j})}{m^*}}) \quad (7.13)$$

eq. (7.11) simplifies to

$$I(E) = \frac{eT}{\pi\hbar} \left[f(E - \mu_S) - f(E - \mu_D) \right] \quad (7.14)$$

For zero temperature,[1] the Fermi functions become Heaviside step functions $\theta(\mu_j - E)$, and we obtain a total current of

$$I = \frac{eT}{\pi\hbar} \int_{E=0}^{\infty} [\theta(\mu_S - E) - \theta(\mu_D - E)] dE = \frac{2e^2}{h} TV \quad (7.15)$$

---

[1] The effect of $\Theta > 0$ is considered in exercise E 7.2.

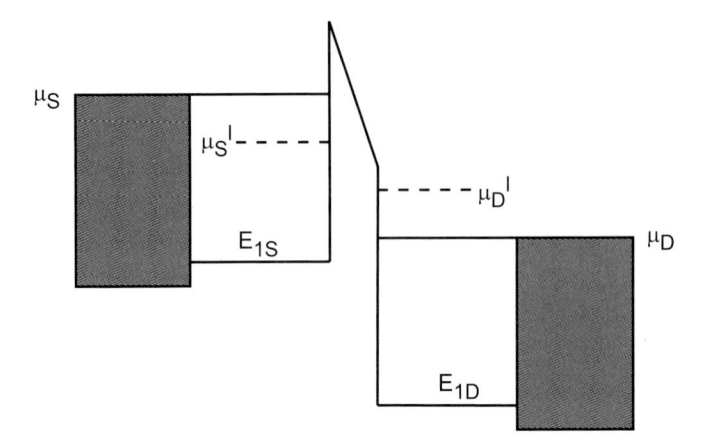

Figure 7.8: Electrochemical potentials of the system of Fig. 7.7, both at the reservoirs and close to the barrier (dashed lines).

Therefore, the conductance for a mode with $T = 1$ equals

$$G = \frac{2e^2}{h} \tag{7.16}$$

It is quantized because the energy dependence of the one-dimensional density of states and that one of the electron velocity cancel each other!

Apparently, the quantized conductance follows quite naturally from this simple consideration. The result is nevertheless quite surprising. For $T = 1$, the electrons suffer no scattering at the QPC. A finite conductance for such a scenario is counter-intuitive. The answer can be found in the subtlety that the voltage drop across the QPC is *not* the difference between the source and the drain potentials, divided by e. Within the mean free path around the barrier, transport is ballistic, and therefore, states moving away from the barrier are only occupied if an electron has been scattered into them by reflection at the barrier. Clearly, the occupation probability of the states does not have the form of a Fermi-Dirac distribution. This point has been quantified by [Buttiker1985].

In order to explain how this idea can shed light on the issue, we again allow $0 \leq T \leq 1$ and consider the open channel as a special case later on. For the sake of simplicity, let us assume that the source-drain bias voltage is sufficiently small[2], which means that the density of states can be set constant in this small energy range of interest. Therefore, $\overrightarrow{D}_S(E) = \overleftarrow{D}_S(E) = \overrightarrow{D}_D(E) = \overleftarrow{D}_D(E)$.

Consider the scenario depicted in Fig. 7.8. To the left side of the barrier, all right-moving states are occupied up to μ_S, since they are connected to the source potential. On this side, a left-moving state close to the barrier is occupied only if an electron has scattered into it, which can happen exclusively by backscattering at the barrier. Hence, the electrochemical potential at the left side must be smaller than μ_S. We call this local chemical potential μ_S^ℓ. Likewise,

---
[2] In detail, the condition for the bias voltage reads $V \ll \mu_S - E_{1,S}$.

7.2 Ballistic quantum wires

some right-moving states on the right side with energies within $[\mu_D, \mu_S]$ are occupied by electrons that have been transmitted through the barrier. This results in an local potential at the right side $\mu_D^\ell > \mu_D$.

Recall that the chemical potential μ of a metal can be defined as that energy for which the number of empty states below μ equals the number of empty states above μ. This definition is very convenient here, and we use it to calculate μ_S^ℓ and μ_D^ℓ. On the left side, the only empty states with $E < \mu_S^\ell$ are those left-moving states that have not been occupied by reflection of electrons at the barrier. The density of these states is thus given by

$$n_{S,empty}(E < \mu_S^\ell) = \int_{\mu_D}^{\mu_S^\ell} \overleftarrow{D}_S(E) T \, dE = \overleftarrow{D}_S T (\mu_S^\ell - \mu_D) \tag{7.17}$$

The density of occupied states $n_{S,occ}(E > \mu_S^\ell)$ is the sum of two components. First of all, all right-moving states on the left side in $[\mu_S^\ell, \mu_S]$ are occupied, since they get filled by the source reservoir. Second, there are left-moving states that got populated by backscattering at the barrier. Hence,

$$n_{S,occ}(E > \mu_S^\ell) = \overrightarrow{D}_S(\mu_S - \mu_S^\ell) + (1-T)\overleftarrow{D}_S(\mu_S - \mu_S^\ell) \tag{7.18}$$

Since $\overleftarrow{D}_S = \overrightarrow{D}_S$, we obtain the local chemical potential at the left side via

$$n_{S,empty}(E < \mu_S^\ell) = n_{S,occ}(E > \mu_S^\ell) \Rightarrow$$

$$\mu_S^\ell = \mu_S - \frac{T}{2}(\mu_S - \mu_D) \tag{7.19}$$

A corresponding consideration for the part of the wire attached to drain gives

$$\mu_D^\ell = \mu_D + \frac{T}{2}(\mu_S - \mu_D) \tag{7.20}$$

Question 7.2: Derive eq. (7.20)!

The local voltage drop at the barrier is thus given by

$$V^\ell = \frac{1}{e}(\mu_S^\ell - \mu_D^\ell) = \frac{1}{e}(\mu_S - \mu_D)(1 - T) \tag{7.21}$$

and we find a conductance of the barrier of

$$G_{barrier} = \frac{eI}{V^\ell} = \frac{2e^2}{h} \frac{T}{1-T} \tag{7.22}$$

which certainly makes a lot of sense: as T approaches one, the barrier conductance approaches infinity. Since the overall conductance between source and drain is given by

$$G_{SD} = \frac{2e^2}{h} T \tag{7.23}$$

there must be a resistance in series with the barrier given by

$$R_{contact} = \frac{1}{G_{SD}} - \frac{1}{G_{barrier}} = \frac{h}{2e^2} \qquad (7.24)$$

It is called a *contact resistance*, since it occurs at the interface between the reservoirs and the one-dimensional wire. In a sense, it manifests the difficulty of feeding the reservoir electrons into the one-dimensional wire, which has conceptually nothing to to with feeding the electrons from the wire in the QPC. Experimentally, however, both regions are not well separated spatially. Suppose now we have $T = 1$. In that case, the right-moving states in the wire are occupied up to μ_S, while the left-moving states are populated up to μ_D only. This means that the chemical potential inside the wire is $(\mu_S + \mu_D)/2$, since for this potential, the number of occupied states above equals that one of empty states below. The voltage thus drops by $V/2$ at the entrance as well as at the exit of the wire.

Our purpose here was to shed some light on the conductance quantization in quasi-one dimensional systems. Along the way, many questions remain, and a lot of assumptions have certainly been made. To work out all these details is a formidable task; we briefly outline some of them below.

First of all, we have considered only a single mode. The Landauer formula ([Landauer1957], see also [Buttiker1986])

$$G = \frac{2e^2}{h} \sum_{\alpha,\beta} |t_{\alpha\beta}|^2 \qquad (7.25)$$

gives the conductance of a multi-mode QPC. The transmission amplitude from mode β to mode α is $t_{\alpha\beta}$. The partial conductances of the individual modes are thus additive for negligible coupling between different modes. This assumption implies that the electrons remain in their initial mode throughout their trip across the QPC. This kind of transport is called adiabatic. It requires that the width of the channel changes smoothly on the scale of the Fermi wavelength, which is usually the case in the experiments under discussion. The properties of adiabatic constrictions and the consequences of non-adiabaticity in experimental realizations are discussed in [Yacoby1990].

Another effect not considered are reflections of the electronic wave functions, which may take place at the entrance and the exit of the wire. Multiple reflections may lead to transmission resonances, as we will discuss in more detail in chapter 8. It is intuitively clear that a smooth potential shape in the above sense suppresses backscattering of electrons at the exit, and interference effects are absent. Extensive numerical simulations have quantified how the QPC conductance depends on the potential shape. An example is shown in Fig. 7.9. In some experiments, weakly pronounced oscillations, superimposed on conductance steps, have been observed, but the agreement with theoretical considerations is poor. It has turned out to be very difficult to fabricate samples with sufficiently sharp contact regions. Thermal smearing further hampers a clear observation.

The conductance steps vanish at a characteristic temperature given by the energy spacing of the modes. As can be seen in exercise E 7.2, a sharp transmission step at $\Theta = 0$ is thermally smeared at higher temperatures according to

$$G = \frac{2e^2}{h} f(E_1 - \mu) \qquad (7.26)$$

7.2 Ballistic quantum wires

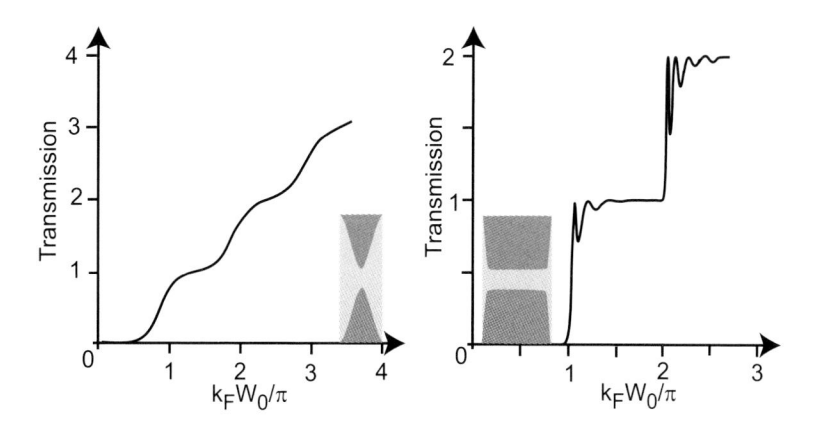

Figure 7.9: Transmission as a function of the Fermi wavenumber k_F and the electronic width at its narrowest point, W_0, calculated with the recursive Green's function technique [Fisher1981], for zero temperature. A QPC with a smooth transition region, left (the geometry is shown in the inset), shows smooth transmission steps. In longer wires with sharp contacts to the reservoirs, resonances due to interference effects are found (right). After [Maao1994].

which gives a characteristic temperature of

$$\Theta_{char} = \frac{\Delta}{2k_B ln(3+2\sqrt{2})} \approx \frac{\Delta}{3.52 k_B} \quad (7.27)$$

In the sample shown in Fig.1.2, for example, the temperature for which the steps vanish is roughly 20 K, and the mode spacing is therefore of the order of 6 meV. One might thus expect that the steps become infinitely sharp as the temperature approaches zero. This is not observed experimentally. Rather, the steepness of the conductance steps tends to saturate as Θ is reduced below a few hundred mK. In [Buttiker1990], the potential step of our model is replaced by a saddle-shaped potential,

$$V(x,y) = \frac{1}{2}m^*(\omega_y^2 y^2 - \omega_x^2 x^2) \quad (7.28)$$

which should represent an excellent approximation for typical QPC geometries. The transmission probability of this potential is given by

$$T(\epsilon) = \frac{1}{1+\pi\epsilon} \quad (7.29)$$

In this model, the energy separation between the modes equals $\hbar\omega_y$, while $\hbar\omega_x$ determines the width of the transition range between adjacent conductance steps (Fig.7.10).

7.2.3 Magnetic field effects

As in diffusive quantum wires, the modes of a QPC get of course depleted by perpendicular magnetic fields as well (Fig. 7.11), since the magnetic field increases the confinement and the

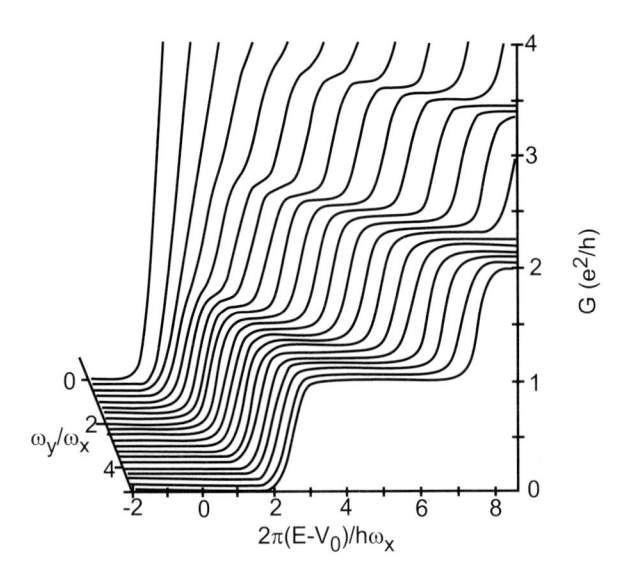

Figure 7.10: Conductance of a QPC with a saddle point potential, characterized by ω_x and ω_y, calculated for zero temperature. The steps vanish as soon as ω_y becomes smaller than ω_x. After [Buttiker1990].

energy separation between the modes becomes larger. Simultaneously, quantum Hall states are formed in the 2DEG, which influences the resistance measurements. We can elegantly explain the data of Fig. 7.11, once the Landauer-Büttiker formalism has been introduced, and we will come back to this issue later. In weak perpendicular magnetic fields, QPCs show a negative magnetoresistance in addition, an effect discussed in [P 7.2]. Diamagnetic shifts of the QPC modes can, however, be conveniently studied in parallel magnetic fields, since here, the effect B exerts on the 2DEG is negligible.

This kind of spectroscopy has been performed by [Salis1999] on a QPC residing in a parabolic quantum well with two occupied subbands in growth direction, Fig. 7.12. Transverse modes in both y- and z- direction contribute to the current. We label the channels in y-direction by ℓ, and in z-direction by m, respectively. Since the confinement in z-direction is significantly larger than in y-direction, we can think of the mode structure as being composed of ladders denoted by m, each with rungs labelled by ℓ. [3] The total number of modes carrying current is given by

$$j = \sum_{m,\ell=1}^{\infty} \theta(\mu - E_{m,\ell}) \qquad (7.30)$$

Here, μ is the chemical potential (we assume that the source-drain voltage is negligibly small), and $E_{m,\ell}$ is the energy of the mode bottom. To keep things simple, we restrict ourselves to the case of two occupied subbands in growth direction. We expect again conductance quantization in units of $2e^2/h$. However, if two degenerate modes (belonging to different ladders)

---
[3] This is possible only if the Hamiltonian is separable.

7.2 Ballistic quantum wires

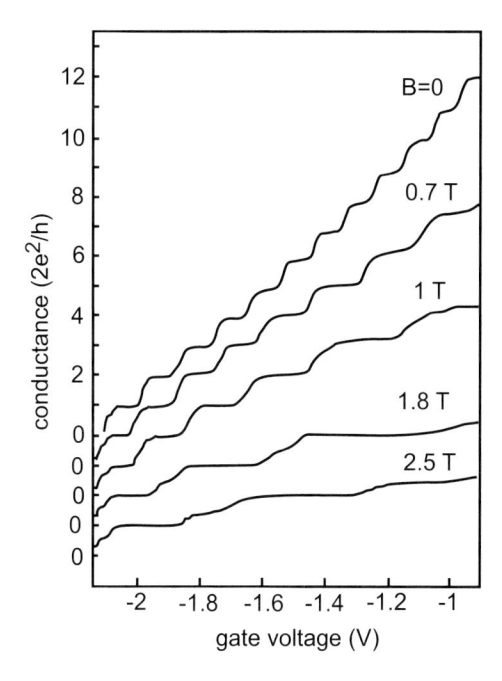

Figure 7.11: Depopulation of QPC modes by a magnetic field. As B is increased, the with of the conductance plateaus increases. This reflects the increasing magnetoelectric confinement in y-direction. At $B = 1.8$ T and above, the spin degeneracy gets lifted and additional plateaux evolve at odd integers of e^2/h. after [Wees1988a].

cross the chemical potential simultaneously, the conductance will change by $4e^2/h$. Such a degeneracy shows up in the experiment as a suppression of conductance steps, Fig. 7.12. Via diamagnetic shifts, the energies of the modes, and thus the degeneracies, can be tuned by in-plane magnetic fields. The transconductance dG/dU_{SG} (U_{SG} is the voltage applied to the split gates) emphasizes the diamagnetic shift of the modes, and is a good representation of the QPC's energy spectrum at the same time. In Fig.7.13 (a) and (b), such measurements are shown for B applied in y- and in x-direction, respectively. Dark regions correspond to low transconductance: here, the chemical potential lies in between two adjacent modes. The bright lines (high transconductance) reflect the mode spectrum.

For both magnetic field directions, the two subband ladders are visible, with each ladder having its own characteristic dispersion. As a function of B_y, the levels of both ladders show a positive dispersion, which is much stronger in the $m = 2$ ladder. For magnetic fields in x-direction, the dispersion of the $m = 1$ states is reversed: their energy decreases as B_x is increased. In both cases, mode crossings are clearly visible. The measurements in Figs. 7.13 (a) and (b) are compared to model calculations in Figs. 7.13 (c) and (d). It is reasonable to assume that the energy in the QPC is proportional to U_{SG}. The lever arm $\alpha = dE/dU_{SG}$ can be estimated by temperature-dependent measurements, which give the energy spacing between adjacent modes according to eq. (7.27). A value of $\alpha \approx 0.02$ eV/V is found.

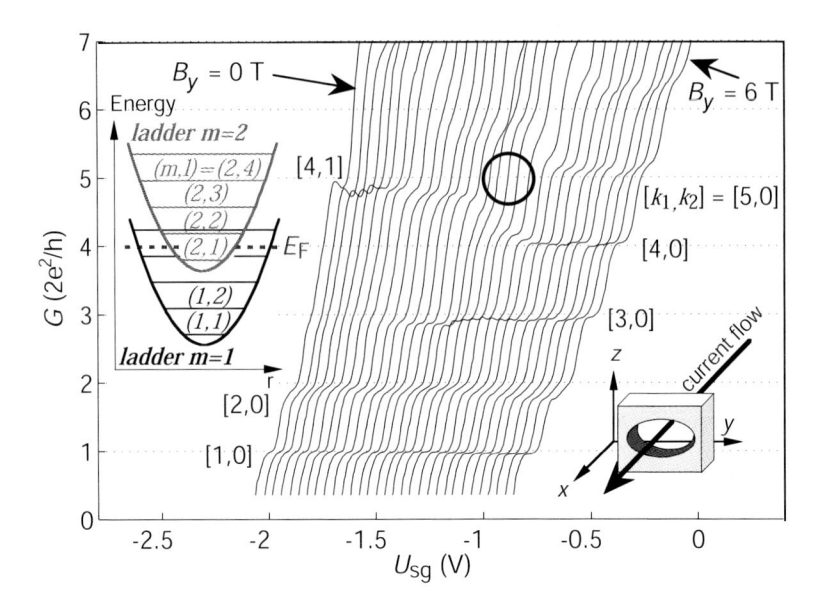

Figure 7.12: The left inset shows a schematic mode spectrum of a QPC, located in a parabolic quantum well with two occupied subbands in growth direction, plotted as a function of the spatial coordinate r. Each subband contains a ladder of equidistant modes. m is the subband index, while l labels the mode within a ladder. The main figure shows the conductance as a function of the QPC gate voltage U_{SG}, measured for magnetic fields $B_\perp = 0...6$ T applied in y-direction (see right inset). Conductance steps can be suppressed and recovered by the magnetic field, as in the encircled region. The labels $[k_1, k_2]$ denote the number of occupied modes in subbands 1 and 2. After [Salis1999].

Furthermore, we assume parabolic confinement in the y-direction as well. The confining potential at the narrowest point of the QPC, which determines the number of transmitted modes, is the given by

$$V(y,z) = \frac{1}{2}m^*(\omega_y^2 y^2 + \omega_z^2 z^2) \tag{7.31}$$

After our previous discussion of parallel magnetic fields in section 6.5, it is intuitively clear that B_y causes magnetic shifts of the modes' z-component and enhances the effective mass in x-direction, while B_x does not modify the effective mass, but causes magnetic shifts of the y- and z- components. In [Sherbakov1996], the Schrödinger equation has been solved for a magnetic field B_y applied in y-direction. The energy eigenvalues are found to be

$$E_{ml}^y = \hbar\omega_y(\ell - \frac{1}{2}) + \hbar\sqrt{\omega_z^2 + \omega_C^2}(m - \frac{1}{2}) \tag{7.32}$$

For a magnetic field B_x in x-direction, the energy levels are given by ([Schuh1985])

$$E_{ml}^x = \hbar\omega_1(\ell - \frac{1}{2}) + \hbar\omega_2(m - \frac{1}{2})$$

7.2 Ballistic quantum wires

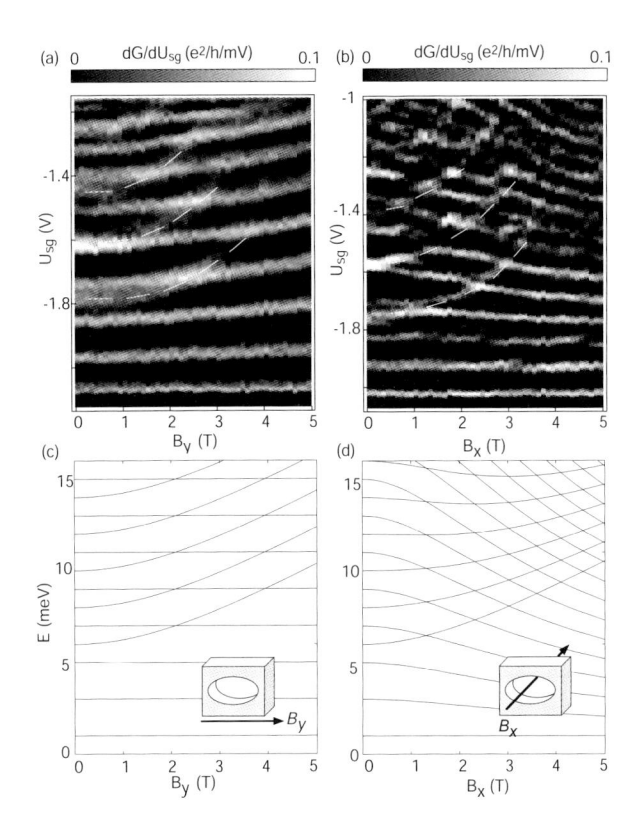

Figure 7.13: Gray scale plot of the transconductance dG/dU_{SG} as a function of B_y (a) and of B_x (b). The bright lines represent the QPC modes. In (c) and (d), the measurements are compared to the energy spectrum of an elliptical parabolic confinement. After [Salis1999].

where

$$\omega_{1,2} = \frac{1}{2}(\omega_C^2 + \omega_y^2 + \omega_z^2) \pm \sqrt{(\omega_C^2 + \omega_y^2 + \omega_z^2)^2 - \omega_y^2\omega_z^2} \quad (7.33)$$

For $\omega_y = \omega_z$, eq. (7.33) becomes the Fock-Darwin spectrum [Fock1928], [Darwin1931]. The calculated spectra agree very well with the experimental result for $\omega_y = 2$ meV and $\omega_z = 5$ meV.

7.2.4 The "0.7 structure"

In 1996, [Thomas1996] reported a very clearly pronounced additional conductance plateau at $G \approx 0.7 \cdot (2e^2/h)$. Subsequent experiments have confirmed this observation and studied its parametric behavior. Not only does this plateau remain unexplained within our model, it has some additional puzzling features (see Fig. 7.14), namely

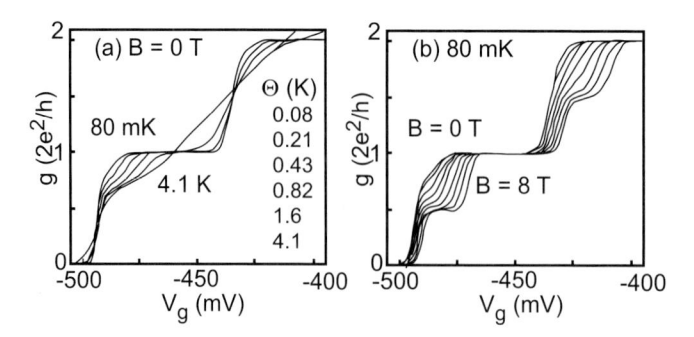

Figure 7.14: Conductance of a QPC as a function of the spit-gate voltage. (a) The "0.7 feature" shows a different temperature dependence than the conventional conductance steps. (b) In strong parallel magnetic fields, the spin degeneracy is removed and additional plateaus are visible at odd integers of e^2/h. As B is reduced, the spin-split plateau at $G = 0.5e^2/h$ evolves into the 0.7 feature. After [Cronenwett2002].

- as the temperature is increased, it does not suffer thermal smearing, but becomes more prominent instead;
- it emerges from the spin-split plateau at $G = e^2/h$ as a strong parallel magnetic field is reduced;
- its presence or absence seems to depend randomly on the sample.

In fact, the origin of this plateau has not yet been clarified unambiguously, although there is strong evidence that it is caused by electronic correlation effects known as Kondo correlations [Cronenwett2002]. This example illustrates that we have not yet reached a complete understanding of even very simple ballistic systems.

7.2.5 Four-probe measurements on ballistic quantum wires

Measuring the resistance of a ballistic QWR without the contact resistance clearly requires to attach voltage probes in between the two contacts. This has to be done without disturbing the current flow. For example, if the electrons get backscattered at the probes along their trip from source to drain, the resistance will be increased. In other words, the resistance R_{wp} between the wire and the probe has to be large compared to R_{SD}. Such an experiment has been performed on quantum wires defined by cleaved edge overgrowth (see chapter 5). The sample layout is reproduced in Fig. 7.15. With all gates grounded, a wire extends along the cleaved edge, which is coupled to the 2DEG that resides in the quantum well. The length scale for scattering between the wire modes and and the states in the 2DEG is $\ell_{2D-1D} \approx 6$ μm.

Negative voltages applied to the tungsten gate stripes deplete the 2DEG underneath. The ballistic wire of length $L = 2$ μm can be tuned by activating gate 2, while gates 1 and 3 remain grounded. In this operation mode, the 2DEG areas to the left and right of gate 2 serve as source and drain. Clear resistance quantization is observed (dashed line in Fig. 7.16; this

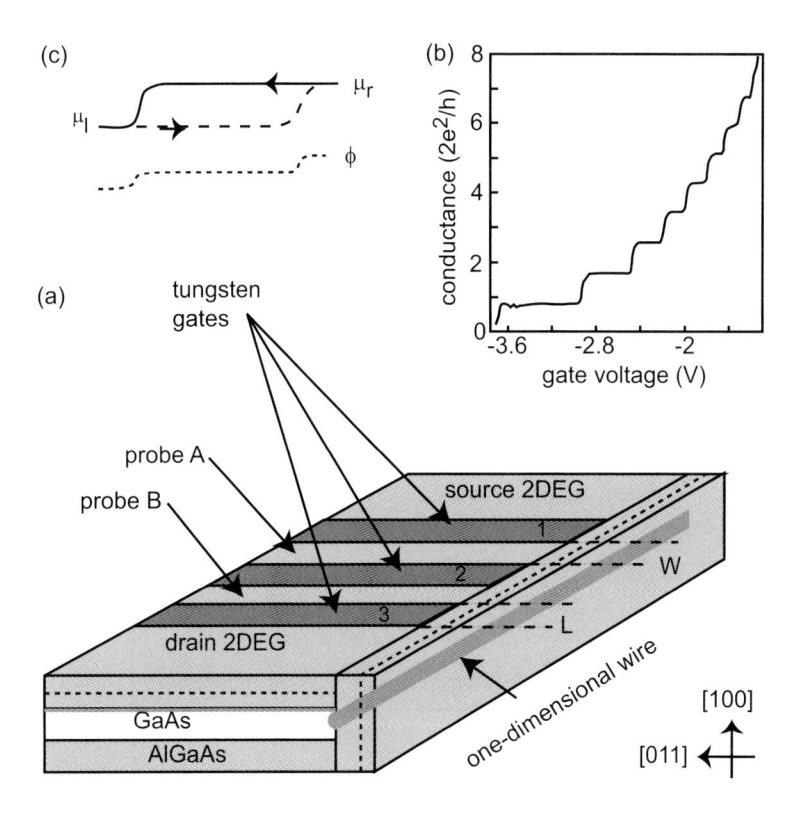

Figure 7.15: (a): sample geometry for four-terminal resistance measurements on a ballistic quantum wire. A quantum well is grown in [100] direction, and three tungsten gate stripes of width $W = 2\ \mu\mathrm{m}$, separated by $L = 2\ \mu\mathrm{m}$, are evaporated on top. The wafer is then cleaved, and a modulation-doped layer of $\mathrm{Al}_{0.3}\mathrm{Ga}_{0.7}\mathrm{As}$ is grown along the [110] direction. The wire extends along the cleave. (b): conductance quantization as a function of the voltage at gate 2, with the other gates grounded. (c): spatial variation of the electrochemical potentials of left- and right-moving electrons along the wire, as discussed in the text. In addition, the corresponding potential drop ϕ is shown. After [Picciotto2001]

trace is equivalent to the conductance trace in Fig. 7.15). The plateaux deviate from the expected $G = je^2/h$ by up to 25 %, an effect that has its origin in non-ideal contacts to source and drain in this particular sample geometry. [4] By additionally activating gates 1 and 3 to appropriate voltages, strictly one-dimensional leads are generated along the corresponding regions of the wire. Also, the 2DEG regions between the gates now form two voltage probes that couple weakly to the wire, since their width is smaller than ℓ_{2D-1D}. Clearly, the voltage probes are located in between the contact regions of the wire to source and drain, and the measured voltage difference between A and B in Fig. 7.16 should be zero. This is in fact the case for all plateaux, except close to pinch-off around a gate voltage of -4.6 V.

These measurements have confirmed in a beautiful way the model of contact resistances

---

[4] This interesting effect is the topic of [P 7.2], but is of no relevance for the present discussion.

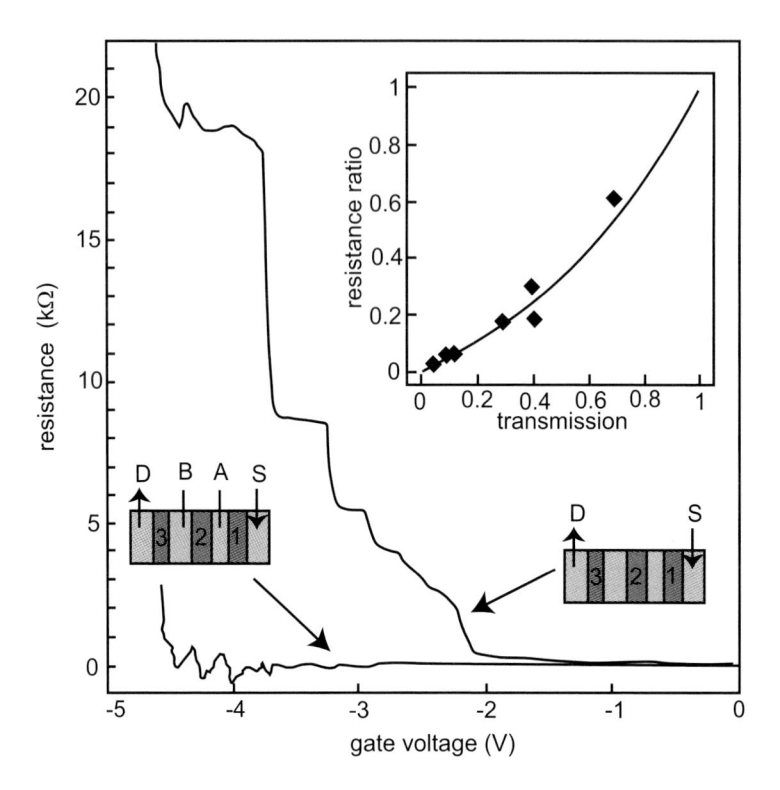

Figure 7.16: Main figure: the two-terminal resistance of the wire (dashed line) along gate 2 and its four-terminal resistance (full line) are compared. The inset shows the ratio between the four-terminal resistance and the two-terminal resistance as a function of the probability T for electrons to be transmitted between the wire and the probe. The solid trace in the main figure has been performed at $R_{wp} = 250$ kΩ, i.e. $T \approx 0.05$ in the inset. After [Picciotto2001]

and ballistic transport in one-dimensional systems. There is actually a quite different kind of experiment which conceptually proves this as well, namely four-terminal resistance measurements on 2DEGs in the quantum Hall regime. This is the topic of the following sections.

7.3 The Landauer-Büttiker formalism

In the previous section, we have argued that the quantized conductance of ballistic quantum wires stems from contact resistances. We have also seen that four-probe measurements give a resistance of zero, as expected from its interpretation in terms of contact resistances. In fact, a conceptually very similar system is a 2DEG in the Quantum Hall regime. As already indicated in section 6.2, the electrons skip along the edge of the Hall bar in strong magnetic fields. The origin of this dynamics is illustrated in Fig. 7.17.

7.3 The Landauer-Büttiker formalism

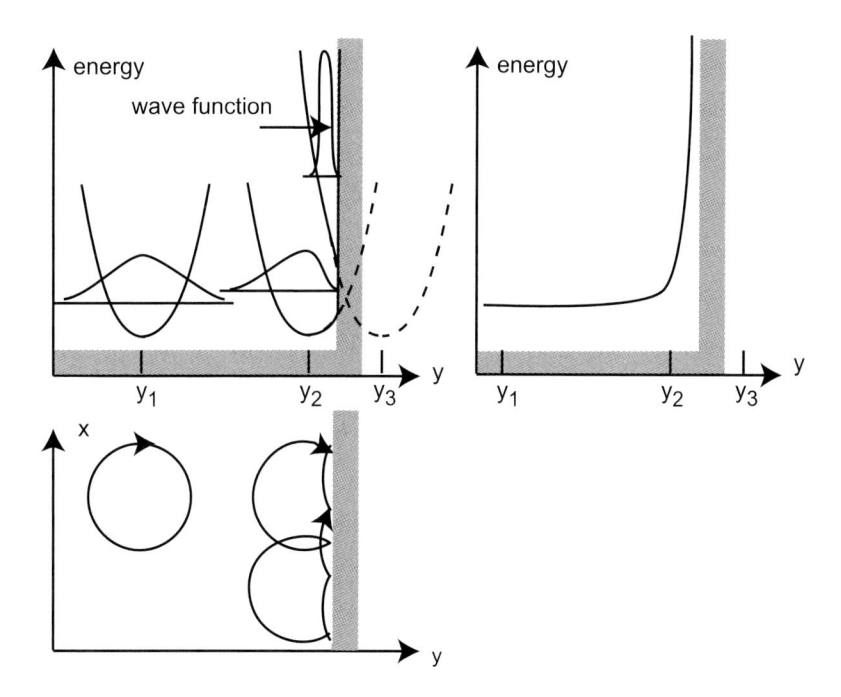

Figure 7.17: Modification of the magnetoelectric confinement of the electrons as they approach the edge of the 2DEG (left). The undisturbed cyclotron motion at y_1 is increasingly squeezed as the guiding center approaches the edge (positions y_2 and y_3). As a consequence, the energy of the Landau level increases (right), while the electrons delocalize along the x-direction (bottom).

7.3.1 Edge states

At the edge of the 2DEG, the conduction band bottom increases sharply and modifies the combined potential of the Landau harmonic oscillators and the electrostatic confinement. The increased confinement shifts the Landau levels to higher energies. Each LL crosses the Fermi level at some point, and consequently, the density of states at the Fermi level is always larger than zero. As sketched in Fig. 7.17 as well, the electrons skip along the edge. Therefore, we speak of "skipping orbits" and "edge states". Edge states have several peculiar features, which become immediately apparent. First of all, they are one-dimensional: the electron motion is confined perpendicular to the sample edge, but free in the direction parallel to it. Second, all the electrons at one sample edge move in the same direction, while the electrons at the opposite edge move in the opposite direction. In the bulk, all electrons are localized at potential modulations, expect for special filling factors, as already shown in section 6.2. The resulting edge state configuration with the directions of current flow is shown in Fig. 7.18.[5] There is no backscattering in edge states, i.e. the elastic mean free path approaches infinity: suppose an electron in an edge state hits a scatterer close to the edge. Its momentum right

---
[5]Since the electrons circulate around the edge, one speaks of a "chiral" Fermi-liquid.

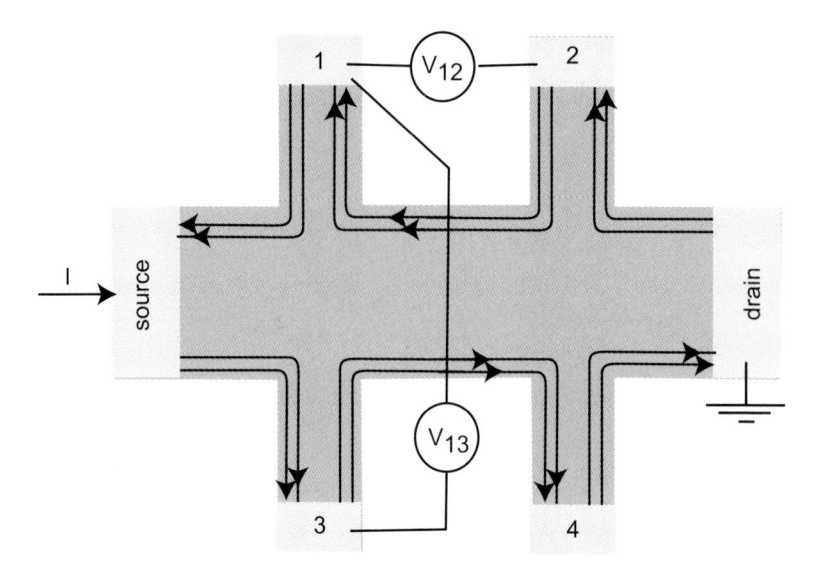

Figure 7.18: Top view of a Hall bar in a strong magnetic field. Current flows in one-dimensional edge states only, in the directions indicated by the arrows. Here, two Landau levels are occupied.

after the scattering event may be reversed, but the strong magnetic field bends the momentum back in forward direction. In order to be backscattered, the electron has to traverse the whole Hall bar and reach the opposite edge! Hence, backscattering is greatly reduced. It follows that a 2DEG in the quantum Hall regime comes very close to an ideal ballistic quantum wire: it is one-dimensional and backscattering is absent. We can even attach voltage probes inside the quantum wires without inducing backscattering. Therefore, the voltage drop between, e.g., contacts 1 and 2 in Fig. 7.18 should be zero. You have realized, of course, that this is exactly what we measure in a Shubnikov-de Haas experiment. In [Buttiker1988], the Landauer formula has been generalized to an arbitrary number of contacts, such that circuits of ballistic quantum wires can be treated. The concept is known as Landauer-Büttiker formalism. It should be noted that this formalism works for all networks of ballistic quantum wires.

Consider a circuit of ballistic quantum wires, like, e.g., the system of Fig. 7.18. We define the *direct* transmission probability of contact p into contact q as $T_{q \leftarrow p} = T_{qp}$. $T_{qp} > 1$ is possible, since more than one mode may connect the two contacts. Note that T_{qp} does not have to be an integer. Note further that within this notation, T_{pp} is a backscattering probability. The *total current emitted by contact p* is denoted by I_p, while μ_p is the electrochemical potential of contact p. Again, an "ideal" contact absorbs all incoming electrons and distributes the emitted electrons equally among all outgoing modes, such that they are filled up to μ_p, assuming zero temperature. In this notation, the Landauer formula generalizes to the Büttiker formula

$$I_p = \frac{2e}{h} \sum_q (T_{qp}\mu_p - T_{pq}\mu_q) \qquad (7.34)$$

which is a direct consequence of current conservation. We proceed by applying the Büttiker

7.3 The Landauer-Büttiker formalism

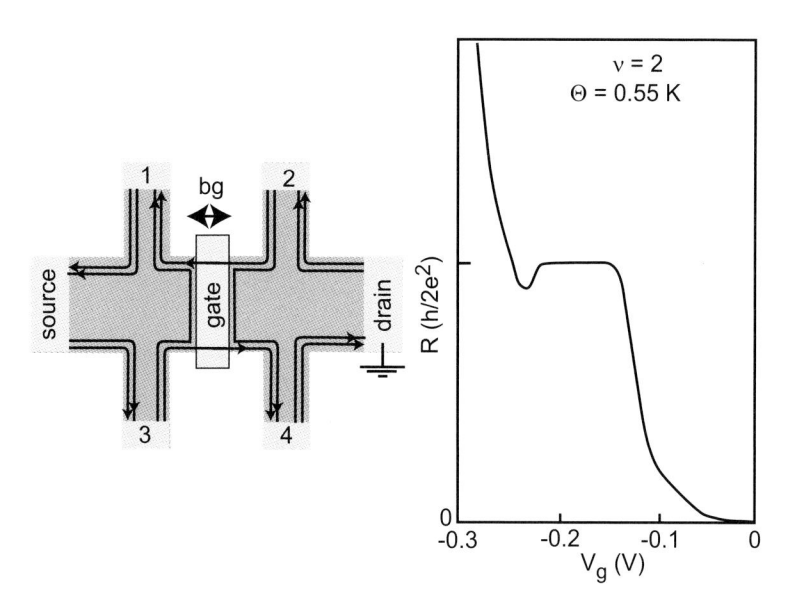

Figure 7.19: Left: sample geometry to control backscattering between edge states. A top gate covers the Hall bar in between 4 voltage probes. At suitable gate voltages, the inner one of the two edge states gets reflected. Right (after [Haug1989]): for a 2DEG in the regime of filling vector 2, with spin-split edge states, a plateau at $R_{xx} = h/2e^2$ is observed as a function of the gate voltage, once the reflection of the inner edge state at the gate is complete. The two traces correspond to different gate lengths bg in x-direction. The dip around a gate voltage of -0.2 V can be explained within a trajectory network formed below the gate, as explained in [Haug1989].

formula to the sample shown in Fig. 7.18. It gives a system of 6 linear dependent equations, one for each contact:

$$\begin{pmatrix} I_s \\ I_d \\ I_1 \\ I_2 \\ I_3 \\ I_4 \end{pmatrix} = \begin{pmatrix} G & 0 & -G & 0 & 0 & 0 \\ 0 & G & 0 & 0 & 0 & -G \\ 0 & 0 & G & -G & 0 & 0 \\ 0 & -G & 0 & 1 & 0 & 0 \\ -G & 0 & 0 & 0 & G & 0 \\ 0 & 0 & 0 & 0 & -G & G \end{pmatrix} \cdot \begin{pmatrix} V_s \\ V_d \\ V_1 \\ V_2 \\ V_3 \\ V_4 \end{pmatrix}$$

with $G = j\frac{2e^2}{h}$. By choosing $\mu_d = 0$ as a reference potential, and after eliminating the drain current as a consequence of current conservation (remember that the voltage probes measure the potentials without drawing current), we can eliminate the drain row and column, and the matrix equation

$$\begin{pmatrix} I_s \\ I_1 \\ I_2 \\ I_3 \\ I_4 \end{pmatrix} = \begin{pmatrix} G & 0 & 0 & -G & 0 \\ -G & G & 0 & 0 & 0 \\ 0 & -G & G & 0 & 0 \\ 0 & 0 & 0 & -G & G \\ 0 & 0 & 0 & 0 & G \end{pmatrix} \cdot \begin{pmatrix} V_s \\ V_1 \\ V_2 \\ V_3 \\ V_4 \end{pmatrix}$$

results. Its solution gives

$$V_s = V_1 = V_2; V_3 = V_4 = 0; I_s = G \cdot V_s \tag{7.35}$$

Therefore, we find the longitudinal resistance

$$R_{xx} = \frac{V_1 - V_2}{I_s} = \frac{V_3 - V_4}{I_s} = 0 \tag{7.36}$$

and the Hall resistance

$$R_{xy} = \frac{V_1 - V_3}{I_s} = \frac{V_2 - V_4}{I_s} = \frac{h}{2je^2} \tag{7.37}$$

Within the edge state picture, the quantized Hall resistance is obtained, and the longitudinal resistance vanishes. The accuracy of the quantization is so much more accurate than in a QPC, since backscattering is greatly suppressed. Let us now consider what happens as we increase the magnetic field, such that the uppermost occupied LL gets depleted. The corresponding edge state, which is the innermost occupied one, is depopulated as well. Since the velocity in x-direction of the electrons in edge state j is given by

$$v_j(k_x) = \frac{1}{\hbar}\frac{\partial E_j(k_x)}{\partial k_x} = \frac{1}{\hbar}\frac{\partial V(y(k_x))}{\partial y}\frac{\partial y}{\partial k_x} = \frac{1}{eB}\frac{\partial V(y)}{\partial y} \tag{7.38}$$

where we have used $y(k_x) = \frac{\hbar k_x}{eB}$, the velocity of the electrons approaches zero as the edge state gets depleted. As a consequence, the edge state begins to soften and the electron trajectories penetrate into the bulk. Finally, the electrons can percolate all the way to the opposite edge, backscattering sets in, and the conductance quantization vanishes. [Haug1989] have performed an instructive experiment related to this picture, see Fig.7.19. A gate stripe extends across a Hall bar inside an area that can be measured by 4 voltage probes. Biasing the gate tunes the electron density, and thus the number of occupied Landau levels, underneath. If the filling factor under the gate is smaller than outside the gated area, edge states get redirected at the gate. This changes the transmission probabilities in eq.(7.34).

In Exercise [E 7.3] the resistances of this system will be calculated. The result for filling factor N in the ungated region and M in the gated region is

$$R_{12} = R_{34} = \frac{h}{e^2}(\frac{1}{M} - \frac{1}{N})$$

$$R_{13} = R_{24} = \frac{h}{e^2}\frac{1}{N}$$

$$R_{14} = \frac{h}{e^2}(\frac{1}{M} - \frac{2}{N})$$

$$R_{23} = \frac{h}{e^2}\frac{1}{M} \tag{7.39}$$

Note that the results of some measurements now depend on the direction of the magnetic field.

The Landauer-Büttiker formalism is a powerful tool, which allows to treat a variety of problems very elegantly. Further examples are treated in the exercises.

7.3 *The Landauer-Büttiker formalism* 189

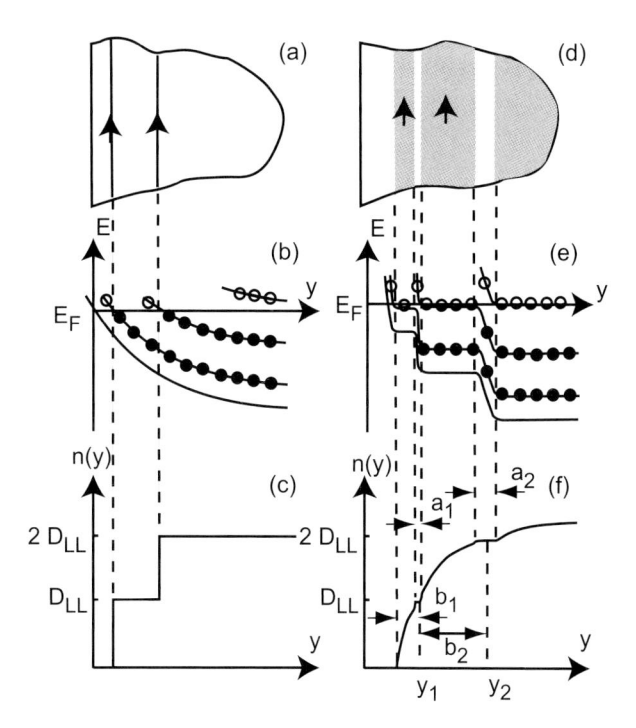

Figure 7.20: (a): guiding center trajectories within the edge state picture, for the case of filling factor 2. Along these lines, the system is metallic, while it is insulating everywhere else. In (b), the corresponding energies of the Landau levels are shown. Full circles denote occupied states. The resulting spatial variation of the electron density is shown in (c). In the edge channel picture, the potential gets screened in the metallic regions, and the potential drop concentrates within the insulating regions, (d) and (e). The edge states evolve into metallic stripes of non-zero width, separated by insulating stripes. The stripe width is determined by the electrostatics of the configuration. The resulting electron density close to the edge is shown in (f). After [Chklovskii1992].

7.3.2 Edge channels

So far, we have interpreted edge states as guiding centers of electron trajectories in strong magnetic fields. Within this picture, the trajectories of electrons moving in different edge states intersect, and we may expect a strong inter-edge state scattering rate. In fact, the edge states are spatially separated in sufficiently strong magnetic fields, and inter-edge state scattering is suppressed.[6] This can be understood by studying the effects of screening at the edge, see Fig.7.20. At points where the edge states intersect the Fermi level, the system has a metallic character. Here, the electrons in the edge state are able to screen the confining potential, and edge channels are formed. The potential drop is concentrated in the insulating regions.

---

[6]Note that this kind of scattering does not show up in the resistance, unless special geometries are considered, like that one of [E7.5].

The electrostatics of edge states, which is the topic of [P7.3], has been considered first by [Chklovskii1992]. Please note that just before the innermost occupied edge channel gets emptied, its width approaches infinity and extends all the way to the opposite edge. As within our picture of the previous section, backscattering becomes possible under these circumstances.

7.4 Further examples of quantum wires

The conductance quantization has been detected first in QPCs defined in 2DEGs by gate voltages, and the vast majority of experiments have been performed on such systems. But these results have also triggered the search for similar effects in other materials, in particular in conventional metals and in carbon nanotubes. Even quantized transmission of light through a pinhole has been discovered afterwards [Montie1991].

7.4.1 Conductance quantization in conventional metals

Once the behavior of QPCs in semiconductors was known, observing conductance quantization turned out to be possible in a very simple experiment: just pull a metallic wire and measure its resistance simultaneously. Right before it breaks, you will observe quantized conductance. Meanwhile, this experiment is performed in several undergraduate lab courses,

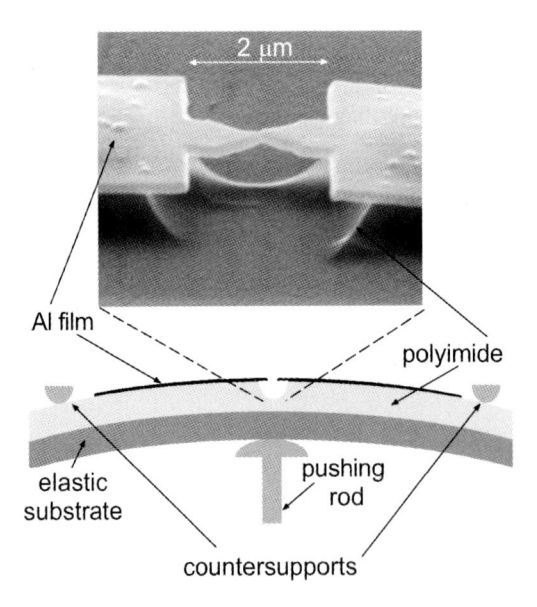

Figure 7.21: A mechanically controlled break junction for observing conductance quantization in an Al QPC. The elastic substrate is bent by pushing rod with a piezoelectric element. The thin Al bridge, fabricated by electron beam lithography, can be broken and reconnected for many cycles. Taken from [Scheer1997].

7.4 Further examples of quantum wires

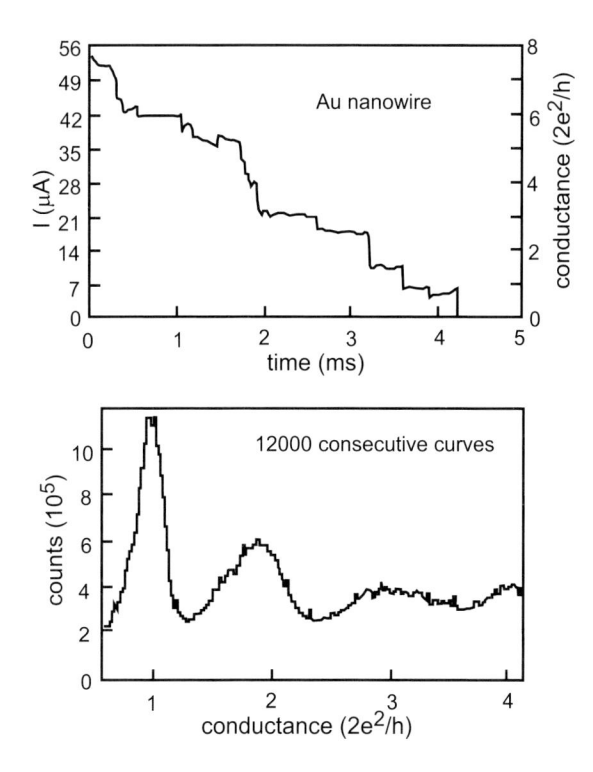

Figure 7.22: Top: conductance as a function of the deformation time of a gold junction (a STM setup was used). The observed steps are usually not quantized in units of $2e^2/h$. Bottom: a histogram of many consecutive sweeps of the upper type, however, reveals that steps of the expected height dominate. Adapted from [CostaKramer1997].

and you can even do it on your kitchen table using a household wire [CostaKramer1996]. Most of these experiments, however, are done in a more controlled setup. For example, a thin metal wire is mounted as the tip of an STM and pushed towards a metal surface. Fig. 7.21 shows a mechanically controlled break junction, another widely used setup. Fig. 7.22 reproduces a typical experimental current trace as a function of time over which the junction is deformed. Clear conductance steps can be observed, although they do not necessarily have the "right" values. This is attributed to the details of the breaking process: possibly, several QPCs are generated in parallel. Disorder in the junction may modify the plateau values as well. Conductance histograms taken over many cycles of breaking and reconnecting the junction, however, show that the conductance is predominantly quantized in units of $2e^2/h$. The details of these experimental results contain a lot of information. To a crude approximation, it may be assumed that right before the wires break, the current is carried via a single atom. The degeneracies of the conducting modes of the atomic junction can be material specific, as has been demonstrated, e.g., by [Ludoph2000].

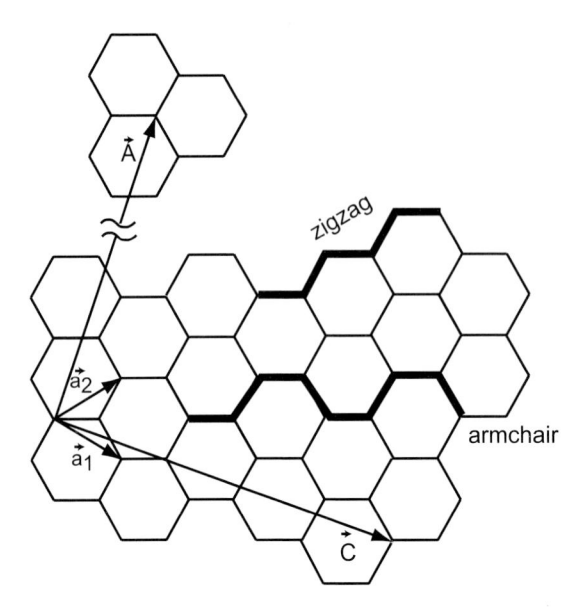

Figure 7.23: Scheme of a sheet of graphite used to roll up a CN.

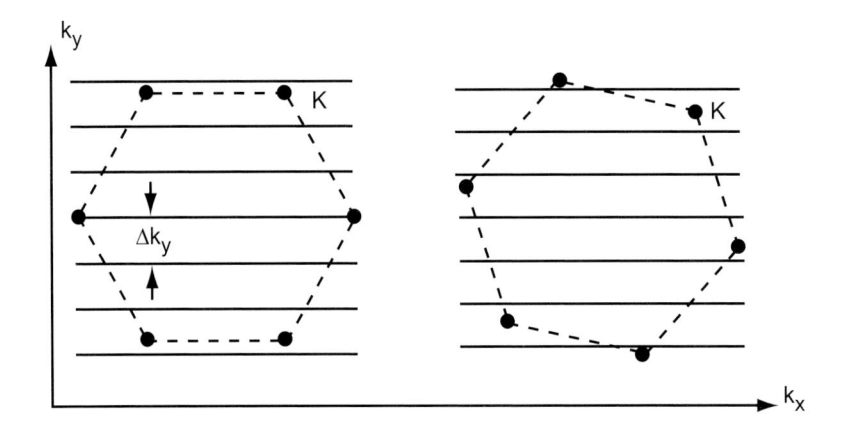

Figure 7.24: By periodic boundary conditions in y-direction, the Brillouin zone of the graphite sheet (dashed hexagon) gets partly quantized, and one-dimensional modes (lines) in x - direction, i.e. along the CN axis, result. Left: the 1D modes miss the K - points. The CN is semiconducting. Right: if some K - points fall on the modes, the CN is metallic.

7.4.2 Carbon nanotubes

Carbon nanotubes (CNs) have enjoyed a wide popularity since 1992. They have unique structural, mechanical and electronic properties, which are treated in several excellent books, as well as in review articles (for references, see the end of this chapter). Before we look at the

7.4 Further examples of quantum wires

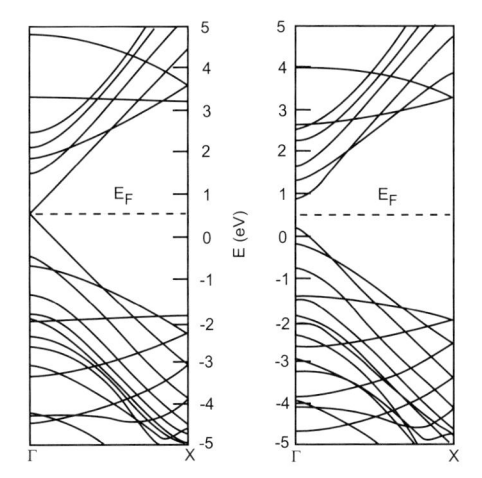

Figure 7.25: Band structures of metallic (a) and and semiconducting (b) CNs. After [Hamada1992].

quantum wire aspects of CNs, it should be defined what a CN actually is. We have seen in chapter 2 that graphite consists of two-dimensional, weakly coupled sheets of honeycomb lattices. By, e.g., laser ablation, it is quite easy to produce individual graphite sheets. In 1991, it has been discovered that under certain experimental conditions, these sheets roll up and form hollow carbon cylinders (CNs) with diameters of a few nanometers only [Iijima1991]. Their length, however, can be many microns. One distinguishes between single-walled and multi-walled CNs, depending on the number of concentric carbon cylinders. Of particular interest are single-walled CNs, since these well-defined systems can be treated theoretically to a considerable depth. In Fig. 7.23, a graphite sheet used to form a CN is shown. The "chirality vector" $\vec{C}$ defines a stripe that is cut out of the 2D lattice. The CN is formed by connecting the edges of this stripe, located at the bottom and the top of $\vec{C}$. Hence, $|\vec{C}|$ is the diameter of the CN. It has become common practice to characterize the structure of a CN by the coordinates (n_1, n_2) of $\vec{C}$ with respect to the lattice vectors $\vec{a}_1$ and $\vec{a}_2$ of the graphite sheet. For example, the CN shown in Fig.1.6 is a $(10, 5)$ tube, which means that $\vec{C} = 10\vec{a}_1 + 5\vec{a}_2$. Note that it suffices to consider tubes with $0 < n_2 < n_1$. For apparent reasons, CNs with cross sections along the bold lines are called "zigzag" and "armchair" tubes, respectively. The elementary lattice vector of the resulting CN, which is perpendicular to $\vec{C}$ in the graphite sheet, is denoted by $\vec{A}$. Its length is the minimal distance for which top and bottom and the top see identical environments. The additional periodic boundary condition modifies the electronic band structure. Clearly, a CN is a quasi one-dimensional system. It is furthermore obvious that the wave vector k_y perpendicular to the tube axis gets quantized, such that one-dimensional modes emerge from the two-dimensional energy dispersion of the graphite sheet (Fig. 2.3). Depending on the direction of the CN axis (parallel to $\vec{A}$ and the circumference of the CN), k_x of one mode may or may not hit a K point of the graphite sheet's Brillouin zone, see Fig. 7.24, and hence, a metallic or a semiconducting CN results. It can be shown that CNs are metallic if $(2n_1 + n_2)$ is a multiple integer of 3 (see, e.g., the literature given at the end of this chapter). The energy dispersion of these two classes of CNs are reproduced in Fig. 7.25.

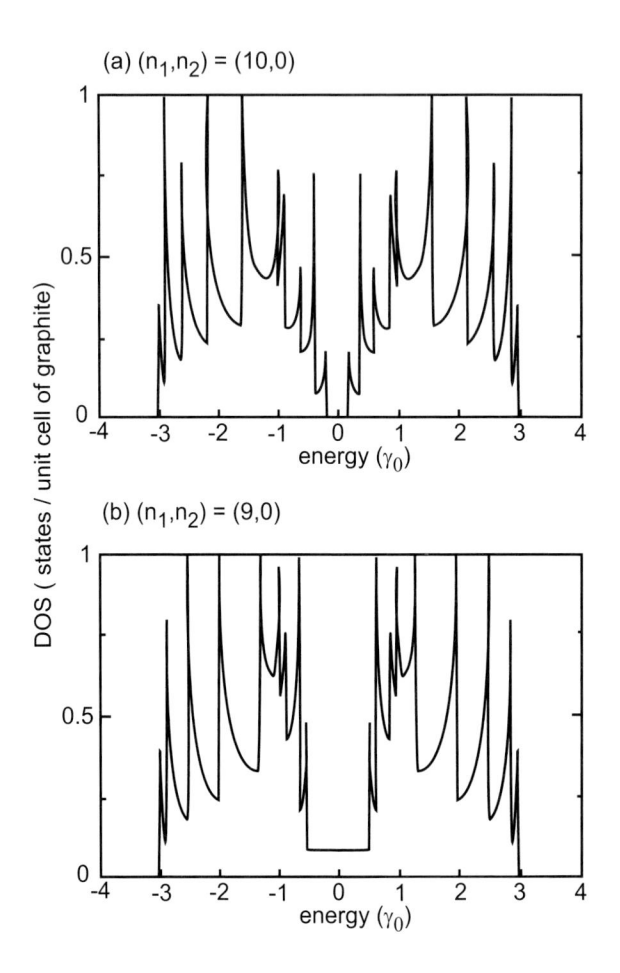

Figure 7.26: The full lines are the calculated density of states of a semiconducting CN (top) and of a metallic CN, bottom. The Fermi energy is at $E = 0$. The band gap equals 0.7 eV in the semiconducting CN. Adapted from [Saito1992].

The calculated densities of states (Fig. 7.26 gives an example) the theoretically obtained densities of states agree very well with experimental results [Wildoer1998], [Odom1998] obtained by scanning tunnelling spectroscopy. The quasi one-dimensional character is apparent. The mode separation is of the order of 100 meV, i.e., much larger than in lithographically patterned QWRs.

Making low resistance contacts to CNs has turned out to be very difficult. Such samples are typically prepared by depositing CNs on an insulating substrate that contains some metallic electrodes. By chance, a CN will make contacts with two or more electrodes, and multi-terminal transport measurements become possible. Usually, the contact between, say, a gold electrode and a CN in such a sample is a tunnel barrier. It has been shown that the contact resistance can be reduced by electron beam irradiation of the contact region [Bachthold1998];

7.5 Quantum point contact circuits

also, the contact becomes much better when the electrodes are patterned *on top of* the deposited CNs. In neither of these setups, however, conductance steps could be observed. An experiments that demonstrates conductance quantization in multiwalled CNs has been performed by [Frank1998]: the authors immersed a CN attached to the tip of a scanning tunneling microscope in liquid mercury, and observed conductance steps as a function of the tip position at room temperature.

In metallic CNs, there are two spin degenerate modes at the Fermi level, and we expect a quantized conductance of $G = 4e^2/h$, in case a single cylinder of the multiwalled CN couples to the reservoir. Instead, steps of height $2e^2/h$ are observed. The origin of this discrepancy has remained unexplained. Possibly, the spin degeneracy is lifted by electron-electron interactions.

7.5 Quantum point contact circuits

7.5.1 Non-ohmic behavior of collinear QPCs

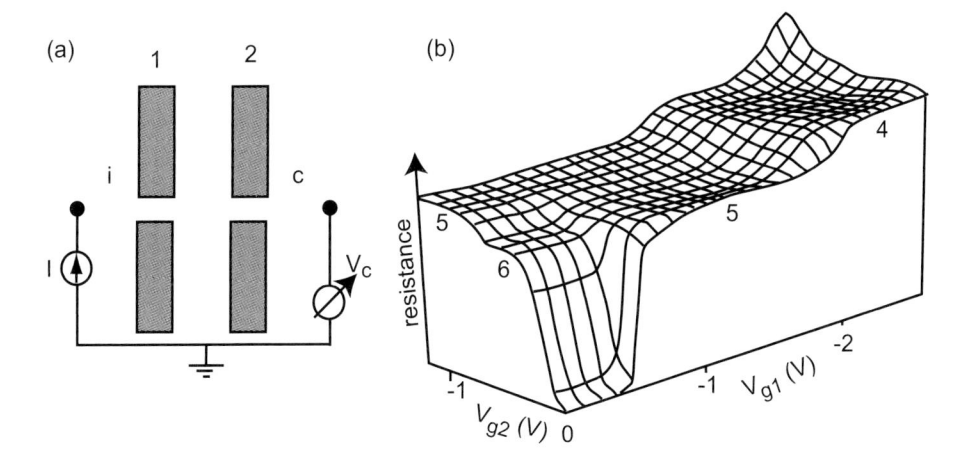

Figure 7.27: Sketch of two QPCs in series (a). The resistance as a function of the voltages V_{g1} and V_{g1} applied to the split gates 1 and 2 is shown in (b). The behavior is non-Ohmic, and indicates that the total resistance equals roughly the individual resistance of the QPC with fewer occupied modes. The temperature was about 300 mK. Adapted from [Wharam1988a].

Combinations of QPCs offer a variety of experimental options. Even the most elementary one, namely just two QPCs, raises already interesting questions. In Fig. 7.27, the resistance of two QPCs in series, with a separation much smaller than the elastic mean free path, is shown as a function of the voltages applied to the two split gates. It is immediately apparent that Ohm's law is violated: the series resistance is much smaller than the sum of the individual QPC resistances. We denote the individual conductances of QPC_1 and QPC_2 by G_1 and G_2, respectively. The series resistance of both QPCs between the injector and the collector is denoted by G_{ic}. Experimentally, $G_{ic} \approx min\{G_1, G_2\}$ was found in this experiment. Apparently, the QPC which is more narrow determines the total resistance, while the electrons pass the second one with no or little further resistance. One is therefore tempted to guess that the contact resistance of the second QPC is strongly reduced. This behavior can be explained by the ballistic character of the electron motion in between the QPCs, in combination with the adiabatic coupling mentioned in section 7.2.2. Suppose the electrons exit the QPC in an adiabatical fashion, i.e., the remain in the mode they used to pass through the QPC. As the width of the constriction gets wider, the energy of the mode, and with it the transverse momentum p_y, gets reduced. Consequently, the electrons in that mode get *collimated* as they exit the first QPC: the ejection angle is smaller than $\pm \pi/2$, while the angular probability distribution of the electrons in the 2DEG reflects the sum of the probability densities of the QPC modes that carry the current, see Fig. 7.28. This spatial distribution has been verified experimentally by an ingenious experiment of [Topinka2000]. Thus, the electrons more or less already have the

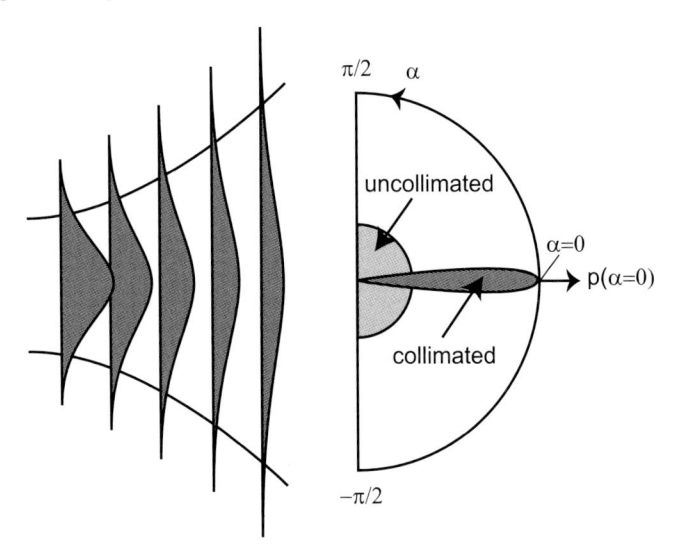

Figure 7.28: Adiabatic spreading of the wave function belonging to the lowest mode of a QPC at its exit. As the channel gets wider, the transverse electron momentum k_y is reduced. The result is sketched to the right: as compared to the uncollimated beam, the probability density $p(\alpha)$ for the electrons to be emitted in x - direction is peaked around $\alpha = 0$.

correct transverse momentum needed for entering the second QPC. To be more quantitative, we once again use the Landauer-Büttiker formula. There is a direct transmission of electrons from the injector to the collector T_{ic}. However, since in the setup of Fig. 7.27, the central region in between the QPCs floats and the collector is grounded, current conservation tells us that the injected flow of electrons will will either escape back into the injector, or it will finally make it into the collector, after some scattering events that will take place far away from the QPC region. If we assume that the middle area is large enough for equilibrating these electrons, we can speak of a chemical potential of the middle μ_m. Thus, electrons are either transmitted directly from the injector into the collector, or they get absorbed and re-emitted by

7.5 Quantum point contact circuits

an effective reservoir at potential μ_m. For the sake of simplicity, let us study the case of equal QPC conductances, i.e. $G_1 = G_2 = N \cdot 2e^2/h$. The Landauer-Büttiker equations then read

$$\frac{h}{2e^2} I_i = NV_i - T_{mi}V_m - T_{ci}V_C$$

$$\frac{h}{2e^2} I_c = NV_c - T_{mc}V_m - T_{ci}V_i$$

$$0 = (N_m - R_{mm})V_m - T_{mi}V_i - T_{mc}V_C \qquad (7.40)$$

If we assume $T_{mi} = T_{mc}$ and use $I_i = -I_c$, addition of the first two equations of (7.40) gives

$$R_{ic} = \frac{V_i - V_s}{I_i} = 2\frac{h}{2e^2}\frac{1}{(N + T_{ic})} \qquad (7.41)$$

This tells us that the series resistance is just the resistance of one QPC if $T_{ic} = N$, when all the electrons are directly transmitted. If $T_{ic} = 0$, the resistances follow Ohm's law.

If the the electrons were ejected from the first QPC with equal probability into all directions, only a fraction of $\approx \frac{w}{2\pi L}$ would be directly transmitted for $w \ll L$. Here, L is the separation of the QPCs, and w_c is the effective QPC width.

That the collimation effect actually determines the non-Ohmic addition of QPC resistances in series can be seen by studying T_{ic} as a function of a magnetic field. We expect that the collimated electron beam pattern gets deflected due to the Lorentz force, and T_{ic} should show a strong peak at $B = 0$, haloed by smaller side peaks at nonzero magnetic fields for QPCs with more than just one conducting channel. Such experiments have been performed by [Molenkamp1990] and by [Shepard1992].

7.5.2 QPCs in parallel

The resistance of parallel QPCs in a ballistic circuit is a periodic function of the magnetic field, see Fig. 7.29(a). The electrons ejected from the first QPC are forced on cyclotron orbits. For the correct polarity of the magnetic field, the deflected electron beam is directly injected into the second QPC, provided the separation between the QPCs is a multiple integer j of the cyclotron diameter. A fraction of the electrons thus gets caught at the QPCs and circulates around the separating barrier. Consequently, the resistance shows maxima for $jr_C = s$, where s is the separation. This can be nicely seen in Fig. 7.29 (b), where these resistance maxima have been measured in a one-dimensional array of 43 QPCs in parallel, with a spacing of $s = 4.6 \mu m$. Since the electron density was $n = 2.2 \times 10^{15}$ m^{-2}, a peak spacing of

$$\Delta B = \sqrt{8\pi n}\frac{\hbar}{es} \approx 34 \text{ mT}$$

is observed. Note that the peak resistance is significantly larger than the resistance at $B = 0$, while it is known that the resistance of an individual QPC drops as B is increased, see [Houten1988], which is [P 7.2.]. Due to the arrangement of many QPCs in series, additional resonances are found in 7.29, which are labelled by m and correspond to trajectories which obey the resonance condition for next-nearest neighbor QPCs.

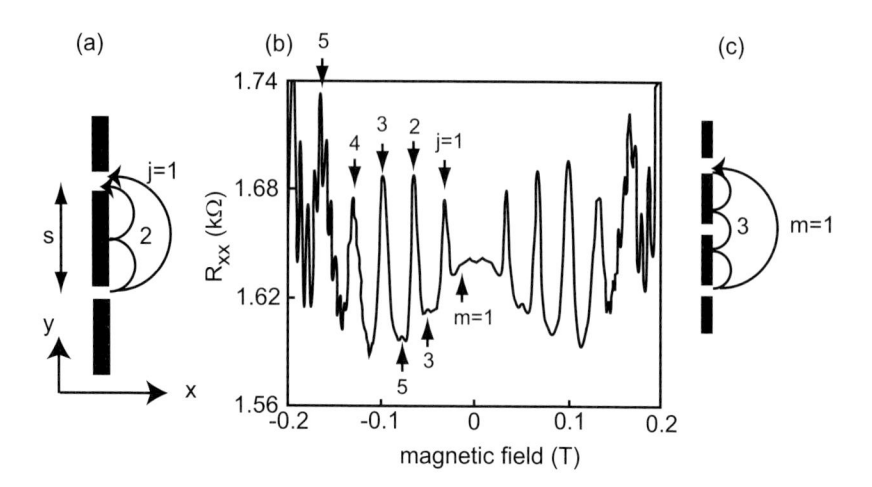

Figure 7.29: The magnetoresistance of QPCs in series (a) shows periodic peaks (b), due to commensurability between the cyclotron orbits and the QPC spacing s (a,c). Adapted from [Moon1997].

In this apparently simple explanation, two crucial assumptions have been implicitly made. First of all, the scattering at the separating barrier must have a large specularity. Here, the barriers have been defined by a shallow etch of the Ga[Al]As surface, which give a specularity of 1, to a good approximation [Moon1997]. Second, despite the collimation effect, the electrons are not strictly ejected in x-direction from the first QPC. However, a simple geometric consideration shows that electrons that exit the QPC at an arbitrary angle in x-direction all get focused at separations $\delta y = j\sqrt{8\pi n}/(eB)$, where δy denotes the distance of the focus from the QPC in y-direction. Further details of this *magnetic electron focusing*, which is well-known from normal metals as well as from charged particles in vacuum tubes, can be found in [P 7.4].

7.6 Concluding remarks

As the feature size of our electronic devices keeps decreasing, the quantum wire aspect of an electrical connection will become more and more important. Conductance quantization in quantum wires is one of the milestones in mesoscopic physics, and its detailed explanation is not trivial. Throughout this chapter, we have treated the electrons as non-interacting. It is known, however, that the Landauer formula remains valid for an interacting region connected to non-interacting leads, which host the eigenstates of the incoming and outgoing waves.

Both non-interacting and interacting electron gases in one dimension can be mapped onto non-interacting bosons, as long as the energy of the excitations is small, i.e., the energy dispersion can be assumed as linear. The resulting system is a Luttinger liquid. Luttinger liquids are distinctly different from Fermi liquids. For example, they show spin-charge separation, which means that the spin and the charge are decoupled and have different group velocities.

7.6 Concluding remarks

The Luttinger liquid aspects of one dimensional electron systems are currently an active field of research. The interested reader is referred to [Schonhammer1996] as a starting point for further studies.

Papers and Exercises

P 7.1: In [Glazman1988], the authors derive the QPC conductance quantization by assuming that the narrowest point of the constriction determines the resistance. Discuss what determines the conductance in this model, and why it is quantized!

P 7.2: At weak magnetic fields, QPCs show a negative magnetoresistance, as reported first by [Houten1988]. Develop a picture of this effect!

P 7.3: In [Chklovskii1992], the electrostatics of edge channels is developed. Quantify the edge channel geometry illustrated in Fig. 7.20!

P 7.4: The article [Houten1989a] discusses in detail the effect of magnetic electron focusing in 2DEGs.
Start by studying the formation of caustics (appendix C). Next, follow the instructive derivation of the expressions for the four-terminal experiments of Section IV, as elaborated in appendix D and Section V of this paper.

E 7.1: An infinitely long quantum wire extends along the x-direction. In z-direction, only the lowest subband is occupied. The confinement in y-direction is parabolic: $V(y) = \frac{1}{2}m^*\omega_0^2$. A magnetic field is applied in z-direction.

   (i) Write down the two-dimensional Schrödinger equation (use the Landau gauge $\vec{A} = (By, 0, 0)$, and ignore the z-direction). Solve the equation using the ansatz $\Phi(x,y) = e^{ik_x x}\psi(y)$. Show that, by suitable substitution, the problem is equivalent to a harmonic oscillator.

   (ii) Interpret the above results! Focus in particular on the evolution of the energy levels as a function of B, as well as on the energy dispersion in x-direction. Discuss further the limits $\omega_0 \to 0$ and $\omega_c \to 0$.

   (iii) Derive eq. (7.10). Hint: use the density of states of a parabolic quantum wire and approximate the sum by an integral.

E 7.2: Consider the transmission of the lowest mode of a QPC, modelled by the transmission function $T(E) = \Theta(E - E_1)$. Show that the conductance G as a function of the chemical potential in the reservoirs and of E_1 has the form of eq. (7.17). Calculate also the characteristic temperature for which the conductance quantization is thermally smeared. Assume that adjacent modes are separated by an energy difference Δ. Assume that the source-drain bias voltage is infinitely small.

E 7.3: In this exercise, the resistances measured on the sample shown in Fig. 7.19 shall be calculated. We assume that the filling factor in the ungated region is N; below the gate, the filling factor is given by M $\leq$ N. Both filling factors are integers. In all edge states, the spin degeneracy is lifted.

   (a) Set up the matrix equation obtained within the Landauer-Büttiker formalism.

   (b) Calculate the resistances R_{ij}, i,j=1...4 ! Explain the plateau at $R_{xx} = h/2e^2$ in Fig. 7.19.

7.6 Concluding remarks

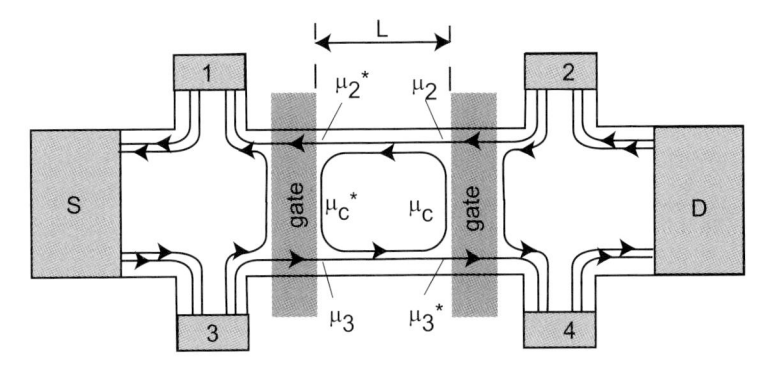

Figure 7.30: Sample geometry for [E 7.4]

E 7.4: Conventional Hall geometries are insensitive with respect to electron scattering between edge states at the same edge. Nevertheless, such scattering exists, and can be characterized by a coupling p between adjacent edge states along a distance L. p is defined as $p = (\Delta\mu - \Delta\mu^*)/\Delta\mu$, where $\Delta\mu$ and $\Delta\mu^*$ denote the potential differences between the edge states at the beginning, and at the end, respectively, of the distance L (is this a meaningful definition?). We study the coupling between the two spin-split edge states of the first Landau level. Experimentally, a Hall bar is adjusted in the regime of filling factor 2. The sample contains two gates across the Hall bar (see Fig. 7.28). These gates are biased to a regime where edge state 1 is transmitted, while edge state 2 is reflected. The distance between the gates is our length L.

(i) Assume that p is equal for both edges. Write down the matrix equation and determine μ_i (i=1...4, 2*,3*,c, and c*), as well as the current $I_S = I_D$. Note that due to charge conservation, $\mu_3 + \mu_c^* = \mu_3^* + \mu_c$, and similarly, $\mu_2 + \mu_c = \mu_2^* + \mu_c^*$.

(ii) The equilibration length L_{eq} is defined as the distance, the electrons have to travel before the potential difference $\Delta\mu$ between two adjacent edge states has dropped to $1/e = 0.368$ of its initial value. Determine L_{eq} from p! How large is L_{eq} for typical experimental numbers of $R_{12} = 0.53$ h/e² and L = 50 μm?

E 7.5: (a) Express for the $\vec{C}$ shown in Fig. 7.23 in terms of the lattice vectors $\vec{a}_1$ and $\vec{a}_2$ of the graphite sheet. Determine n_1 and n_2.
Give the general condition for (n_1, n_2) for armchair and zigzag tubes!

(b) Calculate $A(\vec{n_1, n_2})$ for the $\vec{C}$ drawn in the figure!

(c) Calculate the fundamental reciprocal lattice vector $\vec{B}$ of the CN! Draw the first Brillouin zone of the CN in the Brillouin zone of a graphite sheet. What is the mode spacing Δk_y due to quantization along the CN circumference? Illustrate these modes in the Brillouin zone!

(d) Is this particular CN metallic or insulating?

(e) Consider the energy dispersions of Fig. 7.25. The metallic and semiconducting tubes are $(12, 0)$ and $(13, 0)$, respectively. Estimate the effective mass in the conduction/valence band of the semiconducting tube close to the band extremal points. What is the effective mass at the Fermi level of the metallic tube? Calculate the density of states around the Fermi level for the metallic CN. Why is is constant, although the system is one-dimensional? Does the chemical potential depend on temperature?

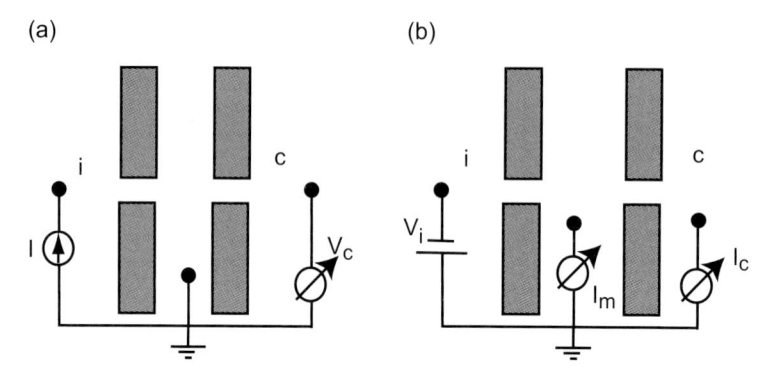

Figure 7.31: Measurement configurations considered in E 7.6.

E 7.6: Use the Landauer-Büttiker formalism to calculate V_c/I in 7.31(a), and I_c/V_i in (b). Assume identical QPCs in both cases. Discuss the pluses and minuses of these setups!

Further Reading

More on quantum wires in general can be found in [Beenakker1991]. The reader is also referred to [Datta1997], as well as to [Ferry1997]. An excellent review on QPCs is the article [Houten1992]. Finally, Carbon nanotubes are the topic of two recent books, namely [Saito1998] or [Harris1999].

8 Electronic Phase Coherence

In the previous chapters, we have implicitly assumed that the electrons are phase coherent in the confined directions, but have not considered possible consequences of phase coherence in the extended directions. Do such effects exist at all? The answer is yes, as you may have guessed. Electronic phase coherence manifests itself in electronic interferences, which can take place within the time scale τ_ϕ and the corresponding length scale ℓ_ϕ. Interference effects remind us of wave optics, and there are in fact many analogies. The most important signatures of electronic phase coherence in diffusive systems are Aharonov-Bohm type effects, weak localization, and universal conductance fluctuations, which are the topic of sections 1, 2, and 3, respectively. In section 4, we have a look at phase coherence in ballistic systems. The final section 5 of this chapter introduces the resonant tunnelling effect, i.e. the transmission barriers of tunnel bariers in series with a spacing smaller than the phase coherence length.

8.1 The Aharonov-Bohm effect in mesoscopic conductors

In 1959, [Aharonov1959] published a seminal *gedanken* experiment. The authors predicted that the partial waves of a charged particle enclosing an electrostatic or magnetic potential experience a phase shift, even if the electric and magnetic field vanishes in the regions of non-zero probability density. Interferences as a function of the relative phase shift occur, which are known as the electrostatic and magnetic Aharonov-Bohm (AB) effect, respectively. In Fig. 8.1, an experimental setup suited to test this prediction is shown. A ring is patterned out of a metal or a 2DEG, with a circumference smaller than the phase coherence length. Suppose

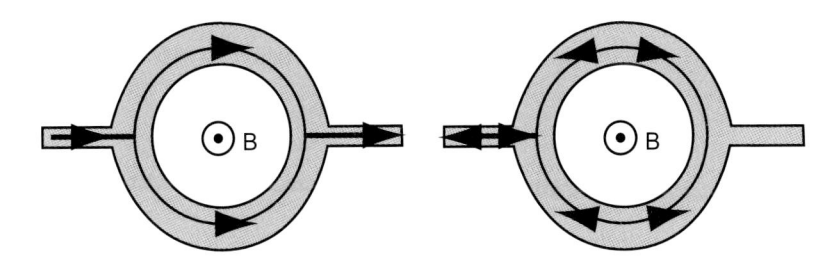

Figure 8.1: Sketch of a sample used to study the Aharonov-Bohm effect. Interfering trajectories that cause h/e and $h/2e$ oscillations are drawn to the left and to the right, respectively.

a conducting ring encloses a magnetic vector potential $\vec{A}$ that generates a constant magnetic field perpendicular the the plane of the ring.[1] The phase collected by the electrons during their passage through branch j of the ring (j denotes the upper or the lower semicircle) is given by

$$\phi_j = \frac{e}{\hbar} \int_j \vec{A} d\Gamma$$

$$\phi_{upper} = \frac{eBS}{2\hbar}, \qquad \phi_{lower} = -\frac{eBS}{2\hbar} \qquad (8.1)$$

Here, S denotes the area enclosed by the ring, and Γ is the parametrized trajectory. The

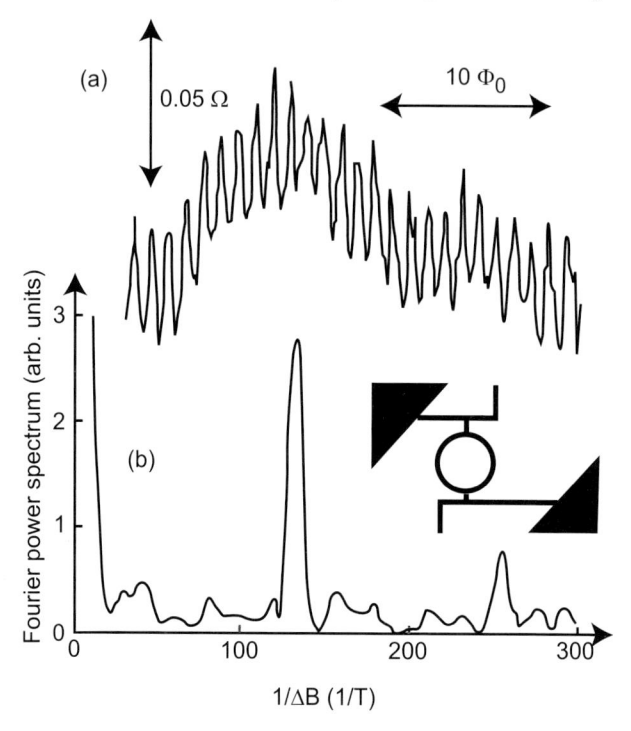

Figure 8.2: AB effect as observed in a gold ring (the sample is sketched in the inset, the ring diameter is about 1 μm). In (b), the Fourier spectrum of the raw data of (a) is shown. A strong eS/h frequency is observed, while the second order effect at $2eS/h$ is much weaker. After [Webb1985].

total transmission probability T is obtained by summing up all the probability amplitudes and calculation of the absolute value of the square. We neglect multiple reflections at the entrance and exit of the ring for now. Let us further assume that for $\vec{A} = 0$, both branches have identical transmission amplitudes t_0. T is obtained from

$$t_j = t_0 \cdot e^{\pm i \phi_j} \Rightarrow$$

---

[1] This is not exactly the original proposal by [Aharonov1959], since the branches of the ring are penetrated by $\vec{B}$. For very narrow arms compared to the ring diameter, however, this modification is irrelevant.

8.1 The Aharonov-Bohm effect in mesoscopic conductors

$$T = (t_{upper} + t_{lower})^* \cdot (t_{upper} + t_{lower}) = 2T_0(1 + \cos(\frac{eBS}{\hbar} + \phi_0)) \tag{8.2}$$

Apparently, T oscillates as a function of B, with a period of one magnetic flux quantum $\Phi_0 = \frac{h}{e}$ that penetrates the ring. Experimentally, the AB effect on metallic loops was first observed in gold rings by [Webb1985], see Fig. 8.2. It was later reproduced in rings defined in a Ga[Al]As heterostructure by [Timp1988], see Fig. 1.4 for such data. In these experiments,

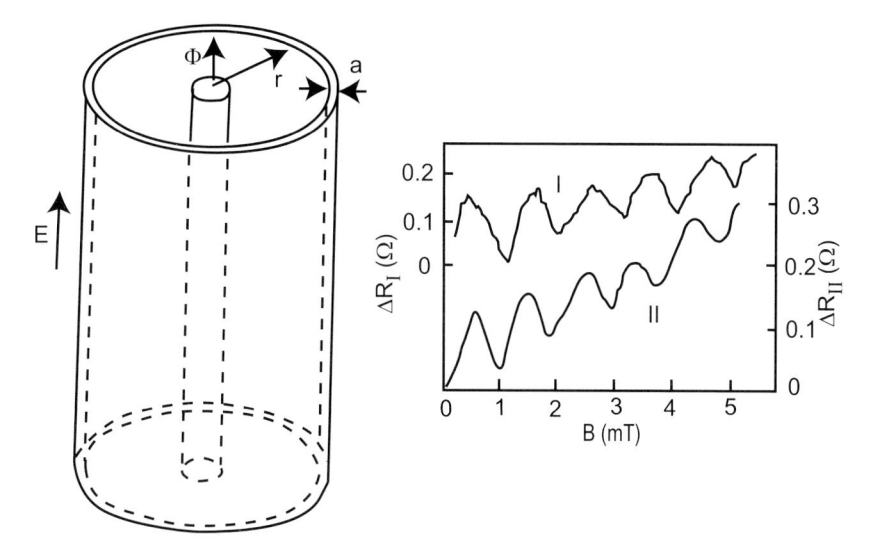

Figure 8.3: The experiment of [Sharvin1981]. The resistance as a function of the magnetic field along the cylinder axis oscillates with a period of $h/2eS$, i.e. it shows AAS oscillations. AB oscillations are absent. After [Sharvin1981].

the amplitudes were much smaller than predicted by eq. (8.1), since ℓ_ϕ was smaller than the ring circumference, and thus only a fraction of the electrons could pass the ring coherently.

So far, interferences have been taken into account to first order only. For sufficiently large ℓ_ϕ, there are of course also higher order interferences, for example, between two partial waves that have traversed both arms clockwise and counterclockwise, respectively, see Fig. 8.1. Due to their interfere at the ring's entrance, the reflection probability is aperiodic function of B. The period of these oscillations is half the AB period, $h/2eS$. They are known as *Altshuler-Aronov-Spivak (AAS)* oscillations [Altshuler1981] and are of particular relevance in mesoscopic physics. Since both trajectories traverse exactly the same path, their phase difference at $\vec{A} = 0$ is always zero, independent of the size and shape of the loop. The interference is constructive, which leads to a backscattering probability which is enhanced, compared to the classical value. We speak of coherent backscattering. In an ensemble of AB rings, all the initial phases ϕ_0 are randomly distributed, while they are always zero for the AAS oscillations. We therefore expect that in ensembles of rings, the AB oscillations average to zero, while the AAS oscillations survive ensemble averaging. This has been demonstrated first by [Sharvin1981], long before the AB oscillations could be detected in individual rings. This experiments was carried out with a insulating cylinder (diameter about 1.5 μm), coated with a thin Mg film. Measuring the resistance between the top and the bottom of the cylinder in a magnetic field along the cylinder axis can be thought of ensemble averaging over many rings in parallel. Due to the small variations in the diameter, several oscillation periods can be observed.

Ensemble averaging was systematically investigated by [Umbach1986]. Different numbers ($j = 1, 3, 10$, and 30) of silver rings were patterned in series, and the amplitudes of the various oscillation periods were investigated. It was found that the amplitude of the AB oscillations drops with $1/\sqrt{j}$ while the amplitude of the AAS oscillations remains constant, Fig. 8.4.

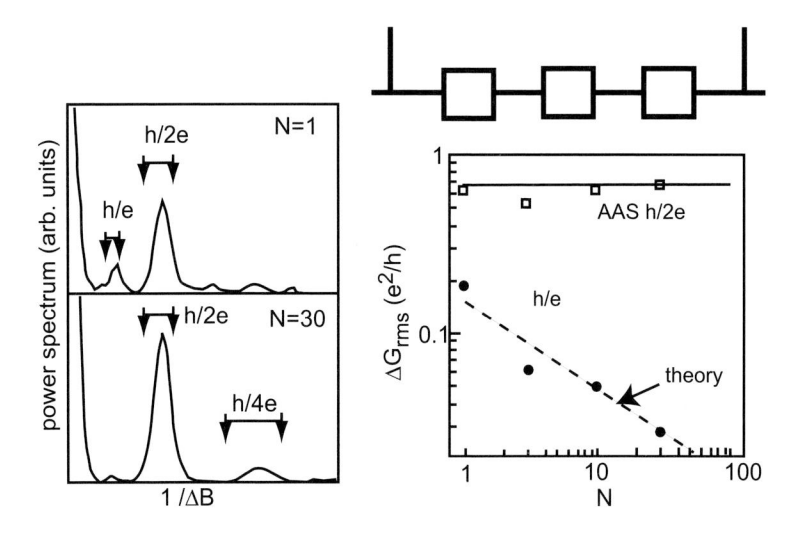

Figure 8.4: Top: three Ag loops in series. The size of the loops is 940 nm × 940 nm. The width of the Ag wires is about 80 nm. Left: the Fourier spectra of the magnetoresistance oscillations observed in one ring (top) shows both a h/e and a $h/2e$ component. For 30 loops in series (bottom), however, the h/e component is absent, while a weak $h/4e$ component has emerged. To the right, the relative strengths of the Fourier peaks are plotted vs. the number N of rings in series. Adapted from [Umbach1986].

8.2 Weak localization

Imagine an experiment in which a lot of rings with a broad size distribution are measured simultaneously. What will remain of the magneto-oscillations?[2] We surely can no longer expect to observe them, since each loop area has its own period, which will ensemble average. Note, however, that all the resonant backscattering waves are in phase at $B = 0$, and the resistance is enhanced as compared to the incoherent case. To be a bit more specific, we follow the line of arguing presented in [Beenakker1991] and consider the probability $P(\vec{r}_1, \vec{r}_2, t)$ for

---
[2]Small magnetic fields are assumed, such that Shubnikov - de Haas oscillations are absent.

8.2 Weak localization

the electron to move from $\vec{r_1}$ to $\vec{r_2}$ within the time t. $P(\vec{r_1}, \vec{r_2}, t)$ is thus the squared sum of all probability amplitudes A_i for this propagation within t:

$$P(\vec{r_1}, \vec{r_2}, t) = |\sum_i A_i|^2 = \sum_i |A_i|^2 + \sum_{i \neq j} A_i A_j \qquad (8.3)$$

The first term on the right hand side is the classical probability for the electron to propagate between the two points along any path within t. The second term results from interferences. Since the phases of A_i are uncorrelated, the interference term averages to zero, with one exception: for $\vec{r_1} = \vec{r_2}$, we can form pairs of trajectories that correspond to identical paths, traversed in opposite directions. In other words, the two paired propagators can be mapped on each other by time inversion. At $B = 0$, their phases are identical, and interference is constructive. Such pairs thus give a non-vanishing contribution to the interference term. Since $|A_i + A_i^{time\ reversed}|^2 = 4 |A_i|^2$, we find an enhancement of the backscattering probability by a factor of 2, compared to the classical value, for such propagators. As the magnetic field is increased, the contribution the largest rings in the ensemble to the resistivity will oscillate rapidly, while the phase difference in the smallest rings will remain essentially unchanged. Hence, the larger the magnetic field, the fewer rings will contribute to the constructive interference, and the resistance should drop to its classical value, once the phase shift in the smallest rings is of the order of π. This is exactly the situation encountered in a diffusive electron gas. The ensemble of loops is formed by the elastic scatterers, see Fig. 8.5. This coherent

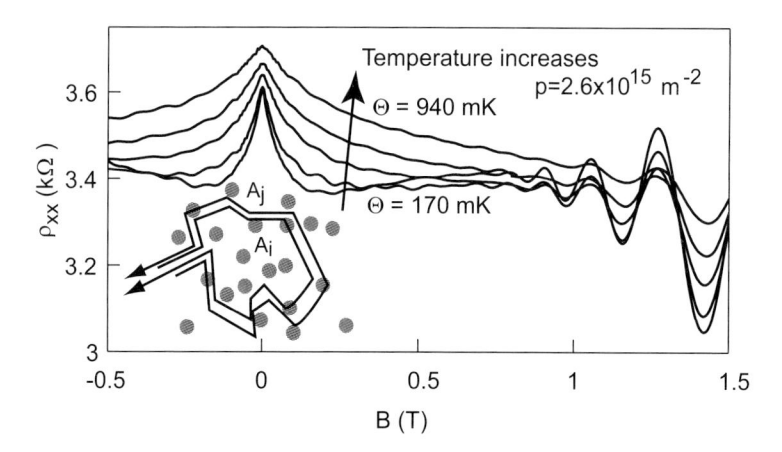

Figure 8.5: Inset: a fraction of the electronic trajectories in a diffusive 2DEG forms closed loops and lead to coherent backscattering. Main figure: WL peak as a function of B, and for various temperatures between 170 mK and 940 mK. The sample was a Si/SiGe quantum well containing a hole gas. Adapted from [Senz2002].

backscattering on on elastic, randomly distributed scatterers is called *weak localization* (WL). The localization of the electrons due to coherent backscattering is thereby distinguished from strong localization, which takes place in highly disordered samples. Experimentally, we can

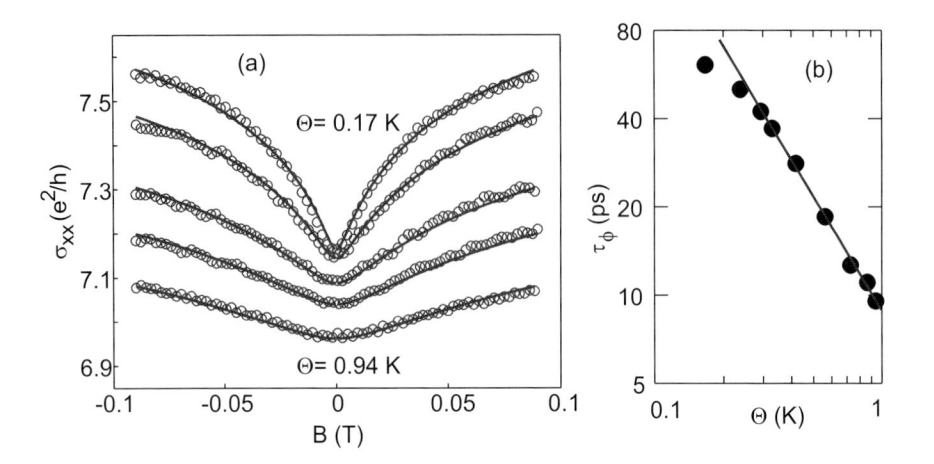

Figure 8.6: (a): the data of Fig. 8.5 (circles), translated into longitudinal conductivity $\sigma_{xx}(B)$. The lines are least square fits according to eq. (8.4). (b): temperature dependence of τ_ϕ, as determined from these data. The dashed line represents the theoretically expected $1/\Theta$ dependence. Adapted from [Senz2002].

thus expect an increased resistivity in diffusive samples due to WL, which is reduced to its classical value as the magnetic field increases. Of course, there will be no magneto-oscillation because of the averaging. This is in fact what is observed in experiments, see Fig. 8.5. The functional form of this WL peak depends on many parameters, in particular the elastic mean free path, the phase coherence length, as well as on the dimensionality of the electron gas. For a two-dimensional system, i.e. for a sample width larger than ℓ_{phi}, [Altshuler1982] have derived the magnetic field dependence of the WL correction to the classical conductivity :

$$\Delta\sigma_{xx}^{WL}(B) = \frac{e^2}{2\pi^2\hbar}[\Psi(\frac{1}{2} + \frac{\tau_B}{2\tau_\phi}) - \Psi(\frac{1}{2} + \frac{\tau_B}{2\tau_D}) + ln(\frac{\tau_\phi}{\tau_D})] \quad (8.4)$$

which at $B = 0$ reduces to

$$\delta\sigma_{xx}^{WL} = -g_s\frac{e^2}{4\pi^2\hbar}ln(1 + \frac{\tau_\phi}{\tau_D}) \quad (8.5)$$

Here, $\Psi(x)$ is the Digamma function. For large arguments, it can be approximated by

$$\Psi(x) \approx ln(x) - \frac{1}{x}$$

Fig. 8.6 shows that typical data can be fitted very well to the Altshuler formula.[3] The fit parameter is the phase coherence time.

It is widely accepted that at low temperatures, the electronic dephasing occurs via quasi-elastic electron-electron collisions [Altshuler1982]. For such a type of dephasing, one expects

---
[3] Note that in order to transform ρ_{xx} into σ_{xx}, the Hall resistivity has to be measured as well.

a characteristic $\tau_\phi \propto \frac{1}{\Theta}$ dependence, which is usually found in experimental data, see Fig. 8.6, except at very low temperatures. Here, τ_ϕ saturates in most experiments. Although it is not entirely clear what the reason for this saturation is, a plausible explanation here would be that the electron temperature deviates from the bath temperature of the helium mixture in dilution refrigerators at very low temperatures; see however [P 8.2].

8.3 Universal conductance fluctuations

Let us take another look the quantum wire presented already in Figs. 7.3 and 7.4. In Fig. 8.7, its two-probe conductance as a function of the electron density and of magnetic fields in the regime $\omega_C < 0.45\omega_0$, where Landau quantization can be neglected, is shown. Around $B = 0$, a pronounced dip in the conductance can be seen, which is due to weak localization. In addition, conductance fluctuations are observed as a function of both parameters outside the weak localization peak. Cross sections of the conductance as a function of only one parameter with the second parameter fixed show this more clearly.

Note that these fluctuations are *not* noise: the features are fairly symmetric with respect to magnetic field inversion, and the features shift towards larger electron densities as $|B|$ is increased. In addition, the fluctuations smear out rapidly as the temperature is increased to about 1 K. The typical fluctuation amplitude is of the order of $0.2 e^2/h$. Characteristic fluctuation periods are ≈ 20 mT, and 5×10^6 m^{-1}, respectively. These fluctuations are *parametric*, i.e., they are perfectly reproducible as a function of the parameters, but they are nevertheless fluctuations, since there is no way in controlling their individual appearance. Also, the fluctuations look different in other samples with identical macroscopic properties, and they change in an individual sample when we warm it to room temperature and cool it down again.

Apparently, the fluctuations depend on the mesoscopic structure of the sample. It is known that thermal cycling essentially changes the impurity configuration of the sample, but not their macroscopic features, like the impurity density or the scattering times. It can thus be assumed that the fluctuations somehow characterize the specific impurity configuration in the sample during a particular cooldown. Note that the amplitude and the width of the weak localization peak fluctuates as well. This is an indication that the measurement does not average over a huge number of weak localization loops, but just a few of them. Furthermore, the fluctuations show a very strong temperature dependence at ultra-low temperatures below 1 K, while the elastic scattering times are essentially independent of temperature in this regime. This suggests that these fluctuations are again a manifestation of electronic phase coherence; we have just seen in the section on weak localization, that $\tau_\phi \propto 1/\Theta$ in this temperature range. This very qualitative line of arguing essentially sketches the generally accepted interpretation of these conductance fluctuations.

Similar fluctuations can be observed in many mesoscopic samples where the sample size is comparable to the phase coherence length, but larger than the elastic mean free path. The quantum wire under consideration here, for example, has a length of $L = 40$ μm, while $\tau_D = 5.7$ μmm and $\tau_q = 460$ nm for the 2DEG the wire was made of. The phase coherence length inside the wire is $\ell_\phi \approx 7$ μm. In most samples with $\ell_e < L < \ell_\phi$ and for negligible temperature, it is found that the average fluctuation amplitude does neither depend

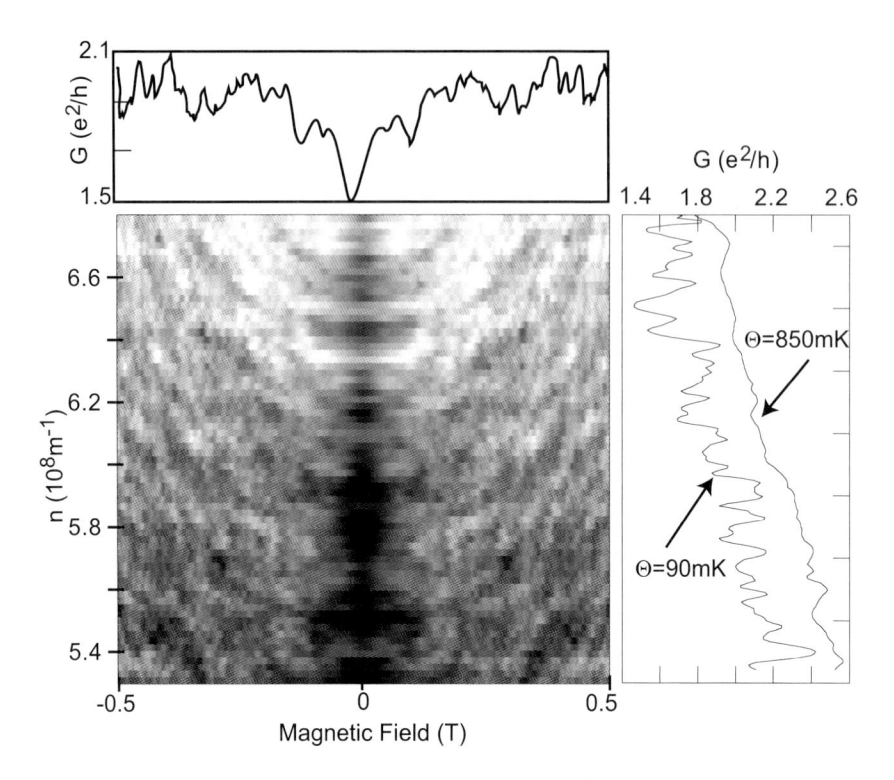

Figure 8.7: Main figure: conductance of the quantum wire shown in Figs.7.3 and 7.4, as a function of the one-dimensional electron density n and the magnetic field B. The gray scale ranges from $G = 1.4e^2/h$ (dark) to $2.5e^2/h$ (bright). Cross sections as a function of B for constant $n = 6 \cdot 10^{-8}$ m^{-1}, and as a function of n at $B = 0$ are shown to the top and to the right, respectively. The temperature has been 90 mK. The right figure shows the thermal smearing of the fluctuations as the temperature is moderately increased to 850 mK.

on the sample size, nor on the strength or the configuration of the elastic scatterers. This is quite remarkable; apparently, more scatterers does not imply more pronounced smearing of the fluctuations. Therefore, they are often referred to as *universal conductance fluctuations - UCF*. Unfortunately, there is no simple, intuitive picture for UCF, in contrast to weak localization. Furthermore, the quantitative description of the fluctuations strongly depends on many length scales, in particular the width and length of the sample, as well as on ℓ_e, ℓ_ϕ and ℓ_T. We therefore refrain from a discussion of UCF in all these regimes, and exemplify it by some qualitative arguments for our wire. Theoretical models for parametric UCF are based on the *ergodicity theorem*. Suppose we measure the resistivity of various samples with identical macroscopic parameters and sample sizes of a few elastic mean free paths only. Furthermore, the samples are cooled to a regime where $\ell_\phi > \ell_e$. Even though the resistivities are identical, we can no longer expect to measure identical resistances in different samples, since the exact number of scatterers will be sample-dependent. In addition, the microscopic configuration of

8.3 Universal conductance fluctuations

elastic scatterers is sample-specific, even if their number is identical, and consequently, the interference pattern of the electron waves, and with it the transmission and reflection probabilities, will be unique for each sample. As in weak localization, the interferences of the electronic wave functions generate localization and reduce the conductance.

Experimentally, it is of course a rather tedious task to fabricate and measure sufficiently many samples for a good statistics. It is, however, generally accepted that the systems under consideration here behave *ergodic*. Suppose we measure a quantity q in an ensemble of samples with identical macroscopic parameters, and determine the variance of q. In a second experiment, we measure the same quantity in only one member of the ensemble as a function of a parameter p (which can be, e.g., the electron density or the magnetic field), and average the quantity over p. Per definition, a system is called ergodic (with respect to q) if the two averaging procedures give the same mean value, and the same variances. This definition assumes that both the number of samples in the ensemble, and the scan range of the parameter, are infinite. In a simple picture, we can imagine that in both experiments, we average over many microstates of the system. Non-ergodic samples are thus samples where tuning the external parameter does not induce sufficient transitions between microstates. This can be due to metastable states, for example. For a more detailed discussion of ergodicity, see [Palmer1982].

By which amount do we have to change p, before the microstate of the system can be regarded as different, or before the values $q(p)$ of the measured quantity can be regarded as statistically independent, respectively? The answer is given by the autocorrelation function $C_q(\Delta p)$.[4] At $\Delta p = 0$, $C_q(0) = C_0$. For a fluctuating q as a function of p, $C_q(\Delta p)$ typically drops to zero within a certain range of Δp. The autocorrelation value p_C is the parameter value for which the autocorrelation function has dropped to $0.5 C_0$, and the data can be considered as statistically independent as soon as p differs by at least p_C.

As long as $\ell_\phi > L$, the average UCF amplitude has been found to be of the order of e^2/h, independent of the number of elastic scatterers and of L. This is quite surprising, since it means that averaging over disorder does not weaken the fluctuations, as long as the sample is phase coherent. This result can be derived if one assumes strong correlations between the transmission amplitudes of different paths across the whole sample, while the reflection amplitudes are uncorrelated. An intuitive argument for such a scenario would be that electrons that manage to cross the sample do so via identical sections of trajectories which contain many scatterers, while a single backscattering event suffices for reflection [Lee1986].

In many circumstances, however, the non-zero temperature cannot be neglected. Its effect on the length scales is twofold: first of all, it may reduce ℓ_ϕ to the regime of $\ell_e < \ell_\phi < L$. Second, the thermal length comes into play. In [Beenakker1988], it has been argued that for a quantum wire with $w \ll \ell_\phi$, the fluctuation amplitude can be approximated by

$$\delta G = \frac{1}{\beta} \frac{e^2}{h} \sqrt{\frac{12(\ell_\phi/L)^3}{1 + \frac{9}{2\pi}\frac{\ell_\phi}{\ell_T}}} \tag{8.6}$$

which, for the QWR under study here, agrees fairly well with the experiment if a phase coherence length inside the wire of $\ell_{phi} \approx 7$ µm and a thermal length of $\ell_T \approx 11$ µm is assumed.

---

[4]For a brief introduction to correlation functions, see appendix 2.

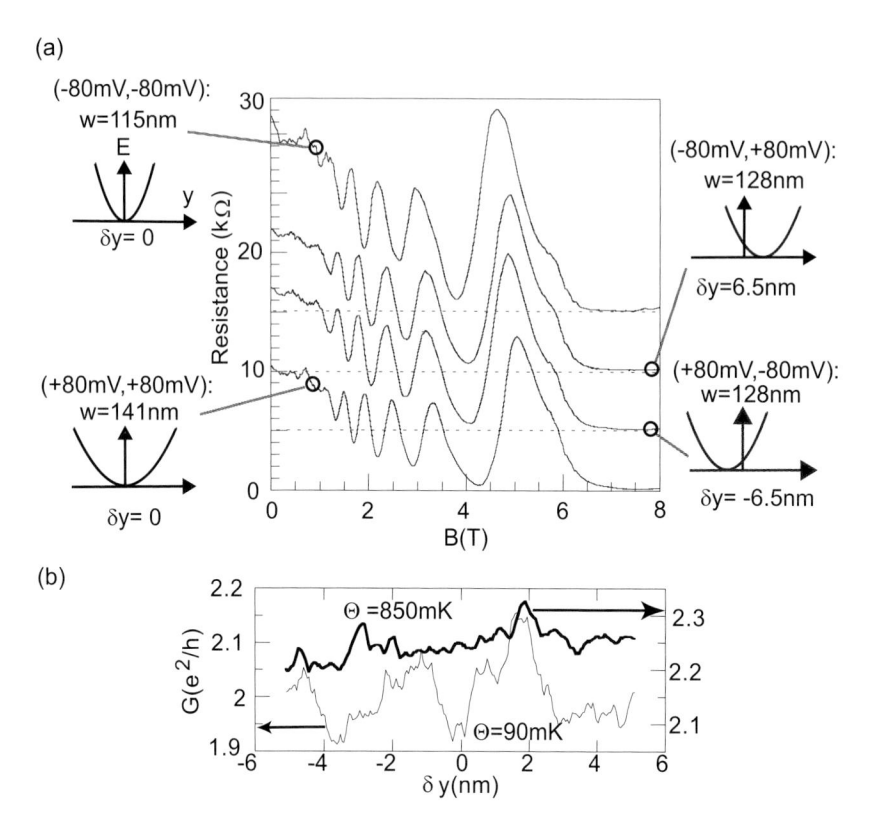

Figure 8.8: (a): magnetoresistance of the QWR (the sample is shown in Fig. 7.3) for different combinations of in-plane gate voltages (V_1, V_2). Adjacent traces are offset by 5 kΩ, and the dashed lines denote the corresponding zeroes. Fitting the oscillations to eq. 7.10 shows that the wire width w and ω_0 do not alter for antisymmetric gate voltage changes, i.e. for $V_1 + V_2 =$ *constant*. Potential shape and position of the QWR are indicated by the sketches outside the main figure. (b): measured conductance fluctuations as a function of the wire displacement δy, shown for two temperatures.

We conclude this section by discussing an experiment in which the impurity configuration has been changed parametrically [Heinzel2000]. The QWR under discussion was tuned by varying the two in-plane gates (see Fig. 7.3) such that $\delta V_{as} = \delta V_1 = -\delta V_2$. In Exercise 6.3, it has been shown that a constant electric field displaces a parabolic potential without changing its shape. By analyzing the magnetoconductance oscillations in large magnetic fields, it has been made sure that ω_0 remains constant in such an experiment. Within the approximation of a parabolic confinement, the QWR is thus spatially displaced in y-direction as δV_{as} is scanned, Fig. 8.8(a). The displacement δy equals the change of the wire width as a function of one in-plane gate voltage with the second gate voltage fixed. Shifts up to $\delta y = 12$ nm have been possible before a leakage current between wire and in-plane gates sets in. Furthermore, δy is linear in δV_{pgi} (i=1,2) within experimental accuracy, and a lever arm $\delta y / \delta V_{as} = 80$ nm/V is measured.

In Fig. 8.8 (b), the conductance of the wire is shown as a function of its displacement. Again, reproducible conductance fluctuations are observed, with a temperature dependence similar to the fluctuations as a function of B and n. The average period and amplitude is $\delta y \approx 2$ nm and $\delta G \approx 0.15\, e^2/h$, respectively. Apparently, the interference pattern can also be changed by shifting the wire through the crystal. Intuitively, this is quite clear, since the interference pattern depends on the potential landscape at which the electronic waves scatter. As the wire is shifted through the crystal, the scattering potential is scanned, and the interference pattern changes accordingly. But what determines the fluctuation period of $\delta y \approx 2$ nm?

We denote the relevant density of bumps in the scattering potential by $1/d^2$ and try to relate d to a characteristic length scale of the sample. On average, the number of bumps inside the wire should change by one as the wire is displaced by $\delta y = d^2/2L$. The factor of 1/2 enters because the bumps may enter the wire on one side as well as exit it on the other. One finds $\delta y = 2.7$ nm for $d = \ell_q$, the quantum scattering time. This indicates that the conductance fluctuations are caused by all scatterers, and not just by the large-angle scatterers that determine ℓ_e, which enter or leave the wire region as it is displaced. It should be mentioned that shifting the impurity configuration within the wire also generates conductance fluctuations, which, however, have a characteristic fluctuation period of the order of the Fermi wavelength $\lambda_{F,1}$ of the electrons in the lowest one-dimensional subband, as has been numerically demonstrated by [Cahay1988] for the case of displacing a single scatterer. In the experiment under discussion here, however, this length scale cannot be seen, since $\lambda_{F,1}$ equals ≈ 30 nm, which is larger than the displacement range.

Question 8.1: How do you explain that in Fig. 8.7, the feutues in the conductance shift to larger electron densities as $|B|$ is increased?

8.4 Phase coherence in ballistic 2DEGs

Electronic phase coherence in ballistic 2DEGs is a relatively unexplored territory. All the experiments discussed so far in this chapter relied on the diffusive character of the sample; scattering at impurities was essential for obtaining information on τ_ϕ. An important experiment addressing the issue of phase coherence in the ballistic regime has been performed by [Yacoby1991], see Fig. 8.9. Electrons were injected into a ballistic 2DEG via a quantum point contact acting as an emitter (E), and collected using a second QPC (C). Since the sample in between emitter and collector is ballistic, only those electrons that move very close to the straight line that connects E and C contribute to the voltage buildup. The upper half plane of the 2DEG was partly covered by a gate of length L, which could be used to tune the electron density, and hence the phase shift, of the electrons underneath. Hence, approximately 50% of the electrons entering the collector have passed below the gate. Assuming a linear relation between the gate voltage V_G and the Fermi energy E_F in the 2DEG, the phase shift the electrons

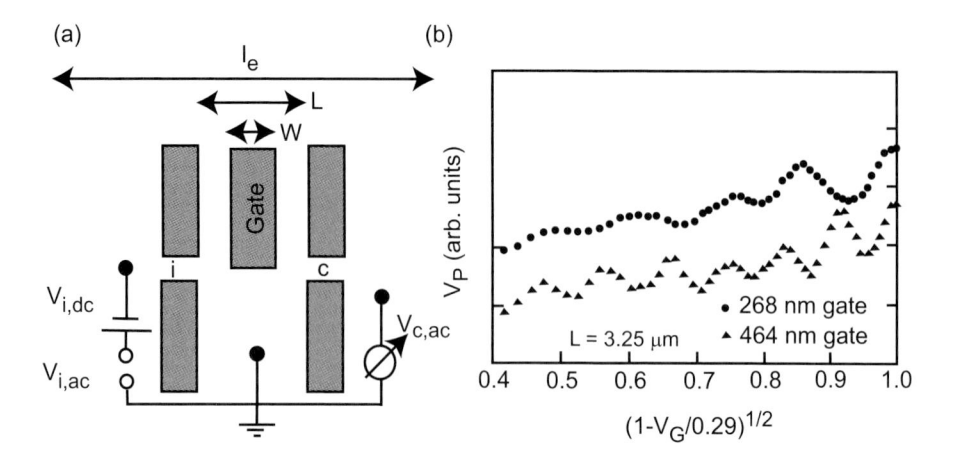

Figure 8.9: (a: sketch of the experimental setup used to investigate the dephasing in ballistic 2DEGs. The 2DEG is enclosed by two split gates acting as emitter (E) and collector (C), separated by a distance $L < \ell_e$. Note that the ballistic 2DEG is grounded, and the ac voltage buildup is measured behind the collector QPC. (b): voltage buildup at C, as a function of $\sqrt{1 - V_G/V_T}$. Periodic oscillations are observed, with a period that drops as the gate width W is increased. The measurements were performed at $\Theta = 1.4$ K. (Adapted from [Yacoby1991]).

acquire underneath the gate is given by

$$\delta\phi = Lk_{F,0}\delta(\sqrt{1 - V_G/V_T}) \tag{8.7}$$

where $k_{F,0}$ denotes the Fermi wave vector below the gate for $V_G = 0$, and $V_D = -290$ mV is the threshold gate voltage, at which the 2DEG below the gate gets depleted.

Question 8.2: Check eq.(8.7), using the assumptions stated in the text!

The phase-shifted electrons that traversed underneath the gate interfere with those that bypass the gated region, and the resulting current should therefore be periodic with a period of

$$\delta\phi = 2\pi \rightarrow \delta(\sqrt{1 - V_G/V_T}) = 2\pi/k_{F,0} \tag{8.8}$$

In fact, the voltage buildup at the collector as a function of the gate voltage showed quasi-periodic oscillations with the expected period, as shown in Fig. 8.9(b). The data shown here have been obtained for a pure ac excitation at E, i.e., for the special case of $V_{E,dc} = 0$. The experiment, however, has been carried out for a variety of dc excitation voltages, in order to compare the experiment with theoretical considerations. Note that only the ac component was measured at the collector!

From the relative oscillation amplitude $a(L, V_{E,dc})$, which is the peak-to-peak amplitude divided by the average voltage buildup, the phase coherence length was calculated via the

8.4 Phase coherence in ballistic 2DEGs

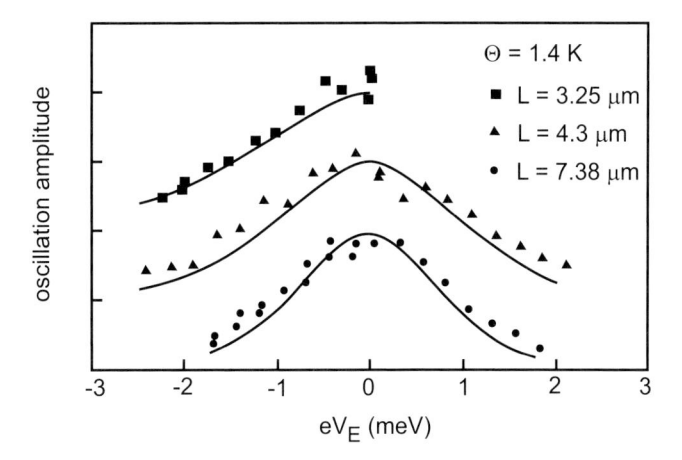

Figure 8.10: The symbols denote the oscillation amplitudes as measured for various gate lengths L and under different dc emitter voltages $V_{E,dc}$. The solid lines are theoretical curves as expected for complete dephasing in single electron-electron scattering events at random locations. The good agreement suggests that this is in fact the actual dephasing mechanism. (Taken from [Yacoby1991]).

relation

$$\ell_\phi = -\frac{L}{ln(a(L, V_{E,dc}))} \qquad (8.9)$$

This equation has its origin in the assumption that complete dephasing takes place via single electron-electron scattering events, which occur randomly.[5] This conclusion is drawn because the theoretical expression for the electron-electron scattering length in ballistic 2DEGs, derived by [Giuliani1982], shown as full lines in Fig. 8.10, agrees very well with the experimental data.

The data imply that $\ell_\phi \propto 1/V_{E,dc}^2$ in this case. If we, somewhat sloppy, identify $V_{E,dc}$ with an effective temperature via $eV_{E,dc} \approx k_B\Theta$, this result indicates that in the ballistic regime, $\ell_\phi \propto 1/\Theta^2$, which is different than in the diffusive regime. Furthermore, $\ell_\phi \approx 100 \ \mu m$ was found for $V_{E,dc} = 0$, which corresponds in the ballistic regime to a dephasing time of $\tau_\phi = \ell_\phi/v_F \approx 37$ ps, which is the same order of magnitude found in diffusive systems.

---

[5]The concept behind this relation is the Poisssson distribution of uncorrelated events, an issue discussed in further detail in exercise E 8.2.

8.5 Resonant tunnelling and S - matrices

Tunnel barriers are an essential part of many nanostructures. They represent small, often tunable resistors with little or no dissipation.[6] The transmission probability T of a rectangular barrier of height V_0 and width a as a function of the energy E of the incident particle is a standard example in elementary quantum mechanics. It is given by

$$T(E) = \begin{cases} \frac{4E(V_0-E)}{4E(V_0-E)+V_0^2 \sinh^2(\sqrt{2m(V_0-E)}a/\hbar)} & E \leq V_0 \\ \frac{4E(E-V_0)}{4E(E-V_0)+V_0^2 \sin^2(\sqrt{2m(E-V_0)}a/\hbar)} & E \geq V_0 \end{cases} \quad (8.10)$$

(see Fig. 8.11). We speak of tunnelling if $E \leq V_0$. In semiconductors, such barriers can

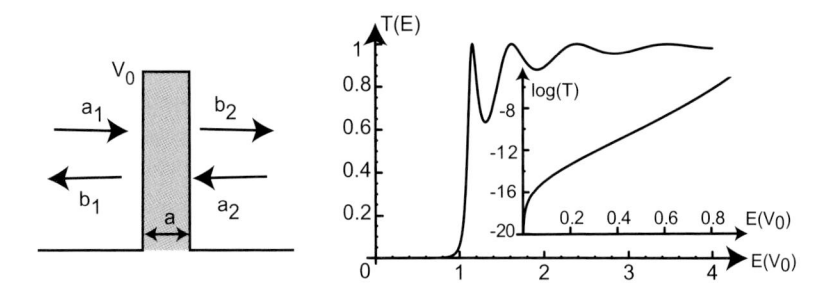

Figure 8.11: Transmission of a single rectangular barrier (sketched to the left), plotted for $a = 8\hbar/\sqrt{2m}$ (m is the mass of the particle). The inset shows a logarithmic plot of the tunnelling regime, showing that in the tunnelling regime, T increases approximately exponentially with energy.

be designed by incorporating a AlAs layer in GaAs during growth. Another widespread experimental realization is by quantum point contacts in the pinch-off regime, as discussed in chapter 7. In metallic nanostructures, a tunnel barrier can be easily formed by depositing a metal layer on top of an oxidized metal.

The S-matrix $\underline{S}$ of a tunnel barrier relates the outgoing wave functions to the incoming wave functions via

$$\vec{b} = \underline{s}\vec{a} \qquad \begin{pmatrix} b_1 \\ b_2 \end{pmatrix} = \begin{pmatrix} r_{11} & t_{12} \\ t_{21} & r_{22} \end{pmatrix} \begin{pmatrix} a_1 \\ a_2 \end{pmatrix} \qquad (8.11)$$

The coefficients of the S-matrix can be calculated from elementary quantum mechanics; how exactly they depend on the barrier parameters is of secondary interest for our purposes. Here, we simply note that the diagonal elements t_{ii} correspond to transmission amplitudes,

---

[6]The electrons that tunnel elastically through a barrier are injected into the collector at an energy eV above the Fermi level. Dissipation occurs inside the collector, within the inelastic scattering length.

8.5 Resonant tunnelling and S - matrices

while the off-diagonal elements r_{ij} represent reflection amplitudes. Due to conservation of probability current density, $\underline{S}$ has to be unitary:

$$\underline{S}^{T*}\underline{S} = \underline{1} \Rightarrow |r_{11}| = |r_{22}|; |t_{12}| = |t_{21}| \;\; r_{11}^*r_{11} + r_{22}^*r_{22} = 1 \;\; r_{11}t_{12}^* + t_{21}r_{22}^* = 0$$

Since the tunnel barrier in Fig. 8.11 is invariant under reflection, the relations

$$r_{11} = r_{22} = r; \qquad t_{12} = t_{21} = t$$

hold in this case. For tunnel barriers of a different shape, the transmission probabilities of course differ from the simple expressions in eq. (8.10). Methods for calculating such transmission coefficients are well established in elementary quantum mechanics. For our purposes, however, the relation between barrier shape and the s-matrix is of no further interest. Any barrier can be characterized by an s-matrix of the type (8.11). We will combine the s-matrices of individual barriers to calculate the transmission probability of more complicated structures, in particular systems with two tunnel barriers in series. The *double barrier* is sketched in Fig.8.12. Each barrier is characterized by its s-matrix. We are interested in the transmis-

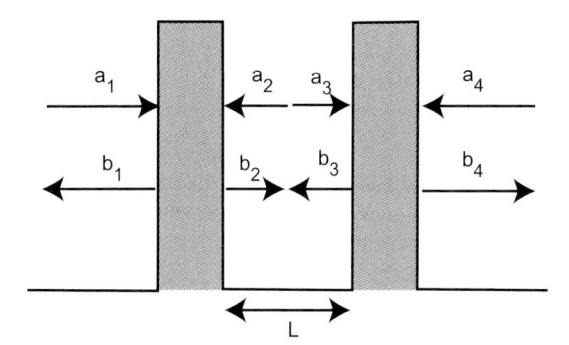

Figure 8.12: Sketch of a resonant tunnelling structure, formed by two tunnel barriers in series separated by a distance L, and the corresponding wave function amplitudes.

sion of the double barrier structure. Suppose the transport between the barriers is completely coherent, i.e., the distance L between the barriers is much smaller than the phase coherence length. As the electrons travel between the barriers, they collect a phase θ. Each time the wave hits barrier j, a fraction $(1 - t_j)$ gets reflected, which leads to interference. The total transmission amplitude from source to drain t_{sd} is obtained by summing up the partial transmission amplitudes along all trajectories the electron wave can take. There are infinitely many of such trajectories, since the electron can experience an arbitrary number of round trips between the barriers before leaving the structure. Therefore,[7]

$$t_{sd} = t_1 e^{i\theta} t_2 + t_1 e^{i\theta} r_2 e^{i\theta} r_1 e^{i\theta} t_2 + \ldots = t_1 t_2 e^{i\theta}\left[1 + \sum_{j=1}^{\infty}(r_1 r_2 e^{2i\theta})^n\right] = \frac{t_1 t_2 e^{i\theta}}{1 - r_1 r_2 e^{2i\theta}}$$

---

[7] The electron waves experience phase shifts during the reflection at the barriers, which are neglected here.

which gives the transmission probability

$$T = t_{sd}^* t_{sd} = \frac{T_1 T_2}{1 + R_1 R_2 - 2\sqrt{R_1 R_2}\cos(\theta)} \tag{8.12}$$

This transmission is plotted as a function of θ in Fig. 8.13. Eq. (8.12) is nothing but the well-known Airy formula describing the transmission of coplanar optical resonators, so-called etalons. The properties of T are therefore extensively discussed in textbooks on wave optics. In optical resonators of this type, the finesse $F^* = \frac{\pi\sqrt{1-T}}{T}$, which essentially measures the average number of partial waves interfering with each other, is an important quantity. For sufficiently small T, the full with at half maximum (FWHM) of a resonance is given by $FWHM = 4arcsin[T/2\sqrt{1-T}]$. For $T_i \ll 1$, the FWHM can be approximated by $FWHM \approx 2T/\sqrt{1-T}$. Furthermore, in this regime, the system can be approximated by a damped harmonic oscillator, where the oscillation is the wave bouncing back and forth between the two barriers, and the damping is provided by tunnelling out of the resonator. Such systems have resonances at an energy E_0 and a homogeneous line width of Lorentzian shape, which can be written as

$$T(E) = \frac{\Gamma_1 \Gamma_2}{\frac{1}{4}(\Gamma_1 + \Gamma_2)^2 + (E - E_0)^2} \tag{8.13}$$

Here, Γ_i denotes the coupling constant of barrier i. It is given by $\Gamma_i = \hbar\nu T_i$, and ν is known as the attempt frequency, i.e. the frequency at which the electron hits barrier i and tries to tunnel. the attempt frequency is given by $\nu = v/2L = \hbar k/2Lm^*$, with v being the velocity of the electron. Hence, $\Gamma_i/\hbar$ represents the tunnel rate, or in other words the number of tunnel events across barrier i per unit time. In addition, the electron phase θ has been mapped onto the electron energy via energy $E(\theta) = \frac{\hbar^2}{8m^* L^2}$ in eq. (8.13). The quality of this approximation for even not so small T_i is demonstrated in Fig. 8.13.

Thus, the double barrier can be thought of an electron interferometer. A resonance occurs when the Fermi wavelength is commensurable with L, i.e. $n \cdot \lambda_F/2 = L$.[8] Within the S-matrix formalism, this equation is easily obtained by multiplication of the s-matrices for the two barriers and the S-matrix describing the electron transfer from one barrier to the other.

With S-matrices, we can treat the double barrier transmission in a more general way. The S-matrices of the two individual barriers read

$$\begin{pmatrix} b_1 \\ b_2 \end{pmatrix} = \begin{pmatrix} r_1 & t_1 \\ t_1 & r_1 \end{pmatrix} \begin{pmatrix} a_1 \\ a_2 \end{pmatrix}; \quad \begin{pmatrix} b_3 \\ b_4 \end{pmatrix} = \begin{pmatrix} r_2 & t_2 \\ t_2 & r_2 \end{pmatrix} \begin{pmatrix} a_3 \\ a_4 \end{pmatrix}$$

while the incoming wave functions are related to the outgoing wave functions via $a_3 = b_2 e^{i\theta}$, $a_2 = b_3 e^{i\theta}$. Let us further assume that a wave is incoming only from the left with amplitude 1, $a_1 = 1$ and no left moving wave exists to the right-hand side of the double barricr, $a_4 = 0$. This results in a vector $\vec{b}$ of outgoing amplitudes as a function of incoming amplitudes $\vec{a} =$

---

[8] Note that this is not exactly true, since the wave function penetrates into the barrier. However, for small transmission amplitudes, this is an excellent approximation.

8.5 Resonant tunnelling and S - matrices

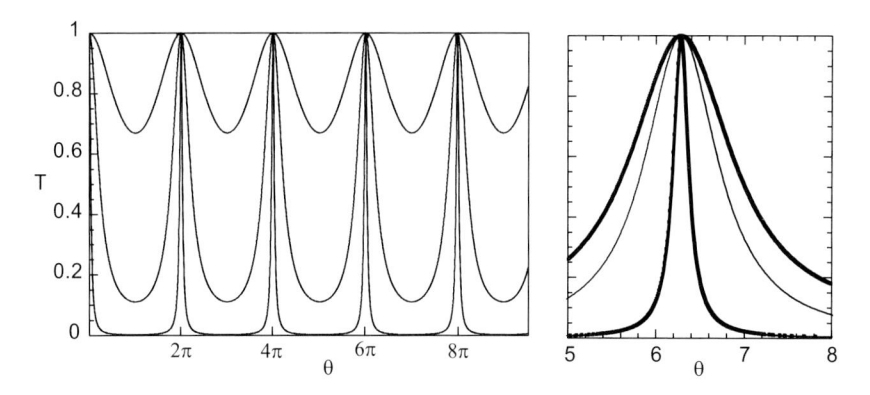

Figure 8.13: Left: coherent transmission of a double barrier as a function of the phase collected during one round trip between the barriers, shown for equal individual barrier transmissions T_b = 0.9, 0.5, and 0.1, respectively. Right: Lorentzian fit (bold lines) for T_i=0.5 and 0.1 (thin lines). For the latter barrier transmission, the fit according to eq. (8.13) is essentially indistinguishable from eq. (8.13).

$(1, b_3 e^{i\theta}, b_2 e^{i\theta}, 0)$, related by

$$\begin{pmatrix} b_1 \\ b_2 \\ b_3 \\ b_4 \end{pmatrix} = \begin{pmatrix} r_1 & t_1 & 0 & 0 \\ t_1 & r_1 & 0 & 0 \\ 0 & 0 & r_2 & t_2 \\ 0 & 0 & t_2 & r_2 \end{pmatrix} \begin{pmatrix} 1 \\ b_3 e^{i\theta} \\ b_2 e^{i\theta} \\ 0 \end{pmatrix}$$

Solving for the transmission amplitude b_4 gives

$$b_4 = \frac{t_1 t_2 e^{i\theta}}{1 - r_1 r_2 e^{2i\theta}}$$

leading to the transmission amplitude $T = b_4^* b_4$ of eq. (8.12). In this particular example, we could easily guess the result by summing up the interference paths. In more complex structures, however, it may not be so easy to do this, and the S-matrices prove to be very useful. We will see an example of this below. Note that thermal smearing has been neglected. It will be discussed in [E 8.4].

Due to inelastic scattering events, electrons may lose their phase coherence as they traverse the double barrier. In case of complete incoherence, we do not have to sum up the transmission amplitudes, but the transmission probabilities of all trajectories. In that case, the result is

$$T_{sd}^{inc} = T_1 T_2 + T_1 R_2 R_1 T_2 + \ldots = \frac{T_1 T_2}{1 - R_1 R_2} \quad (8.14)$$

It should be noted that in real samples, transport is quite often partly coherent. M. Büttiker who found an elegant model for this general situation, described in [Buttiker1986a]. The incoherent part of the transmission is modelled by a reservoir in between the barriers, which absorb and re-eject those electrons whose phase coherence gets lost.

We conclude this section by discussing the transmission of a quantum ring in terms of the S-matrix formalism. Earlier on, we have already studied the transmission of an open ring as a function of the magnetic field, which revealed the Aharonov-Bohm effect. The spectrum of an isolated ring is also well-known: in the simplest model, a one-dimensional wire (length $2\pi R$) is bent into a ring, imposing periodic boundary conditions:

$$\ell\lambda = 2\pi R, \qquad \ell = 0, \pm1, \pm2...$$

where λ is the electronic wavelength. As a consequence, the wave number is quantized in units of $1/R$. A magnetic field perpendicular to the plane of the ring induces a phase shift of $\Delta\phi = 2\pi\Phi/\Phi_0$, where $\Phi = B \cdot A$ is the magnetic flux through the ring (A denotes the ring area), and $\Phi_0 = h/e$ is the magnetic flux quantum. This corresponds to a magnetic wave vector of $k_m = \Delta\phi/2\pi R = (1/R)\Phi/\Phi_0$, and the energy spectrum is given by

$$E_\ell = \frac{\hbar^2}{2m^*R^2}(k_\ell + k_m)^2 = \frac{\hbar^2}{2m^*R^2}(\ell + \Phi/\Phi_0)^2 \tag{8.15}$$

The states are characterized by their angular momentum $\hbar\ell$. This energy spectrum is treated in [E 8.3].

Suppose we now couple the ring to two reservoirs to the left and right via tunable tunnel barriers. How will the spectrum of the isolated ring evolve into the Aharonov-Bohm effect observed in open rings? S-matrices offer a very elegant way to study this evolution. For simplicity, we assume that both tunnel barriers are equal and that the two branches of the ring have the same length (to be more accurate, it is assumed that the phases the electrons collect in each branch are identical), Fig.8.14.

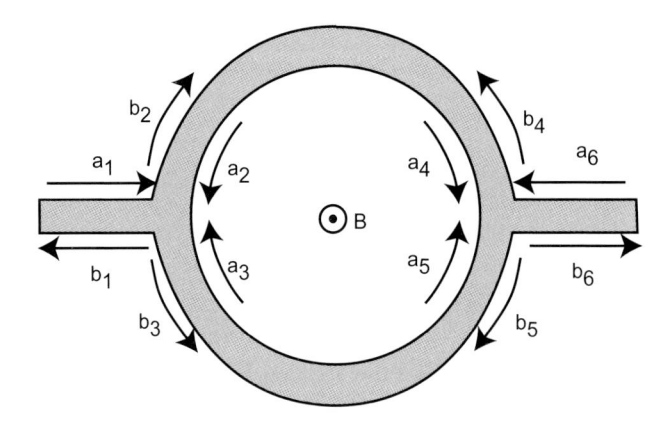

Figure 8.14: Schematic sketch of the quantum ring under study and nomenclature of the partial wave functions.

The junction can be described by the so-called Shapiro matrix

$$(s_{Sh}) = \begin{pmatrix} c & \sqrt{\epsilon} & \sqrt{\epsilon} \\ \sqrt{\epsilon} & a & b \\ \sqrt{\epsilon} & b & a \end{pmatrix}$$

8.5 Resonant tunnelling and S-matrices

c (a) represent the reflection amplitudes for electrons hitting the junction from lead 1, (2 or 3, respectively), while $\sqrt{\epsilon}$ and b are transmission amplitudes. Unitarity of the S-matrix is given for

$$\epsilon = \frac{1}{2}(1-c^2), a = -\frac{1}{2}(1+c), b = \frac{1}{2}(1-c)$$

or

$$\epsilon = \frac{1}{2}(1-c^2), a = \frac{1}{2}(1-c), b = -\frac{1}{2}(1+c)$$

The second set of relations corresponds to two ring branches which become decoupled from each other as c approaches zero. Therefore, the first solution describes the situation of interest. Since c is a measure of the coupling of the ring to the leads, we will express the transmission T_{ring} of the ring as a function of c.

The condition at the entrance of lead 1 an Which means that $\vec{b}_{l,r} = (s_{Sh})\vec{a}_{l,r}$ with $\vec{b}_l = (b_1, b_2, b_3)$, $\vec{b}_r = (b_4, b_5, b_6)$, $\vec{a}_l = (a_1, a_2, a_3)$, and $\vec{a}_r = (a_4, a_5, a_6)$. As above we assume a wave incoming from the left only, with amplitude 1. As above, we denote the phase collected from the vector potential by ϕ, such that the incoming and outgoing waves inside the ring are related via $\vec{a} = (\vec{a}_l, \vec{a}_r) = (1, b_5 e^{i\theta} e^{-i\phi}, b_4 e^{i\theta} e^{i\phi}, b_3 e^{i\theta} e^{-i\phi}, b_2 e^{i\theta} e^{i\phi}, 0)$. This leads to the linear equation

$$\begin{pmatrix} b_1 \\ b_2 \\ b_3 \\ b_4 \\ b_5 \\ b_6 \end{pmatrix} = \begin{pmatrix} c & \sqrt{\epsilon} & \sqrt{\epsilon} & 0 & 0 & 0 \\ \sqrt{\epsilon} & a & b & 0 & 0 & 0 \\ \sqrt{\epsilon} & b & a & 0 & 0 & 0 \\ 0 & 0 & 0 & a & b & \sqrt{\epsilon} \\ 0 & 0 & 0 & b & a & \sqrt{\epsilon} \\ 0 & 0 & 0 & \sqrt{\epsilon} & \sqrt{\epsilon} & c \end{pmatrix} \begin{pmatrix} 1 \\ b_5 e^{i\theta} e^{-i\phi} \\ b_4 e^{i\theta} e^{i\phi} \\ b_3 e^{i\theta} e^{-i\phi} \\ b_2 e^{i\theta} e^{i\phi} \\ 0 \end{pmatrix}$$

The transmission probability is given by $T_{ring}(c, \theta, \phi) = b_6^* b_6$. After solving eq. (8.15) for b_6 and some algebra, one finds the somewhat lengthy expression

$$T_{ring}(c, \theta, \phi) = b_6^* b_6 = \frac{16(1-c^2)^2 \cos^2(\phi) \sin^2(\theta)}{A+B+C+D+E} \quad (8.16)$$

with $A = 5 - 4c + 6c^2 - 4c^3 + 5c^4$, $B = (1+c)^4 \cos^2(2\phi)$, $C = -4(1-c)^2(1+c^2)\cos(2\theta)$, $D = -2(1+c)^2 \cos 2(\phi)[2(1+c^2)\cos(2\theta) - (1-c)^2]$, and $E = 8c^2 \cos(4\theta)$. Fig. 8.15 shows how the transmission as a function of the "electronic phase" θ and the "magnetic phase" ϕ evolves as the reflection amplitude is decreased. Fig. 8.15(a) corresponds to an open ring, showing essentially Aharonov-Bohm oscillations. Note that here, the phase coherence length is infinite. In order to recover the sinusoidal magneto-oscillations typical for the Aharonov-Bohm effect, we would have to expand eq.(8.16) in a two-dimensional Fourier series and plot the first order only. The second order gives the Altshuler-Aronov-Spivak oscillations. Fig.8.15(d) shows the transmission for a reflection amplitude close to 1 (namely $c = 0.99$). Here, the parabolas of eq.(8.15) are found (remember that $E \propto \theta^2$!). In Fig. 8.15(b) and (c), the transmission is plotted for $c = 0.2$ and 0.4, respectively. Hence, as c increases, the transmission gets more and more concentrated at the edges of the ellipsoidal regions of high transmission in Fig. 8.15(a). Simultaneously, the shape of these ellipsoid-like regions evolves into diamond-like structures.

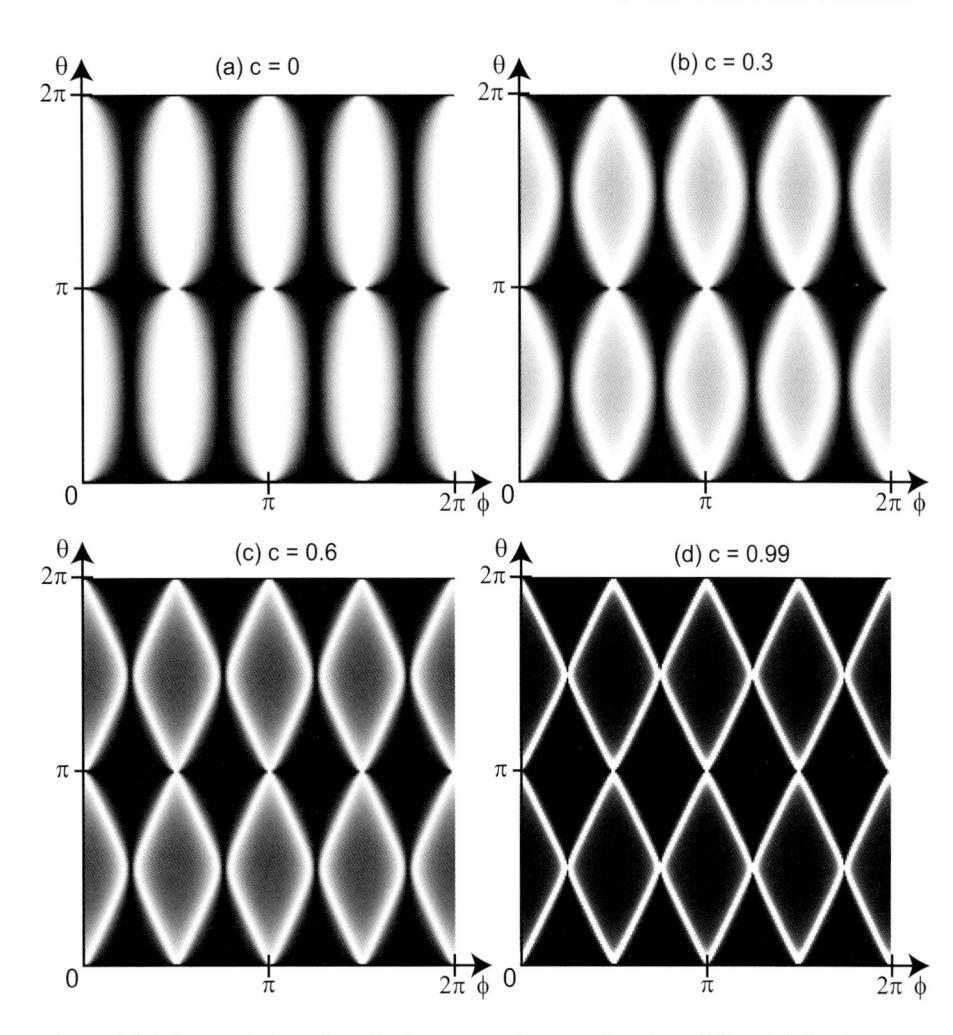

Figure 8.15: Transmission of an ideal quantum ring as a function of θ and ϕ for different reflection amplitudes at the ring entrances. Black corresponds to $T_{ring} = 0$, white to $T_{ring} = 1$

Papers and Exercises

P 8.1: Go through [Beenakker1988]; explain the *flux-cancellation effect* and the theoretical expression for the autocorrelation function of UCF in magnetic fields!

P 8.2: The suggestions made in [Mohanty1997] have been heavily debated since they popped up. What exactly do the authors propose?

P 8.3: What is *weak antilocalization*? To answer this question, consult [Dresselhaus1992]!

E 8.1: Consider the geometry depicted to the left in Fig.8.1. Assume a constant magnetic field is present in z - direction, and calculate the phase difference between partial waves that

8.5 Resonant tunnelling and S - matrices

traverse the upper and the lower branch. Show that $\Phi_{upper} - \Phi_{lower} = 2\pi\Phi/\Phi_0$, where $\Phi_0 = h/e$ denotes the magnetic flux quantum.

E 8.2: The relation between random events and the Poisson distribution is applicable to many situations. Here, it will be discussed using eq. (8.9) as an example: within the assumptions described in the text, random electron-electron (e-e) scattering events determine the amplitude of the Aharonov-Bohm type oscillations of Fig. 8.9.

(i) We denote the average number of e-e scattering events per unit time, i.e., the e-e scattering rate, by γ. Clearly, the exact number of scattering events j within a time t will fluctuate around its average, which equals simply $t \times \gamma$. What is the probability $P(j)$ that exactly j events take place within t ($P(j)$ is the *Poisson distribution*)? Hint: divide the time interval in a large number of sections of equal size, such that more than one event per interval does occur. Count all the possible arrangements of the sections under the constraint that j of them are occupied.

(ii) Use $P(j)$ to define a meaningful e-e scattering length. How does eq. (8.9) emerge from this?

E 8.3: Consider a ring (radius r) with only one radial mode occupied (i.e., the ring has been formed out of a strictly one-dimensional wire).

(i) Calculate the energy spectrum of the ring as a function of a homogeneous magnetic field perpendicular to the ring area. It makes life easier to use cylindrical coordinates and gauge the vector potential as $\vec{A} = (0, r \cdot B/2, 0)$. Use the wave function ansatz $\Psi(\phi) = (2\pi r)^{-1/2} e^{i\ell\phi}$ with ℓ being an integer (what is the physical meaning of ℓ?).

(ii) Calculate the current flowing in the ring as a function of ℓ for zero temperature. How do you interpret this result? Compare the result to a current generated by a single electron circulating in the loop!

(iii) Estimate the current for an odd number of electrons in the ring at B=0. Assume realistic ring diameters and electron densities. How could one measure this current?

E 8.4: Thermal smearing of resonant tunnelling peaks
Calculate, in analogy to our treatment of the thermal smearing of quantized conductance steps in chapter 8, how a transmission resonance of a double barrier is modified by a non-zero temperature.

(i) Consider the limiting case of a purely thermally broadened resonance, i.e. $T(\Theta = 0, E) = \delta(E - E_r)$. Show that the line shape is the derivative of the Fermi function. Calculate its line width (FWHM)!

(ii) How does the line shape look like for a Lorentzian-shaped resonance $T(\Theta = 0, E)$? How would you determine experimentally the thermal and the Lorentzian contribution to the line width?

Further Reading

The reader is encouraged to study the excellent treatment of phase coherent electrons in mesoscopic samples by [Beenakker1991], section 6.

9 Singe Electron Tunnelling

The charge stored on a capacitor is not quantized: it consists of polarization charges generated by displacing the electron gas with respect to the positive lattice ions and can take arbitrary magnitudes. The charge transfer across a tunnel junction, however, is quantized in units of the electron charge (*single electron tunnelling*), and may be suppressed due to the Coulomb interaction. These simple facts lay the foundation of a new type of electronic devices called single electron tunnelling (SET) devices. This effect, also known as Coulomb blockade, was first suggested back to 1951 by Gorter [Gorter1951], who explained earlier experiments [Itterbeek1947] but remained largely unnoticed until almost 40 years later, Fulton and Dolan built a transistor based in single electron tunnelling [Fulton1987]. After introducing the concept of Coulomb blockade in section 1, we will discuss basic single electron circuits, in particular double barrier and the single electron transistor, in section 2. Some examples of applications are given in section 3.

9.1 The principle of Coulomb blockade

Consider a tunnel junction biased by a voltage V. The equivalent circuit of a tunnel junction consists of a "leaky" capacitor, i.e., a resistor R in parallel to a capacitor C (Fig.9.1). For

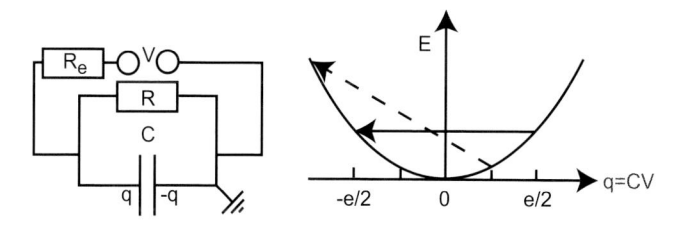

Figure 9.1: Equivalent circuit and current-voltage characteristics of a single tunnel junction. The resistor R_e represents the low-frequency impedance of the environment.

charges $|q| < e/2$, an electron tunnelling across the barrier would increase the energy stored at the capacitor. This effect is known as Coulomb blockade [Likharev1988]. For $|q| > e/2$, the tunnelling event reduces the electrostatic energy, and the differential conductance is given by $dI/dV = 1/R$. Experimentally, it is far from easy to observe Coulomb blockade at a single tunnel barrier, because of two reasons. First of all, in order to avoid thermally activated electron transfers, $\frac{e^2}{8C} \gtrsim k_B \Theta$ is required.

Question 9.1: A typical tunnel junction patterned by angle evaporation is formed by a thin oxide layer (thickness 5 nm, dielectric constant $\epsilon \approx 5$). Estimate the maximum area of the capacitor plates for Coulomb blockade to be observed at (a) 4.2 K, and (b) 300 K!

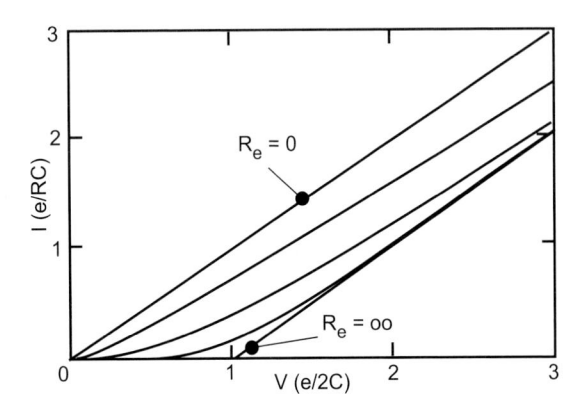

Figure 9.2: Evolution of the I-V characteristics of a single tunnel junction as the resistance of the environment R_e is increased. For $R_e > h/2e^2$, the Coulomb gap becomes clearly visible. The traces are shown for $R_e/R = 0, 0.1, 1, 10$, and ∞ (after [Devoret1990]).

Second, the resistance of the tunnel junction has to be "sufficiently large". We can speak of individual electrons tunnelling through the barrier only if the tunnel events do not overlap, which means that the time between two successive events $\delta t \approx eR/V$ must be large compared to the duration τ of a tunnel event, which can be estimated as $\tau \approx \hbar/eV$ [Korotkov1996]. This leads to the condition $R \gg \hbar/e^2$. Furthermore, quantum fluctuations can destroy the Coulomb blockade as well. So far, we have neglected that the tunnel junction is coupled to its environment, which is modelled by the resistance R_e in Fig. 9.1. More generally, the environment represents a frequency-dependent impedance, although here, we restrict ourselves to very small frequencies, such that the impedance can be replaced by R_e. In fact, our above line of arguing implicitly assumes the so-called *local rule*, which states that the tunnelling rate across the junction is governed by the difference in electrostatic energy right before and right after the tunnel event. According to the *global rule*, on the other hand, the tunnel rate is determined by the electrostatic energy difference of the whole circuit. Since the environment inevitable includes some capacitances much larger than the capacitance of the tunnel junction, we may expect that in this case, the Coulomb blockade vanishes. The influence of the electromagnetic environment on the performance of tunnel junctions is discussed in detail in [Grabert1991]. Here, we just give a simple argument. The local rule holds provided the tunnel junction is sufficiently decoupled from the environment. In the leads, quantum fluctuations of the charge take place. An estimate based on the Heisenberg uncertainty relation tells us what "sufficiently decoupled" actually means: for quantum fluctuations with a characteristic energy amplitude δE, the uncertainty relation $\delta E \cdot \delta t \geq \hbar/2$ holds. Coulomb blockade is only visible for energy fluctuations at the junction much smaller than $e^2/8C$, while the time scale is given by the time constant of the circuit: $\delta t \approx \tau = R_e C$.

Hence, Coulomb blockade can be observed on a single tunnel junction only if the resistance of the environment is of the order of the resistance quantum h/e^2 or higher. The influence of the environmental resistance on the Coulomb blockade has been calculated by [Devoret1990] and is shown in Fig. 9.2. These considerations imply that it is not so easy to observe Coulomb blockade at a single tunnel junction. Since the environment has to be sufficiently decoupled, the resistance of the leads has to be larger than h/e^2. This generates Joule heating, which in turn makes it difficult to keep the electron temperature below $e^2/2Ck_B$. Nevertheless, Coulomb blockade has been observed in single tunnel junctions biased via wires of sufficiently high resistance, Fig. 9.3.

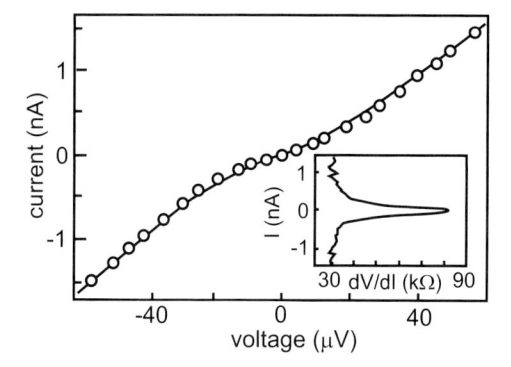

Figure 9.3: I-V characteristics of Al – Al$_2$O$_3$ – Al tunnel barriers, fabricated by angle evaporation. In order to suppress quantum fluctuations, the cross section of the Al wires is only 10 nm × 10 nm. The superconducting state has been destroyed by applying a magnetic field (after [Cleland1990]).

The limitations imposed by the need to decouple the environment and the tunnel junction can be relaxed by using two tunnel junctions in series (Fig. 9.4), since here, quantum fluctuations at the island in between the junctions are strongly suppressed [Grabert1991]. The number of electrons at the enclosed island can change only by tunnelling across one of the barriers, an event essentially free of dissipation. The energy relaxation will take place somewhere in the leads, far away from the island. The resistance of relevance for the suppression of the quantum fluctuations is now that of a tunnel barrier, while the capacitance corresponds to the total capacitance of the island to its environment. Therefore, quantum fluctuations at the island can be suppressed easily without running into heating problems.

9.2 Basic single electron tunnelling circuits

Before we discuss single electron tunnelling in the double barrier system, it is useful to have a look at the problem form a more general point of view, which is then used to analyze specific

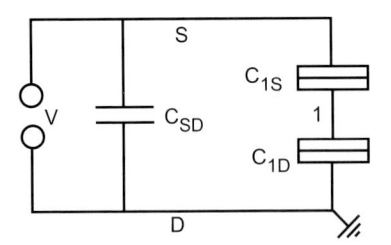

Figure 9.4: A double barrier structure attached to source (S) and drain (D). C_{SD} denotes a residual capacitance between the two leads.

examples including the circuit of Fig. 9.4.

Consider an arrangement of (n+m) conductors embedded in some insulating environment. Each conductor i is at an electrostatic potential V_i, has a charge q_i stored on it, and has a capacitance C_{iD} to drain (ground).[1] Between each pair of conductors i and j, there is a mutual capacitance C_{ij}. Some of these capacitances may belong to tunnel junctions, which allow electron transfers between the corresponding conductors. Furthermore, we assume that m conductors are connected to voltage sources which we call *electrodes*, while the n remaining ones are *islands*.[2] For convenience, we enumerate the n islands form 1 to n, and the m electrodes from n+1 to n+m. The charges and potentials of the islands can be written in terms of an island charge vector $\vec{q}_I$ and potential vector $\vec{V}_I$, respectively. Similarly, charge and potential vectors can be written down for the electrodes, $\vec{q}_E$ and $\vec{V}_E$. The state of the system can be specified by the total charge vector $\vec{q} = (\vec{q}_I, \vec{q}_E)$. Equivalently, this state can also be characterized by the total potential vector defined as $\vec{V} = (\vec{V}_I, \vec{V}_E)$. Charge and potential vectors are related via the capacitance matrix $\underline{C}$

$$\vec{q} = \underline{C}\vec{V} \tag{9.1}$$

We write $\underline{C}$ as

$$\underline{C} = \begin{pmatrix} \underline{C}_{II} & \underline{C}_{IE} \\ \underline{C}_{EI} & \underline{C}_{EE} \end{pmatrix} \tag{9.2}$$

The capacitance sub-matrices between type A and type B conductors (A and B can be electrodes or islands) are denoted by $\underline{C}_{AB}$. Note that the ground is *not* a conductor in terms of our definition, and that $\underline{C}$ is symmetric. The matrix elements of $\underline{C}$ are given by (see Appendix B)

$$(\underline{C})_{ij} = \begin{cases} -C_{ij} & j = 1...n+m; j \neq i \\ C_{iD} + \sum_{k=1; k \neq i}^{n+m} C_{ik} & j = i \end{cases}$$

---

[1] In publications, one frequently encounters an "antisymmetric bias condition", where a voltage of $V_S = +V/2$ is applied to source, and a drain voltage of $V_D = -V/2$. The electrostatics is different in that case.
[2] The electrostatics of such systems in terms of the capacitance matrix is discussed in Appendix B.

9.2 Basic single electron tunnelling circuits

The electrostatic energy E^3 is given by the energy stored at the islands, minus the work done by the voltage sources. Minimizing this energy gives us the ground state.

As we shall see, in single electron circuits, usually the voltages applied to the electrodes are parametrically changed, and the initial island charge vector $\vec{q_I}$ given. As $\vec{V_E}$ is changed, the potential difference between two conductors connected by a tunnel junction may become sufficiently large for electrons to tunnel, resulting in a new charge configuration. Such charge rearrangements will take place as soon as the electrostatic energy of the new configuration is equal to, or smaller than, the energy of the original configuration. The charge transfer can be specified by the change of the charge vector $\Delta\vec{q} = \vec{q}_{new} - \vec{q}$. For a system initially in its ground state, we can find the parametric transition to a new ground state by the condition

$$\Delta E = E_{new} - E \leq 0 \tag{9.3}$$

It may look very cumbersome to calculate the energy differences of all the possible charge transfers and find its minimum. However, as we shall see, their number is usually quite small. ΔE is given by [4]

$$\Delta E[\vec{V_E}, \vec{q_I}, \Delta\vec{q}] = \Delta\vec{q_I}\underline{C}_{II}^{-1}\left[\vec{q_I} + \frac{1}{2}\Delta\vec{q_I} - \underline{C}_{IE}\vec{V_E}\right] + \Delta\vec{q_E}\vec{V_E} \tag{9.4}$$

Eq. (9.4) is an important relation which can be used to analyze Coulomb blockade in all systems that can be characterized by a capacitance matrix. Note that it cannot be used to study Coulomb blockade at the single junction, since the crucial time scale involved there does not enter the formalism leading to eq. (9.4). We are now ready to study the double barrier shown in Fig. 9.4.

9.2.1 Coulomb blockade at the double barrier

The system consists of one electrode (source S) and one island (1). In the following, islands will be labelled by arabic numbers and electrodes by capital letters. The capacitance matrix reads

$$\underline{C} = \begin{pmatrix} C_{11} & -C_{1S} \\ -C_{1S} & C_{SS} \end{pmatrix}$$

with $C_{11} = C_{1S} + C_{1D}$ and $C_{SS} = C_{1S} + C_{SD}$. The charge on the island is given by the number n of electrons tunnelled onto it, plus an arbitrary background charge q_0, induced by the environment: $q = q_0 - ne$. Four different charge transfers are relevant. An electron can hop in both directions across C_{1S} or C_{1D}. For electron transfers across C_{1S}, we have $\vec{V} = (V_1, V_S)$, $\vec{q} = (q_0 - ne, q_S)$, and $\Delta\vec{q} = \pm e(-1, 1)$. Here "+ (-)" corresponds to a transfer of one electron from S (1) to 1 (S). Consequently, the energy difference reads according to eq. (9.4)

$$\Delta E[V, q_0 - ne, \pm e(-1,1)] = \frac{e}{C_{11}}[\frac{e}{2} \pm (ne - q_0 + C_{1D}V)] \tag{9.5}$$

---

[3] The electrostatic energy is the free energy $E = U - \mu N$, where U is the total energy, μ the electrochemical potential, and N the number of electrons. This free energy does not have a particular name.

[4] For a derivation of eq. (9.4), see Appendix B.

For tunnel events across C_{1D}, $\Delta \vec{q} = \pm e(-1,0)$. Again "+(-)" corresponds to a transfer of one electron from D (1) onto 1 (D). This gives

$$\Delta E[V, q_0 - ne, \pm e(-1,0)] = \frac{e}{C_{11}}[\frac{e}{2} \pm (ne - q_0 - C_S V)] \quad (9.6)$$

Question 9.2: C_{ii} of such a metallic grain is sometimes estimated by its self-capacitance $C_{self} = V/q$, where V denotes the potential of the grain and q the charge transferred onto it from infinity (at zero potential). For a sphere, C_{self} equals $4\pi\epsilon\epsilon_0 r$, while for a circular disk, $C_{self} = 8\epsilon\epsilon_0 r$ (r denotes the radius of the island). Estimate C_{self} and the charging energy for some reasonable grain radii!

Coulomb blockade is established only if all four energy differences are positive. This defines a voltage interval of vanishing current

$$Max\left\{\frac{1}{C_{1S}}[-q_0 + e(n - \frac{1}{2})], \frac{1}{C_{1D}}[q_0 - e(n + \frac{1}{2})]\right\} < V <$$

$$Min\left\{\frac{1}{C_{1S}}[-q_0 + e(n + \frac{1}{2})], \frac{1}{C_{1D}}[q_0 - e(n - \frac{1}{2})]\right\} \quad (9.7)$$

Let us study some special scenarios:

1. The simplest situation is $n = 0$, no background charges ($q_0 = 0$), and identical junction capacitances $C_{1S} = C_{1D} = C_{11}/2$:
   Eq. (9.7) now reads $-\frac{e}{C_{11}} \leq V \leq \frac{e}{C_{11}}$. For $V = 0$, we get

$$\Delta E[0, 0, e(\mp 1, \pm 1)] = \Delta E[0, 0, e(\pm 1, 0)] = \frac{e^2}{2C_{11}}$$

All four charge transfer processes are suppressed (Fig. 9.5a).
Applying a positive voltage $V = \frac{e}{C_{11}}$ to source means that

$$\Delta E[V, 0, e(-1, 1)] = \frac{e^2}{C_{11}} > 0,$$

$$\Delta E[V, 0, e(1, -1)] = 0 = \Delta E[V, 0, e(-1, 0)],$$

and

$$\Delta E[V, 0, e(1, 0)] = \frac{e^2}{C_{11}} > 0$$

At this voltage, an electron can either tunnel from drain to the island or from the island to source (Fig.9.5b). Both processes have the same probability.

Question 9.3: Suppose an electron has just tunnelled from drain onto the island under these conditions. The system is in the state depicted in Fig. 9.5b. Show that now, an electron will tunnel from the island to source, and a current actually flows! Calculate the energy differences indicated in Fig. 9.5c.

9.2 Basic single electron tunnelling circuits

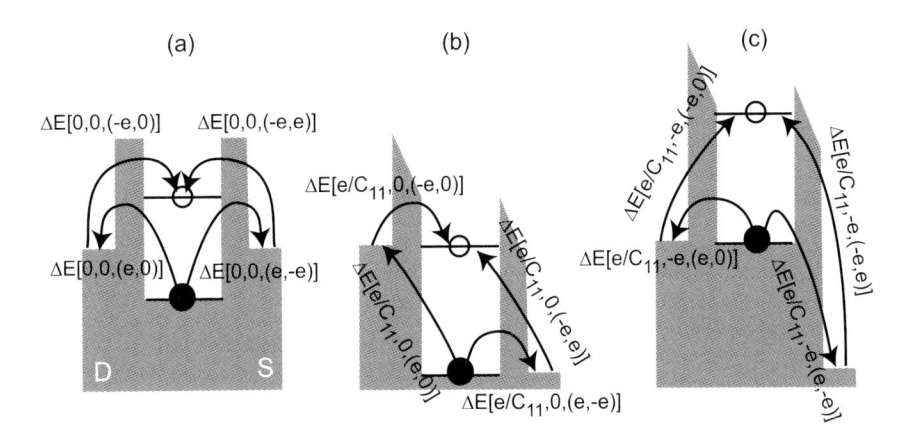

Figure 9.5: Energy differences of the four electron transfers at the double barrier. Empty circles denote empty states, while full circles correspond to occupied states. (a): no voltage is applied ($V = 0$), and Coulomb blockade is established. (b): $V = \frac{e}{C_{11}}$. Electrons can hop from drain onto the island, as well as from the island to source. (c): differences in the electrostatic energy after an electron has, starting from the situation in (b), tunnelled from drain onto the island.

The system thus oscillates between the situations depicted in Fig. 9.5b and c. In each oscillation cycle, a single electron is transferred from drain to source. In addition, the tunnel events show a pair correlation. Shortly after an electron has tunnelled from drain to the island, a tunnelling process from the island to drain will take place, and vice versa.

2. Effect of a background charge q_0:
Let us assume $n = 0$, and $C_{1S} = C_{1D}$, which leads to the condition for Coulomb blockade

$$Max\{\frac{2}{C_{11}}(-q_0 - \frac{e}{2}), \frac{2}{C_{11}}(q_0 - \frac{e}{2})\} < V < Min\{\frac{2}{C_{11}}(-q_0 + \frac{e}{2}), \frac{2}{C_{11}}(q_0 + \frac{e}{2})\}$$

This means that by a non-zero q_0, the Coulomb gap can be reduced, but never be increased. In fact, for $q_0 = (j + 1/2)e$ with j being an integer, the Coulomb gap vanishes completely. Background charges can seriously hamper the observation of the Coulomb blockade, especially when they are time-dependent.

Question 9.4: Draw the energy diagram corresponding to Figs. 9.5a-c for $q_0 = e/4$! Assume equal capacitances.

Question 9.5: Show that for $C_{1S} \neq C_{1D}$, the larger capacitance determines the Coulomb gap, which gets reduced as compared to the Coulomb gap for identical junctions.

Coulomb blockade in metallic islands has been known for a long time. As an example of the early indications, we take a look at an experiment of Giaever und Zeller [Giaever1968]. The authors measured the current-voltage characteristics of a granular Sn film sandwiched between an oxide layer and metallic electrodes (Fig. 9.6). The average diameter of the Sn granules was 11 nm, such that single electron tunnelling is expected to play a role at low temperatures. The system contains an ensemble of double barriers in parallel. Therefore, we expect to observe a gap in the I-V characteristics around $V = 0$ that corresponds to the average single electron charging energy. Leakage currents through the oxide in between the islands are quite small, since the conductance of tunnel barriers decreases exponentially with increasing barrier thickness. At zero magnetic field, both the Al electrodes as well as the Sn granules are in the superconducting state, and the superconducting energy gap strongly influences the transport measurements.[5] However, by applying a magnetic field, the superconducting state is destroyed and our previous model becomes applicable. The Coulomb gap manifests itself in an increased differential resistance around $V = 0$.

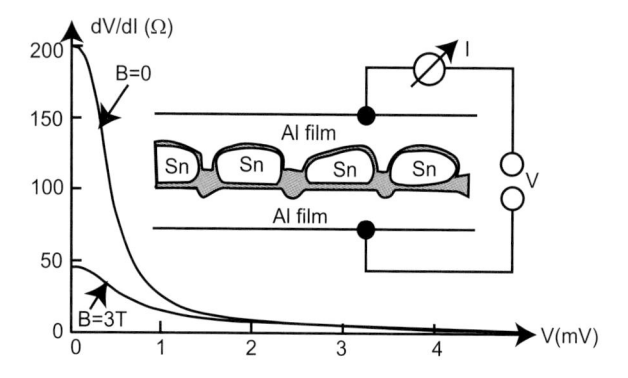

Figure 9.6: The experiment of Giaever and Zeller, after [Giaever1968]. A granular Sn film was embedded in an oxide layer and covered on both sides by Al, which acted as source and drain.

9.2.2 Current-voltage characteristics: the Coulomb staircase

Besides the Coulomb gap around $V = 0$, the Coulomb blockade generates under certain conditions a staircase-like structure in the current-voltage characteristics, known as Coulomb staircase. In contrast to our earlier considerations concerning transport through mesoscopic structures, we study here a system of interacting electrons, and a charge transfer changes the electrostatic energy as well. To include the interaction, we use the so-called transfer Hamiltonian model, which allows us to relate the change in energy ΔE due to a tunnel event with a tunnel rate $\Gamma(\Delta E)$. For the transmission coefficients calculated in earlier chapters, we always assumed that the energy is conserved. Here, however, the electrostatic energy changes as an electron tunnels, and the voltage sources do some work on the system. Such situations can

---
[5]The interplay of superconductivity and single electron tunnelling is a fascinating aspect of single electron tunnelling, which is beyond our scope. See, however, P4.

9.2 Basic single electron tunnelling circuits

be conveniently dealt with by using Fermi's golden rule, which originates in time-dependent perturbation theory. The transfer Hamiltonian model starts from an impenetrable barrier, separating two electron gases. tunnelling is treated as a perturbation and is described by a perturbation Hamiltonian H_t, which is of no further interest to us here. The interested reader is referred to, e.g., [Ferry1997] for details. Applied to a tunnel barrier, Fermi's golden rule states that the transition rate for an an electron in the initial state $|i\rangle$ to a final state $|f\rangle$ on the other side of the tunnel barrier is given by

$$\Gamma_{i\to f} = \frac{2\pi}{\hbar}|\langle i|H_t|f\rangle|^2 \delta(E_f - E_i - \Delta E) \tag{9.8}$$

Here, E_i and E_f denote the energies of the initial and final states with respect to the bottom of the conduction band, and the matrix element $\langle i|H_t|f\rangle$ describes the coupling of the left hand side to the right hand side of the tunnel barrier. This transition rate is just the transmission probability per unit time. In order to determine the total transition rate $\Gamma(\Delta E)$, we have to consider:

1. The tunnelling rate at energy E will be proportional the the spectral electron density $n(e) = D_i(E) \cdot f(E)$. Here the index i denotes the side of the barrier that hosts state i, D_i the relevant density of states, and $f(E)$ denotes the Fermi Dirac distribution function.

2. Since we are dealing with fermions, the electrons can tunnel only into an empty state $|f\rangle$. The transfer rate for an electron in $|i\rangle$ will thus be proportional to $D_f(E + \Delta E) \cdot [1 - f(E + \Delta E)]$.

3. We have to integrate over all energies at which states with non-zero tunnelling probability exist. These are all states above the maximum of the conduction band bottoms at both sides $E_{cb,max}$.

Therefore, the total transition rate is given by

$$\Gamma_{1\to 2}(\Delta E) = \frac{2\pi}{\hbar} \int_{E_{cb,max}}^{\infty} |\langle i|H_t|f\rangle|^2 D_i(E) D_f(E-\Delta E) f(E)[1-f(E-\Delta E)] dE \tag{9.9}$$

Now, 1 and 2 denote the conductors that contain the initial and final states, respectively. For large energy barriers, we can safely assume that the matrix elements of H_t will be approximately independent of energy. Second, we assume that the density of states does not depend on energy, either since the electron gas is two-dimensional, or since the voltage drop is sufficiently small. Furthermore,

$$f(E)[1 - f(E - \Delta E)] = \frac{f(E) - f(E - \Delta E)}{1 - \exp[\Delta E/k_B\Theta]}$$

If we further consider only cases where the temperature is sufficiently low, we can approximate the Fermi functions by step functions, and obtain

$$\Gamma_{1\to 2}(\Delta E) = \frac{1}{Re^2} \frac{\Delta E}{1 - \exp(\Delta E/k_B\Theta)} \tag{9.10}$$

Here, the resistance R of the tunnel barrier has been defined as $R = \hbar/(2\pi e^2 |\langle i|H_t|f\rangle|^2 D^2)$ (see exercise [E 9.2] for a motivation). The current is then obtained from the difference of tunnel rates in both directions, $I = e\left[\Gamma_{1\to 2}(\Delta E_{1\to 2}) - \Gamma_{2\to 1}(\Delta E_{2\to 1})\right]$.

Let us apply this result to the island of Fig. 9.4. For a steady state, the average charge at the island is constant, and the current from source to the island is given by

$$I(V) = e \sum_{n=-\infty}^{\infty} p(n)(\Gamma_{1\to S}(\Delta E_{1\to S}(n)) - \Gamma_{S\to 1}(\Delta E_{S\to 1}(n))) \qquad (9.11)$$

Equivalently, $I(V)$ can be expressed in terms of the drain tunnelling rates. Here, we denote the tunnelling rate from 1 to source by $\Gamma_{1\to S}(\Delta E_{1\to S}(n))$, while the reverse process is denoted accordingly. Of course, the energy differences now depend on the number of excess electrons n stored on the island. The probability of finding n electrons on the island is denoted by $p(n)$. We expect this function to be peaked around one number, which is given by the sample parameters and by V. The steady state condition furthermore requires that the probability for making a transition between two charge states (characterized by n) is zero. This means that the rate of electrons entering the island occupied by n electrons, equals the rate of electrons leaving the island when occupied by $(n+1)$ electrons:

$$p(n) \cdot (\Gamma_{1\to S}(\Delta E_{1\to S}(n)) + \Gamma_{1\to D}(\Delta E_{1\to D}(n))) =$$
$$p(n+1)(\Gamma_{S\to 1}(\Delta E_{S\to 1}(n+1)) + \Gamma_{D\to 1}(\Delta E_{D\to 1}(n+1))) \qquad (9.12)$$

We are now ready to calculate the I(V) - characteristics. Eq. (9.12), together with the normalization condition

$$\sum_{n=-\infty}^{\infty} p(n) = 1$$

allows us to obtain $p(n)$, which we insert in eq. (9.11). This requires some numerics, which is considerably simplified by the fact that only a few occupation numbers have non-vanishing probabilities.

Fig.9.7 shows staircases calculated from (9.11) for different background charges. The staircases are periodic in q_0 with a period of one elementary charge. Qualitatively, the staircase can be understood as follows: suppose the tunnel rate across junction S is much larger than that one across junction D, and the voltage applied is positive. The voltage now drops completely across junction D, i.e. $V_{1D} \approx V$. From eq. (9.4), we calculate $\Delta E[V, -ne, e(-1, 0)] = 0$ the threshold voltages $V(n_0)$ and $V(n_0 + 1)$, which differ by $\Delta V = e/C_{1S} \approx e/C_{11}$. If the voltage is increased by this amount, an additional electron can jump on the island via the drain junction. This increases the current (which is governed by $\Gamma_{1\to D}$ and by $\Gamma_{D\to 1}$) by $\Delta I = e/R_{1D}C_{11}$ for sufficiently low temperatures, as can be seen by inserting $e\Delta V = \Delta E[V, -(n+1)e, e(-1, 0)] - \Delta E[V, -ne, e(-1, 0)]$ in eq.(9.11). The markedness of the staircase steps strongly depends on the sample parameters (Fig. 9.8). The steps become most pronounced if both the resistance and the capacitance of one junction are large compared to those of the second junction. Experimentally, however, this is hard to achieve, since small tunnel resistances tend to correspond to small capacitances as well. An analytical model for the Coulomb staircase in this limit is discussed in [P 9.2].

9.2 Basic single electron tunnelling circuits

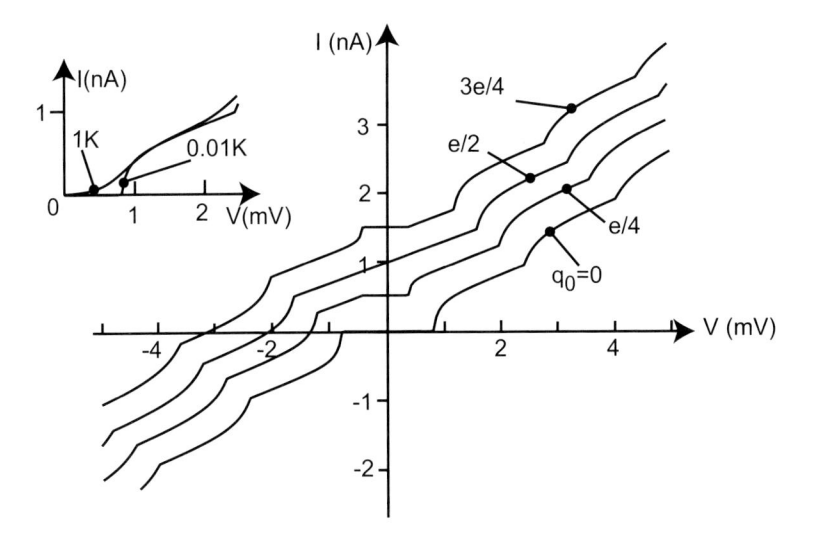

Figure 9.7: Coulomb staircase as calculated from eq.(9.11), for different background charges q_0. The structure is periodic in q_0, with a period of one elementary charge. Typical sample parameters have been assumed, namely $C_{1S} = C_{1D} = 0.1$ fF, $R_{1S} = 20$ MΩ, $R_{1D} = 1$ MΩ, at a temperature of T= 10 mK. The inset shows the thermal smearing of the Coulomb gap (for $q_0 = 0$) as the temperature is increased to 1 K.

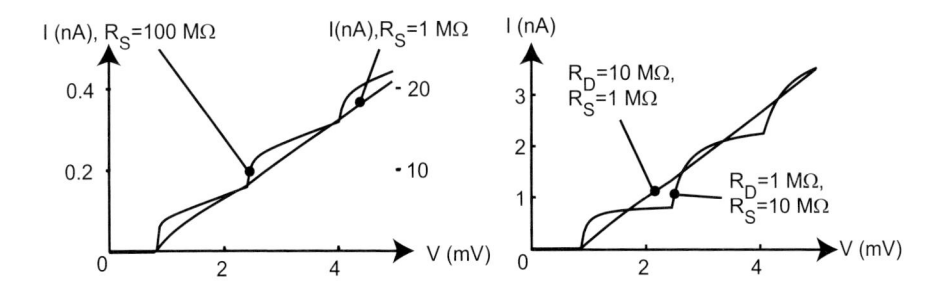

Figure 9.8: Steps of the Coulomb staircase for various sample parameters, as calculated for T=10 mK. Left: $C_{1S} = C_{1D} = 0.1$ fF, $R_{1D} = 1$ MΩ. For $R_{1S} = R_{1D}$, the steps are absent, while for $R_{1S} = 100 R_{1D}$, they are well pronounced. Right: Coulomb staircase of an island with two junctions of both different capacitances and different resistances, i.e., $C_{1S} = 0.1$ fF, $C_{1D} = 1$ aF.

Particularly beautiful Coulomb staircases have been observed in scanning tunnelling experiments on clusters, where the experimental setup consists of a conducting granule or cluster, deposited on an insulating layer on top of a conducting substrate. The tip of a scanning tunnelling microscope (STM) is positioned on top of the cluster (Fig. 9.9(a)), and the current

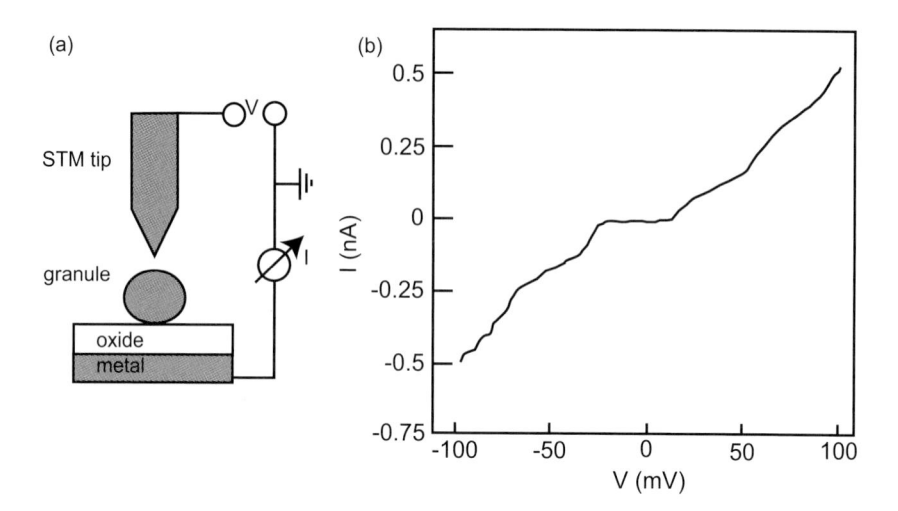

Figure 9.9: (a): one experimental setup for measuring the Coulomb staircase. (b): experimental data (trace A) in comparison to a fit (trace B, shifted by 0.2 nA for clarity), which gives the fit parameters $C_S = 2$ aF, $C_D = 4.14$ aF, $R_S = 34.9$ MΩ, $R_D = 132$ MΩ, and an offset charge of about 1/8 e. Here, the granule was a small indium droplet on top of an oxidized conducting substrate. The temperature was 4.2 K. The measurement is adapted from [Amman1991].

is measured as a function of the voltage applied to the STM tip with respect to the substrate [Amman1991]. In such experiments, the resistance of one barrier is given by the distance between tip and granule, which can be changed in a wide range. Fig. 9.9(b) shows typical experimental data.

9.2.3 The SET transistor

In 1987, Fulton und Dolan [Fulton1987] published a seminal experiment: by angle evaporation, a small metallic island was patterned, coupled to source and drain via tunnel barriers with cross sections in the range of 50 nm by 50 nm. In addition, a third electrode (the *gate*

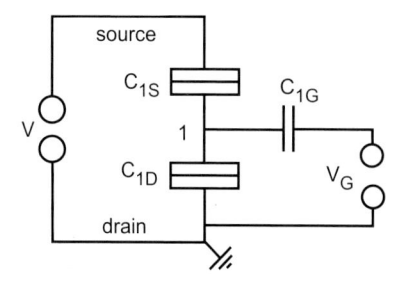

Figure 9.10: Schematic diagram of a SET transistor.

9.2 Basic single electron tunnelling circuits

electrode) was defined such that the gate-island resistance approaches infinity, and thus couples to the island only capacitively. This way, the effective background charge and thus the width of the Coulomb gap can be tuned continuously with the gate voltage, and for sufficiently small source-drain voltages, the current flowing from source to drain can be controlled. The system constitutes a transistor based on the Coulomb blockade and is known as single electron tunnelling (SET) transistor. Its equivalent circuit and an experimental realization are shown in Fig. 9.10. For simplicity, let us assume that the background charge for gate voltage zero vanishes. This is no restriction of generality, since additional background charges can always be compensated for by a gate voltage offset. The inverse capacitance matrix now reads $(\underline{C}^{-1})_{11} = 1/C_{11}$, $(\underline{C}^{-1})_{ij} = 0$ else. Furthermore, $C_{IE} = (-C_{1S}, -C_{1G})$. The electrode voltage vector is given by $\vec{V}_E = (V, V_G)$, while the island charge vector reads $\vec{q}_I = -ne$. The Coulomb gap is given by the onset of the same tunnelling events as for the single is-

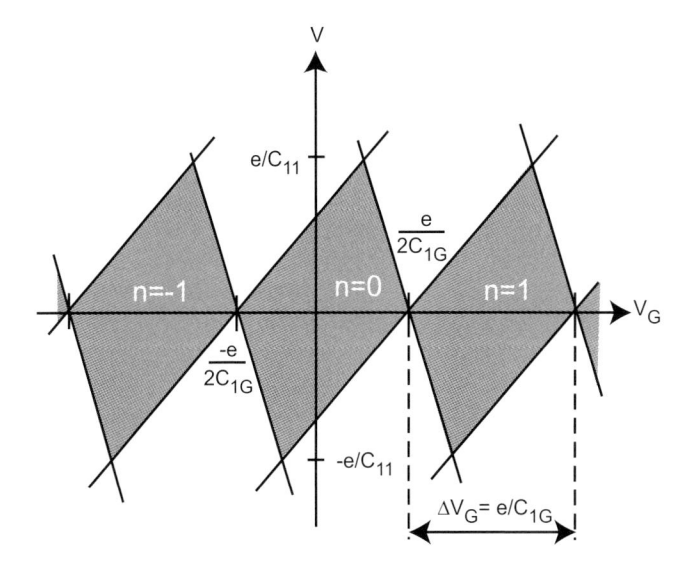

Figure 9.11: Stability diagram of a single electron transistor. Within the diamonds, Coulomb blockade is established, while outside, a current flows between source and drain. The slopes of the boundaries are given by $C_{1G}/(C_{11} - C_{1S})$, and by $-C_{1G}/C_{1S}$, respectively.

land studied above. Now, however, the Coulomb gap depends upon the gate voltage. The corresponding energy differences are

$$\Delta E[(V, V_G), -ne, \pm e(-1, 1)] = \frac{e}{C_{11}} \left[\frac{e}{2} \pm (C_{11} - C_{1S})V \pm ne \mp C_{1G}V_G \right]$$

$$\Delta E[(V, V_G), -ne, \pm e(-1, 0)] = \frac{e}{C_{11}} \left[\frac{e}{2} \mp C_{1S}V \pm ne \mp C_{1G}V_G \right]$$

Coulomb blockade is established if all four energy differences are positive. For each n, this condition defines a stable, diamond-shaped region in the $V_G - V$ - plane, with the four boundaries given by the onset conditions

$$\Delta E[(V, V_G), -ne, \pm e(-1, 1)] = 0 \Rightarrow V(V_G, n) = \frac{C_{1G}}{C_{11} - C_{1S}} V_G - \frac{e(n \pm 1/2)}{C_{11} - C_{1S}}$$

$$\Delta E[(V, V_G), -ne, \pm e(-1, 0)] = 0 \Rightarrow V(V_G, n) = -\frac{C_{1G}}{C_S} V_G + \frac{e(n \pm 1/2)}{C_{1S}} \quad (9.13)$$

These stable regions are known as Coulomb diamonds, and line up along the V_G-axis (Fig. 9.11). Fig. 9.12 shows a measurement of the stability diagram of a Al − Al$_2$O$_3$ single-electron transistor. The experimentally obtained shape of the Coulomb diamonds, as well

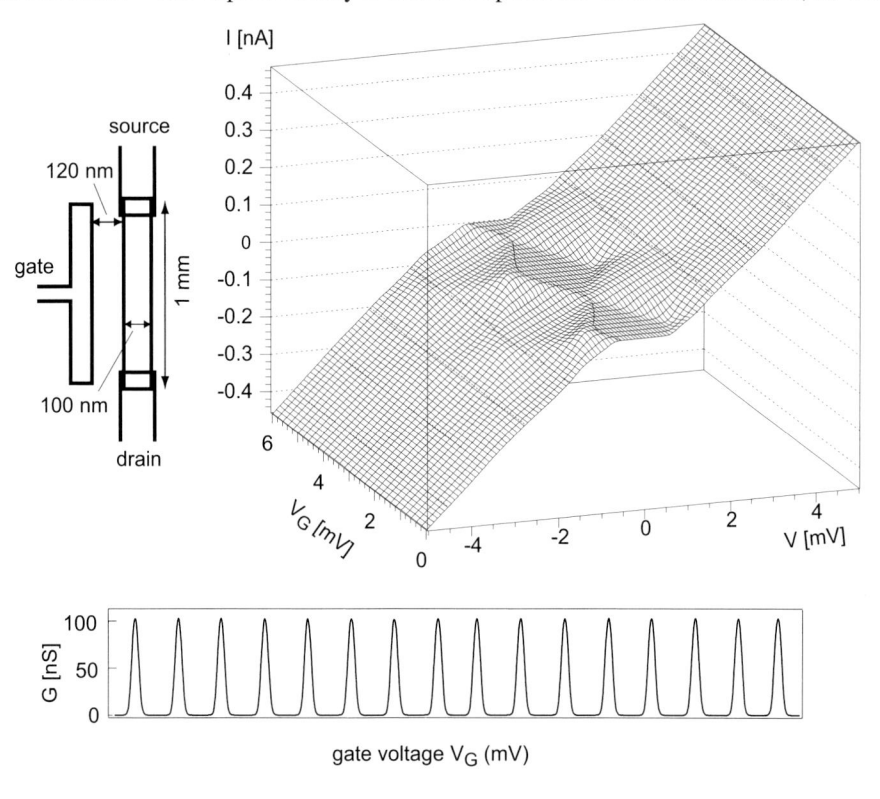

Figure 9.12: Stability diagram of a Al − Al$_2$O$_3$ SET transistor (its dimensions are shown to the left), measured at a temperature of 30 mK. At the bottom, a conductance measurement as a function of the gate voltage, for $V = 10\ \mu$V is shown (Coulomb blockade oscillations).

as the current-voltage characteristics, agree very well with the model just developed. For $|V| < \frac{e}{C_{11}}$, the current oscillates strongly as a function of the gate voltage, an effect known as "Coulomb blockade oscillations". Current peaks occur at $V_G = \frac{e}{C_{1G}}(n + \frac{1}{2})$. In each gate voltage period $\Delta V_G = \frac{e}{C_{1G}}$, n changes by one. It is important to point out that these oscillations have nothing to do with resonant tunnelling. Neither did we assume phase coherence, nor does the nearest-neighbor spacing of the energy levels have to be larger than $k_B \Theta$! In

9.2 Basic single electron tunnelling circuits

fact, for the system shown in Fig. 9.12, the level spacing is well below 1 μeV. We shall see in the following chapter on quantum dots how single electron tunnelling coexists with size quantization. The weak structure outside the diamonds correspond to Coulomb staircases for each gate voltage, telling us that the two tunnel barriers are not identical.

The line shape of the Coulomb blockade resonances in the limit of negligible source-drain voltage has been derived by [Kulik1975] and by [Beenakker1991a]. The typical experimental situation is characterized by $h\Gamma \ll \Delta \ll k_B\Theta \ll E_C$. This is known as the *metallic regime*. Here, Γ denotes the coupling of the island to the leads, while Δ is the spacing of the discrete (kinetic) energy levels of the island. Coulomb blockade is well pronounced in this regime, but many energy levels carry current. The line shape of the conductance resonances is given to a good approximation by

$$G(E) = \frac{e^2 D_{island}}{2} \frac{\Gamma^S \Gamma^D}{\Gamma^S + \Gamma^D} \cdot \cosh^{-2}\left(\frac{E - E_{max}}{2.5 k_B \Theta}\right) \qquad (9.14)$$

Here, D_{island} is the density of states in the dot, $\Gamma^{S,D}$ denote the couplings of the dot to source and drain, while E_{max} is the energy at the peak amplitude. Note that the gate voltage can be transformed into an energy via $\delta E = eC_{1G}/C_{11}\delta V_G$. Increasing the temperature thus broadens the resonances, but does not change the peak conductance. Since the conductance of an individual energy level of the island scales as $1/\Theta$ (see [E 8.4]), and the number of contributing states is proportional to Θ, the total temperature dependence of the peak conductance just cancels [Beenakker1991a].

It is important to realize that Coulomb oscillations do not measure the density of states of the island, but the addition spectrum. The density of states tells us at how many electrons can be in the system at a particular energy, for a *fixed* number of electrons. The addition spectrum, on the other hand, tells us at which energies electrons can be *added to* the system. If the system is interacting, these two quantities are different, a fact which is clearly demonstrated here. Besides being a somewhat unconventional transistor with an ocillatory transconductance dI/dV_G, this device is extremely sensitive to charges in the vicinity of the island and can thus be used as an electrometer, as used, for example, to study the electrochemical potential in semiconductor heterostructures [Wei1997, Ilani2001]. Particularly appealing is the integration of a SET transistor in the tip of a scanning probe microscope, which results in an electrometer of both high spatial and charge resolution [Yoo1997, Gurevich2000]. The charge resolution is ultimately limited by shot noise; a sensitivity of 10^{-4} electrons has been demonstrated experimentally by [Zimmerli1992].

Question 9.6: Estimate the charge resolution δq achievable with the single electron transistor of Fig. 9.12. Assume the working point is in the wing of a Coulomb blockade resonance, and assume a current resolution of 10 fA. Note that $\delta q = C_{1G}\delta V_G$.

In transistor operation, its advantage is the low power consumption, since for switching, the charge needs to be changed by only a small fraction of e. Schemes for a digital logic based on single electron tunnelling have been developed, and experimental implementations are being investigated [Korotkov1996, Ancona1996]. One problem is to reduce the island size sufficiently in order to operate the devices at room temperature. By now, there are several reports on SET transistors operating at room temperature, see, e.g. [Shirakashi1998], but production of such devices is by no means standard. In addition, the switching is strongly disturbed by fluctuating background charges, although a charge stability of 0.01 elementary charges over weeks has been demonstrated in silicon-based SET transistors [Zimmerman2001]. Furthermore, the voltage gain in such transistors is limited.

These limitations can be overcome, in principle, by using resistively coupled single electron transistors. The circuit is shown in Fig. 9.13a: the gate couples to the island via a gate resistance $R_G \gg \frac{h}{2e^2}$. In describing this device, equation (9.10) has to be modified, since charge can also flow from the gate onto the island:

$$p(q)(\Gamma_{S\to 1}(\Delta E(q)) + \Gamma_{D\to 1}\Delta E(q)) + \frac{1}{R_G C_{11}} \frac{\partial}{\partial q}(q - V_g C_{11} + V C_{1D})) =$$

$$p(q+e)\{\Gamma_{1\to S}(\Delta E(q+e)) + \Gamma_{1\to D}(\Delta E(q+e))\} \quad (9.15)$$

$p(q)$ is now the probability density of finding the total charge q on the island. The corresponding current-voltage characteristics are shown in Fig. 9.12. In this device, the gate voltage

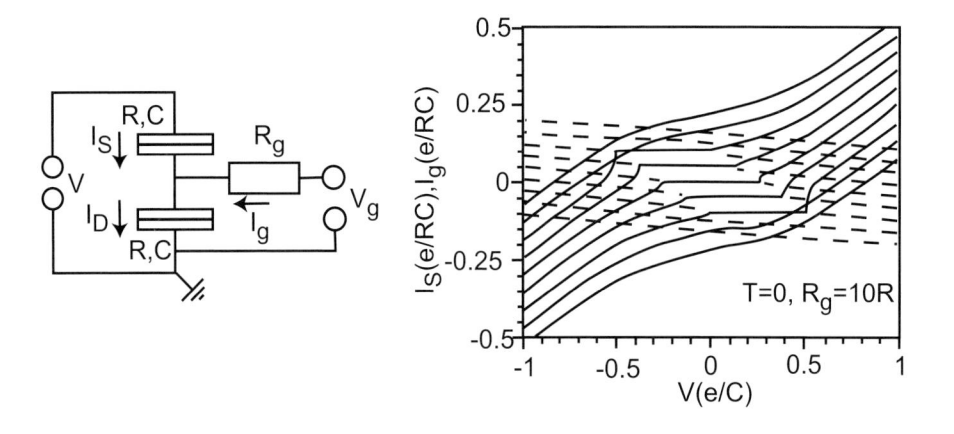

Figure 9.13: Current-voltage characteristics of a resistively coupled single electron transistor. Shown is both the source current (solid lines) and the gate current (dashed lines) for V_g varying from -e/2C to e/2C in steps of e/8C. The traces are offset vertically for clarity. (Adapted from [Korotkov1998]).

keeps the island potential fixed at long time scales ($t \gg 1/R_G C_{11}$). If, however, V is sufficiently large and an electron can tunnel from S into the island, the gate response is too slow to prevent an additional voltage buildup at the drain junction, and the electron is able to tunnel to drain. If $|V_G| > e/C_{1D}$, a gate current starts to flow, and the island is open. Therefore, there is only one Coulomb diamond, centered around $(V, V_G)=(0,0)$. The transconductance is no longer oscillatory in V_G, and the device is much less sensitive to fluctuating background charges. Fabricating such a transistor, however, hits some experimental difficulties that have yet to be overcome: the heating problem is similar to that one in a single tunnel junction, and the stray capacitance between gate and island should be negligible. In addition, this Coulomb diamond is much more sensitive to thermal smearing and noise than those in "conventional" SET transistors [Korotkov1998].

9.3 SET circuits with many islands; the single electron pump

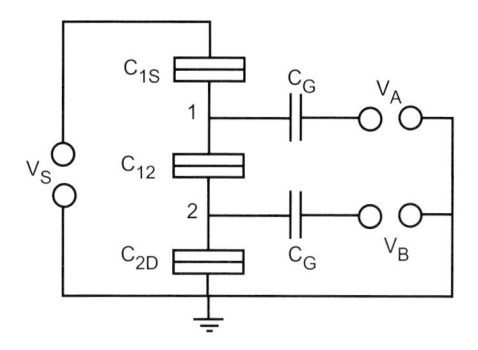

Figure 9.14: Circuit of two islands in series. Each island can be tuned by a nearby gate electrode.

As an example of a more complex SET circuit, we study the system of two islands in series, also known as *single electron pump*, Fig. 9.14. Via a tunnel junction, island 1 is coupled to source and island 2 to drain. The total capacitances C_{11} of both islands are assumed to be equal. Furthermore, we neglect several capacitance matrix elements (except those shown in Fig. 9.14) and assume that electrode A (B) couples only to island 1(2), with equal capacitances. Nevertheless, V_B influences V_1 via the inter-island capacitance C_{12} and vice versa. We will not study the effect of a source drain bias voltage. Rather, we are interested in the ground state of the system as a function of V_A and V_B. We assume that we can probe this state by applying a negligibly small source-drain voltage. Hence, we set $V_S = 0$. The island charge vector is given by $-e(n_1, n_2)$, the electrode voltage vector by $(V_A, V_B, 0)$. The capacitance matrices of interest are

$$\underline{C}_{II} = \begin{pmatrix} C_{11} & -C_{12} \\ -C_{12} & C_{22} \end{pmatrix}$$

$$\underline{C}_{IE} = \begin{pmatrix} -C_G & 0 & -C_{1S} \\ 0 & -C_G & 0 \end{pmatrix}$$

with $C_{11} = C_{22} = C_{1S} + C_{12} + C_G = C_{2D} + C_{12} + C_G$. Six electron transfers are of importance:

1. An electron tunnels between source and 1, $\Delta \vec{q}_I = e(\pm 1, 0)$, $\Delta \vec{q}_E = e(0, 0, \mp 1)$. The onset of this transfer is determined by

$$\Delta E[\vec{V}_E, -e(n_1, n_2), \Delta \vec{q}] = 0 \Rightarrow$$

$$C_{11} \cdot [C_G V_A - (n_1 \mp \frac{1}{2})e] = -C_{12}[C_G V_B - n_2 e] \tag{9.16}$$

2. An electron is transferred between drain and 2, $\Delta \vec{q_I} = e(0, \pm 1)$, $\Delta \vec{q_E} = (0,0,0)$, which gives

$$C_{11} \cdot [C_G V_B - (n_2 \mp \frac{1}{2})e] = -C_{12}[C_G V_A - n_1 e] \tag{9.17}$$

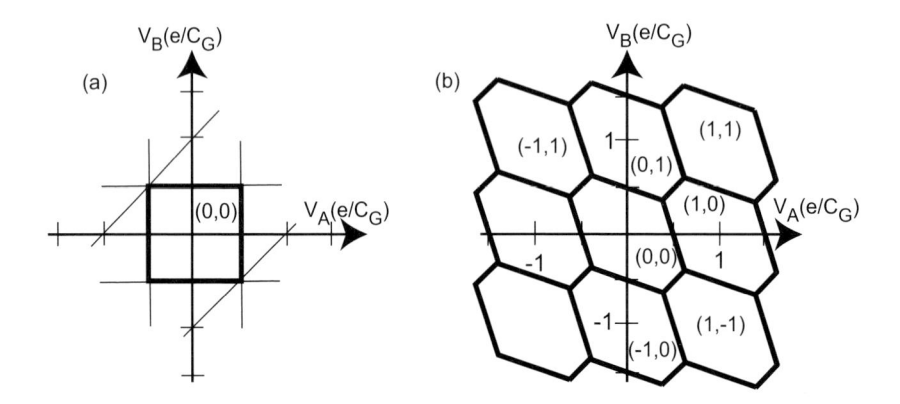

Figure 9.15: Stability diagram of the two-island system of Fig. 9.14, for completely decoupled islands(a) and for an inter-dot capacitance $C_{12} = C_G$ (b).

3. Finally, electrons can be exchanged between 1 and 2: $\Delta \vec{q_I} = e(\pm 1, \mp 1)$, $\Delta \vec{q_E} = (0,0,0)$, leading to

$$V_A - \frac{e}{C_G}(n_1 \mp \frac{1}{2}) = V_B - \frac{e}{C_G}(n_2 \pm \frac{1}{2}) \tag{9.18}$$

These boundaries define regions of stable electron configurations in the (V_A, V_B)-plane, each of which is characterized by the island charge vector that corresponds to the lowest energy. For $C_{12} \to 0$, island 1 and 2 are no longer coupled. It becomes impossible to influence island 1 by V_B and vice versa. In this limit, the stability diagram consists of squares given by conditions 1 and 2. Condition 3 plays no role, since the corresponding lines just touch two corners of the square (Fig. 9.15a).

Question 9.7: Investigate the stability diagram of the double island in the limit of connected islands!

The general situation is shown in Fig. 9.15b: the boundaries (1) and (2) tilt for $C_{12} > 0$, and the stable regions develop a hexagonal shape. A current can pass from source to drain only if electrons can tunnel between the two islands as well as between island 1 (2) and source (drain). This degeneracy exists only at the corners of the elongated hexagons.

9.3 SET circuits with many islands; the single electron pump

Question 9.8: Study the effect of cross capacitances on the stability diagram! Consider equal capacitances between gate A (B) and island 2 (1), which are much smaller than C_G.

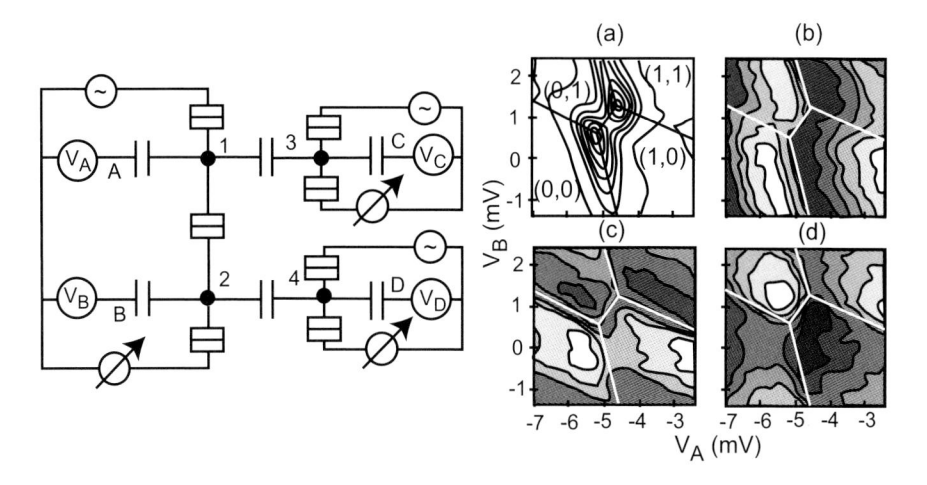

Figure 9.16: Measurement of the stability diagram of the double island system. Left: equivalent circuit of the double island system 1 and 2, with each island coupled to a SET transistor acting as electrometer. Right: (a) conductance of the double island as a function of the gate voltages V_A and V_B in a contour diagram. The conductance of electrometer 3 (4) is shown in (b) and (c), respectively. In (d), the difference signal of the two electrometers is shown Adapted from [Amlani1997].

The charge configuration of the double island system can be directly monitored by coupling a SET transistor to each island, Fig. 9.16. In this setup, the SET transistor labelled by 3 (4) serves as an electrometer to measure the charge on island 1 (2) [Amlani1997]. In Fig. 9.16(a), the current through the double island is shown as a contour plot. As expected, current flows predominantly at the corners of the stable regions. Figs. 9.16(b) and (c) show the conductance of the electrometers 3 and 4, respectively, which is a measure of the charge on island 1 (2). The transition of the island charges is clearly visible as a sharp increase of the electrometer conductance along the direction that corresponds to changing the charge at the measured island. In Fig. 9.16(d), the difference signal of the two electrometers is shown, which emphasizes that in each stable region, the charge configuration is really a different one.

[Pothier1992] were the first ones to demonstrate experimentally that with this device, electrons can be "pumped" by the gate voltages. The current can even be made flowing in the opposite direction of the source-drain bias voltage drop. In order to understand this experiment, we first consider the effect of a non-zero bias voltage: it shifts the boundaries of the stability diagram and generates triangular regions at the corners of the hexagons. Inside the triangles, Coulomb blockade becomes impossible. In order to operate the pump, the DC component of the gate voltages V_A and V_B are adjusted such that the device is located within one of these triangles (Fig.9.17(a)).

Question 9.9: Calculate the shifts of the boundaries given in Fig. 9.17(a)!

In addition, an AC voltage is applied to gates A and B, with a phase shift of (not necessarily exactly) $\pm\pi/2$. For sufficiently large AC amplitudes, the trajectory of the device state is a circle enclosing the triangle. Circling around the triangle labelled "P" in positive direction

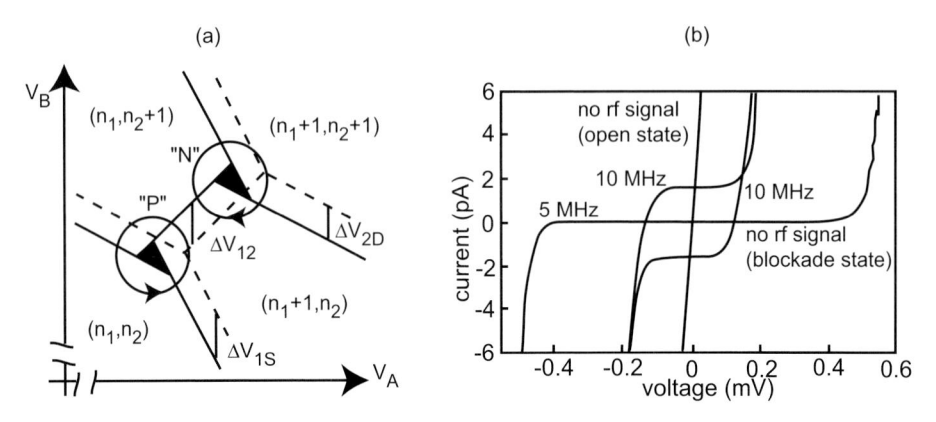

Figure 9.17: Left: a non-zero bias voltage shifts the boundaries of the stability diagram in V_B direction by $\Delta V_{1S} = -\frac{V_S}{C_G}(C + \frac{C^2}{C_{12}} - C_{12})$, $\Delta V_{2D} = -\frac{V_S}{C_G}\frac{C_{12}^2}{C}$, and $\Delta V_{12} = \frac{V_S}{C_G}C_{12}$, respectively. As a result, triangular shaped regions are formed in which Coulomb blockade no longer exists. The circles denote the trajectories of the device as small AC voltages are applied to gates A and B. Right: Operation of the electron pump at different frequencies. The actual phase shift of the AC signal was $\pm 130°$. Also shown are the I-V characteristics in the center and at a corner of a stable region, without an AC voltage applied (adapted from [Lotkhov2001]).

corresponds to a sequence of states $(n_1, n_2) \rightarrow (n_1+1, n_2) \rightarrow (n_1, n_2+1) \rightarrow (n_1, n_2)$. This means that for each round trip, one electron is transferred from source to drain, independent of the direction and magnitude of the bias voltage. The current plateaus of the single electron pump are shown in Fig. 9.17 (b) as a "P" point is encircled with two different frequencies in positive (phase shift $\pi/2$) and negative (phase shift $-\pi/2$) direction. Note that the current is independent of the sign of V_S within a window around $V_S = 0$. Also shown is the current-voltage characteristics when no AC signal is applied. Here, the current plateaux are absent. Provided the trajectory encloses the triangle completely, and that the AC amplitude has to be sufficiently small, such that no other additional electron transfers become possible, the current is coupled to the frequency via

$$I = ef$$

Furthermore, for the system to follow the frequency, f has to be smaller than the inverse time constant $1/\tau$ of the device, given by roughly $\tau = R_{12}C_{12}$. Encircling type "N" points in the same direction, or switching the direction in type "P" points, respectively, reverses the sign of the current.

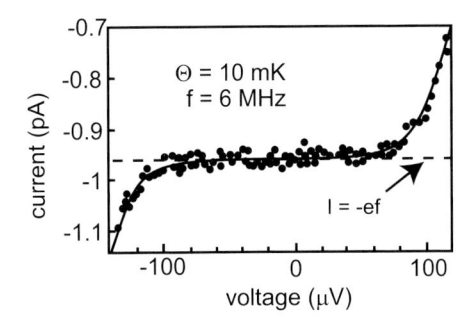

9.3 SET circuits with many islands; the single electron pump

Figure 9.18: Comparison between the observed current plateau of the single electron pump (circles) and the current I expected from $I = e \cdot f$. Close to the center of the plateau, a relative error of 10^{-6} is found. Here, cotunnelling has been suppressed by resistors in series to the single electron pump (Adapted from [Lotkhov2001]).

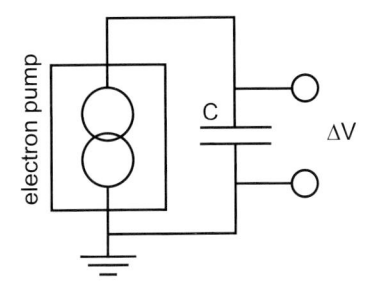

Figure 9.19: Principle of the capacitance standard: the single electron pump, consisting of several SET transistors in series, transfers a well-defined number of electrons onto the plate of a capacitor, and the voltage drop is measured.

Frequencies are the most accurate quantities we have in physics (the "NIST-F1 standard" is currently the frequency standard in the US and has an accuracy of 10^{-15}). This raises the question whether the single electron pump can be used as a current standard, with the current coupled to a frequency (at present, currents can be defined with a relative accuracy of 10^{-6} [Martinis1994]). Here, the low current that can be pumped through a single electron pump constitutes a problem. We may, however, rephrase this question and ask: how accurate is the number of electrons pumped? It turns out that the accuracy is dominated by multi-junction tunnelling events, so-called cotunnelling. Even with Coulomb blockade established, an electron may virtually tunnel onto the island. If this electron, or a second electron tunnels off the island across the second barrier, a real current results. Cotunnelling can be suppressed by increasing the number of tunnel junctions [Martinis1994, Averin1989, Averin1990]. Fig. 9.18 shows an example where the cotunnelling has been suppressed by placing high on-chip resistors in series to the SET device [Lotkhov2001]. [6]

---

[6]The results shown in Figs. 9.16 and 9.17 have actually been obtained with a thin-film Cr resistor located at the entrance and exit of the electron pump, see ref. [Lotkhov2001].

Keller and coworkers [Keller1999] used an electron pump that consists of 6 islands in series to charge a capacitor with an accuracy of 10^{-8}, i.e. the uncertainty is 1 electron for 10^8 pumped electrons. By measuring the voltage drop V across the capacitor after pumping N electrons, the capacitance $C = Ne/V$ could be determined with a standard deviation of $3 \cdot 10^{-7}$.

Papers and Exercises

P 9.1: In [Furlan2000], a single electron transistor is used for detecting charge rearrangements in the substrate. How does this work?

P 9.2: Hanna and Tinkham [Hanna1991] developed an analytical model for the Coulomb staircase in the limit of strongly differing junction couplings. Work out their model and reconstruct the author's "I(V)-phase diagram" in Fig. 1b of this paper!

P 9.3: Geerligs et al. [Geerligs1990] demonstrated the operation of a *single electron turnstile*, a slightly different concept for counting electron than the single electron pump. Explain the pumping mechanism of the single electron turnstile!

P 9.4: Superconductivity adds a new and exciting twist to single electron tunnelling. Work out the basic modifications due to superconductivity! A good starting point is Ref. [Fitzgerald1998]!

E 9.1: The *single electron tunnelling box* consists of an island in between a tunnel barrier and a capacitor with infinite resistance (see Fig. 9.20). The tunnel resistance is sufficiently high to suppress quantum fluctuations. Calculate the number of excess electrons on the

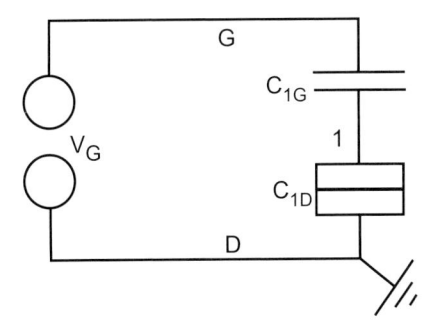

Figure 9.20: Equivalent circuit of the SET box.

island as a function of the voltage!

E 9.2: Calculate the current through a tunnel barrier in the absence of single electron charging effects. Show that our definition of the resistance in (9.8) is reasonable for small voltages applied, since Ohm's law is obtained.

E 9.3: Modify the double island system of Fig. 9.14 such that both source and drain couple to island 1 only. Island 2 "dangles", see Fig. 9.21. How does, in the limit of zero source-drain bias voltage, the phase diagram in the $V_A - V_B$ - plane look like? Discuss the relevance of direct electron transfers between island 2 and the source/drain contacts! Assume identical capacitances.

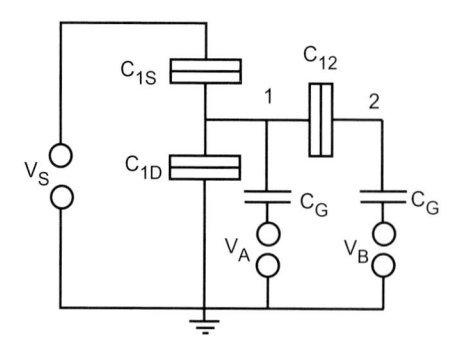

Figure 9.21: Sketch of the double island system of [E 9.3].

Further Reading

A classic review article has been written at the beginning of the "single electron tunnelling age" by [Averin1991]. A stimulating book containing collections of articles on various aspects of single electron tunnelling phenomena is [Grabert1992]. Furthermore, [Korotkov1996] is an article entitled "Coulomb Blockade and digital Single Electron Devices", which focuses on the relevant aspects of a future single electron logic.

10 Quantum Dots

A conducting island of a size comparable to the Fermi wavelength in all spatial directions is called a *quantum dot*. The properties of quantum dots are very similar to those of atoms, and sometimes you hear that "quantum dots are artificial atoms". The differences are essentially the size (0.1 nm for atoms vs. ≈ 100 nm for quantum dots), and the shape and strength of the confining potential. In atoms, the electrons are bound by the attractive forces exerted by the nucleus, while in quantum dots, a mean electric field generated by background charges and gate voltages holds the electrons together. The number of electrons in an atom can be changed by ionizing it, which can be done by irradiating it with electromagnetic waves, or by applying a strong electric field. In quantum dots, the electron number is typically altered by tuning the confinement potential. An equivalent process in atoms would be to replace the nucleus by a neighbor in the periodic table.

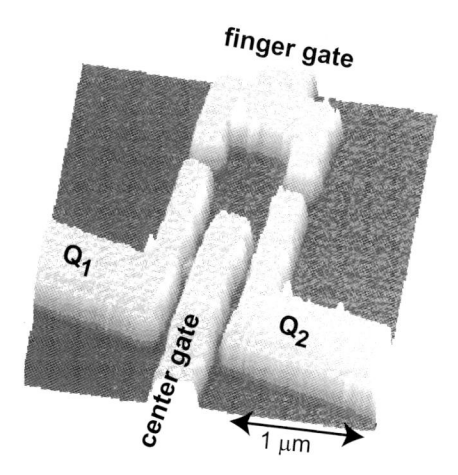

Figure 10.1: Atomic force micrograph of a gate geometry used to generate a quantum dot in a Ga[Al]As heterostructure. The gold electrodes (bright) have a height of 100 nm. The two QPCs formed by the gate pairs $F - Q_1$ and $F - Q_2$ can be tuned into the tunnelling regime, such that a quantum dot forms in between the two barriers. Its electrostatic potential can be varied by changing the voltage applied to the center gate.

The length scales imply characteristic energy scales in quantum dots that differ from those in atoms by roughly 4 orders of magnitude. The energy level spacing in atoms is of the order of 1 eV, while in quantum dots, it is typically 0.1 meV. The ionization energy is in the range of 10 eV for atoms and about 1 meV in quantum dots. Therefore, quantum dots open up novel experimental possibilities not available in atoms. For example, it is easy to break Hund's rules in a quantum dot by applying a magnetic field. Doing this in an atom requires a magnetic field of the order of 10^4 T, two orders of magnitude larger than the strongest magnetic fields available in the laboratory. Such possibilities, combined with the high tunability of quantum dots, have boosted quantum dot research in the past 10 years. The option of tailoring their optical and electronic properties promises a variety of applications as well. Quantum dot lasers with particularly low threshold currents have been built, and it is envisaged to transfer many concepts of quantum optics related to the interaction of photons with atoms into a solid state environment, a goal which would pave the road for novel applications, such as quantum computing, for example.

Here, however, we restrict ourselves to the basic transport properties of quantum dots. From a fundamental point of view, the possibility to probe a small entity of confined, interacting electrons by transport experiments is exciting. Clearly, this aspect is essential for any kind of optoelectronic quantum dot devices as well.

We begin this chapter with a brief survey of the transport phenomenology of quantum dots. The elementary constant interaction model will be introduced in section 2. It offers a crude and simple way to separate the interaction effects from size quantization, and allows to interpret many observations in a straightforward manner. However, many experiments reveal contradictions to this simple model, as demonstrated in section 3. In section 4, we will have a look at the line shapes of the conductance resonances, which offer additional information and complements the information on the discrete energies, which is extracted from the resonance positions. Finally, the chapter is concluded with a look at further experimental realizations of quantum dots, which do not rely on semiconductor heterostructures.

10.1 Phenomenology of quantum dots

The majority of transport experiments on quantum dots has been performed on samples made by the top-down approach, namely by lateral patterning of a semiconductor heterostructure. We pick this realization as an example to introduce the transport phenomenology of quantum dots. Other systems are mentioned in section 5. The gate structure of Fig. 10.1, defined on top of a Ga[Al]As heterostructure, is designed to impose and tune a quantum dot in the 2DEG underneath. In combination with the finger gate F, the gates Q_1 and Q_2 form two quantum point contacts, which can be tuned independently, see Fig. 10.2. Suppose we now adjust the finger gate and the center gate such that the electron gas underneath is depleted. As the conductance of both QPCs is reduced below $2e^2/h$ by tuning the Q gates, the electron puddle in between the gates gets disconnected from the environment, and a closed quantum dot is formed, weakly coupled to source and drain via tunnel barriers. In this regime, conductance oscillations as a function of any of the gate voltages are observed, Fig. 10.3. The voltages applied to F, Q_1 and Q_2, however strongly change the QPC conductances as well, which limits the tuning range of the dot. Therefore, the dot is usually tuned with a center gate voltage, This

10.1 *Phenomenology of quantum dots*   251

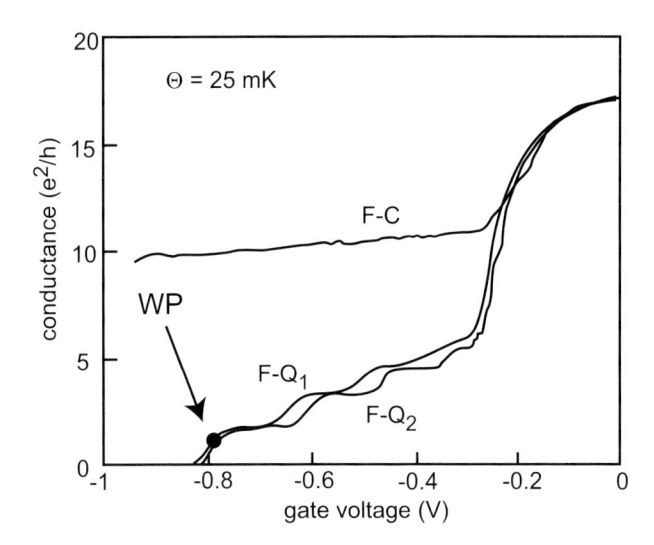

Figure 10.2: Parametric conductances of different pairs of gates of the sample shown in Fig. 10.1. Sweeping the gate pairs $F - Q_1$ and $F - Q_2$ show typical QPC characteristics. Here, all remaining gates have been grounded. Gate pair $F - C$ forms a somewhat poorly defined QPC, which cannot be pinched off. Nevertheless, the formation of the channel due to depletion is clearly seen. "WP" denotes the working point of the gates F, Q_1, and Q_2 used to operate the dot as a single electron transistor.

gate is designed to couple well to the dot, but has a weak influence on the QPC transmission. A typical conductance trace as a function of a center gate voltage is shown in Fig. 1.5.

Qualitatively, such oscillatory behavior is expected from both single electron tunnelling, as well as from the resonant tunnelling discussed in chapter 8. In fact, when this kind of oscillation was first observed, its explanation was not immediately clear. Only additional experimental studies revealed that in fact both of these effects play an important role. For typical experimental parameters, the Coulomb blockade determines the coarse features, while size quantization, i.e., the quantization of the kinetic energy inside the dot, is responsible for the fine structure [Scott-Thomas1989], [Houten1989],[Meirav1990]. A large amount of information on quantum dots has been collected by investigating their conductance as a function of a gate voltage and a second, independent parameter. In Fig. 10.4, the peak positions of the conductance resonances are plotted as a function of the gate voltage and a (perpendicular) magnetic field. While the raw data look rather smooth at first sight, a further investigation reveals a rich fine structure. First of all, the peak spacing in gate voltage is not constant, see the inset in Fig. 10.4(a). There is a general trend towards smaller peak spacings as the gate voltage is increased. This is partly due to a geometrical effect, since the edge of the dot approaches the gate electrode as we fill in electrons. As a consequence, the capacitance between the dot and the gate increases. On top of this effect, fluctuations in the peak spacings are apparent, as we saw already in Fig. 1.5. in the introduction. The fine structure becomes more visible once a constant amount of the peak spacings has been subtracted, Fig. 10.4(b). We divide this pattern into three regimes. At very low magnetic fields, the spacings fluctuate, with a certain tendency

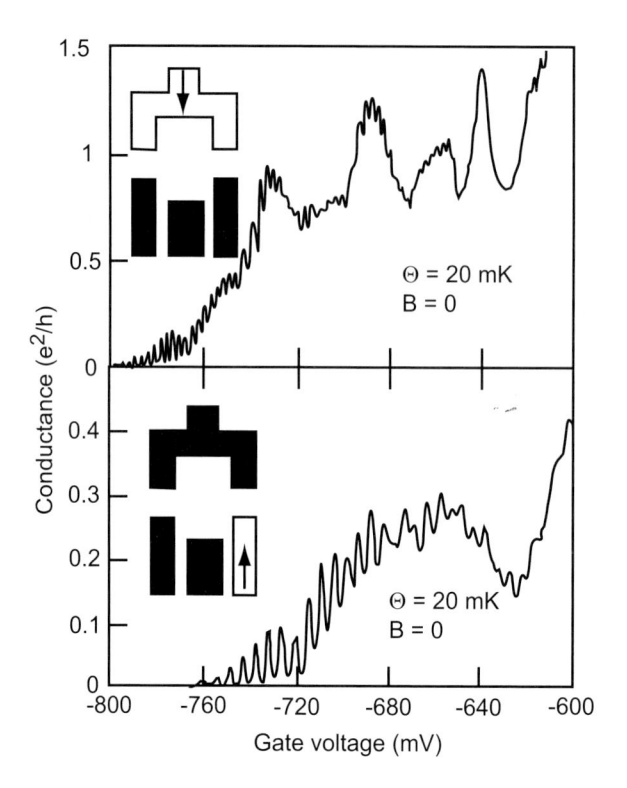

Figure 10.3: Gate voltage characteristics with the dot defined, i.e., all gates are activated. Short-period transmission resonances are seen, which are modulated with a much larger period. Note that the resonances set in at a threshold gate voltage.

to bunch together for small occupation numbers. At intermediate magnetic fields ($B \approx 1$ T and occupation numbers $N > 20$ in this example), the peak positions show quasi-periodic cusps. This pattern changes abruptly as B is increased, with a transition magnetic field that increases with the occupation number.

While the details of this overall pattern depend of the sample, and many additional effects have been found in particularly designed quantum dots and under appropriate experimental conditions, such a phenomenology is a common feature of most samples. In the following, we focus on the interpretation of this overall pattern. Before we begin, however, further experimental results need to be mentioned which contain additional information. Let us first look at the current through a quantum dot as a function of a gate voltage and of the source-drain bias voltage V, Fig. 10.5. Regions of suppressed current are observed, as sketched in Fig. 10.5(a). They resemble the Coulomb diamonds encountered already in chapter 9. Here, however, their sizes fluctuates, while their shapes are essentially identical. At small source-drain voltages, the conductance resonances as a function of the gate voltage look similar to Coulomb block-

10.2 The constant interaction model

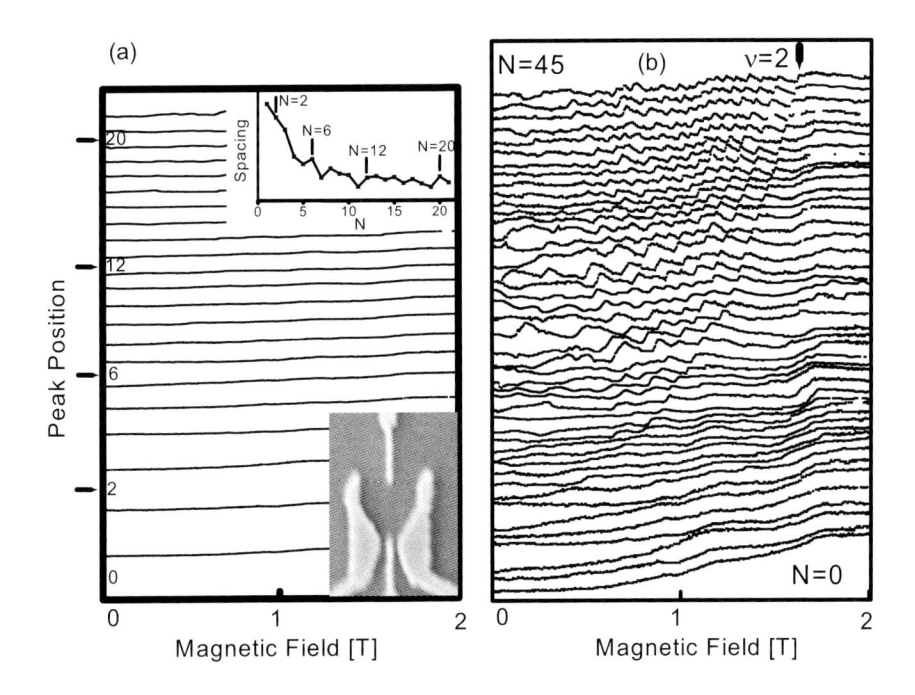

Figure 10.4: (a): positions of 22 consecutive conductance resonances as a function of the gate voltage and the magnetic field. The gate geometry of the sample is shown in the lower inset. The quantum dot has an approximately triangular shape with a width and height of about 450 nm. The center gate is in the center at the bottom. The upper inset shows the peak spacings at $B = 0$. The numbers indicate dot occupations for which particularly large spacings are expected within the Fock-Darwin model (see text). Figure (b) shows the data of (a) up to 45 electron in the dot, with a constant peak spacing removed. $N = 0$ indicates the region where the dot is empty, while $\nu = 2$ denotes the Landau level filling factor inside the dot, see text. After [Ciorga2000].

ade oscillations, although their amplitude fluctuates from resonance to resonance. As the bias voltage is increased, however, a fine structure emerges, Fig. 10.5(b), which is absent in single electron transistors. Finally, the amplitude of the resonances can be tuned by a magnetic field, see Fig. 10.6. In fact, it may change by orders of magnitude and can be suppressed below the noise level.

10.2 The constant interaction model

How shall we interpret these observations? Clearly, a quantum dot is a quasi zero-dimensional system; within a single-particle picture, its density of states consists of a sequence of peaks, with positions determined by the size and shape of the confining potential, as well as by the

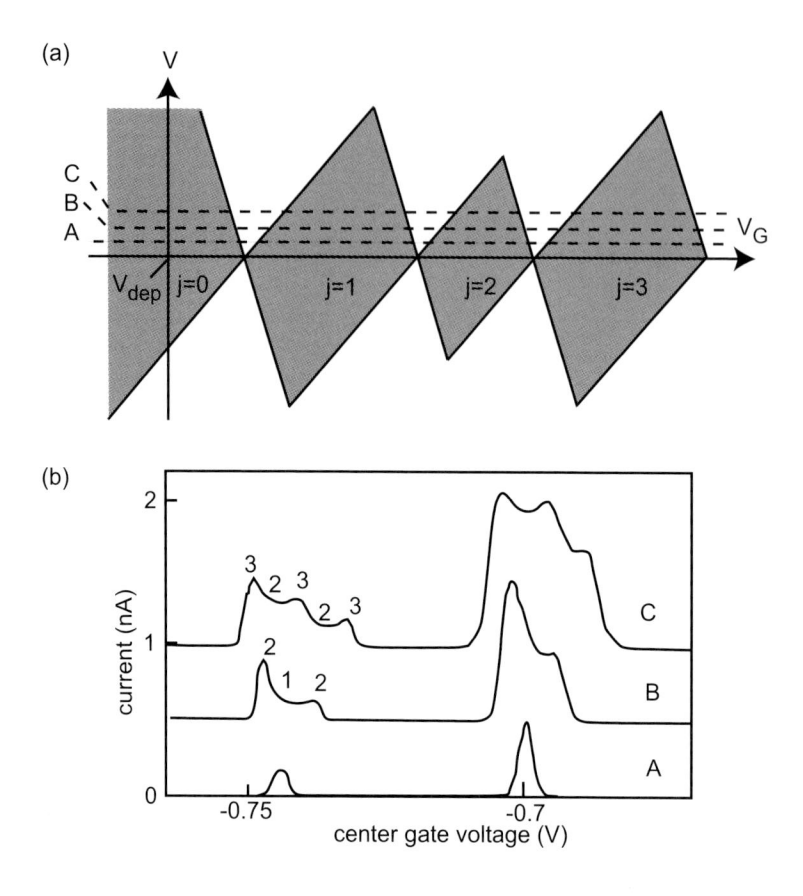

Figure 10.5: (a): schematic sketch of the current through a quantum dot, in a gray scale plot, as a function of the gate voltage and the source-drain bias voltage V. Diamond-shaped regions of suppressed current are observed. As V is increased from A to C, the gate voltage sweeps reveal a fine structure of the conductance resonances (b), (after [Johnson1992]).

effective mass of the host material. A very simple estimation for the average nearest neighbor spacing Δ of these energy levels is obtained by starting from the two-dimensional density of states, $D_2(E) = m^*/\pi\hbar^2$. For a spin degeneracy of 2, there are $m^*A/\pi\hbar^2$ states per energy interval of unit length in an area A, and thus a spin-resolved, average energy level spacing of

$$\Delta \approx \pi\hbar^2/m^*A \tag{10.1}$$

is expected. The second energy scale of relevance is set by the single electron charging energy. For a sufficiently weak coupling to the leads, i.e., when the conductances of the barriers that connect the dot to source and drain are below $2e^2/h$, the electrons of the dot are strongly localized, and Coulomb blockade comes into play.

10.2 The constant interaction model

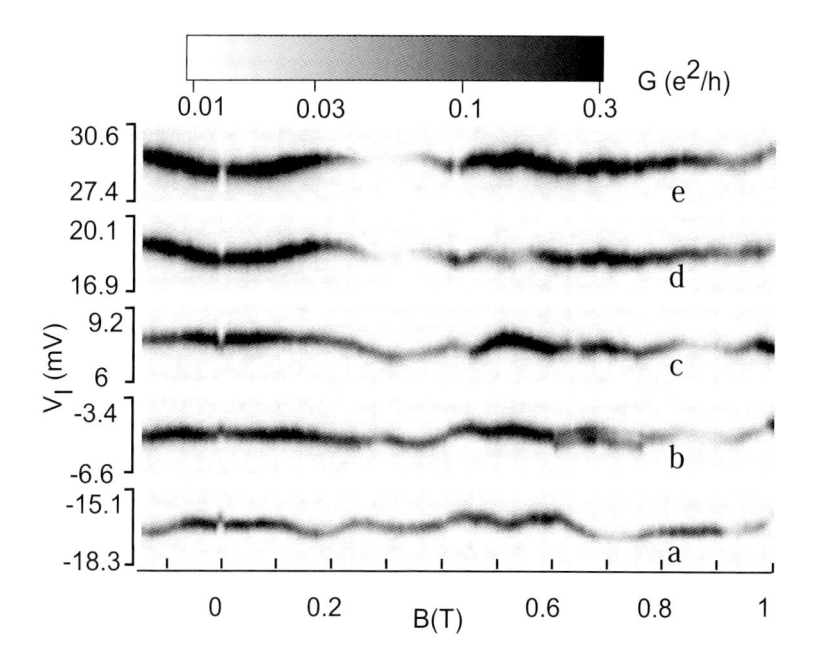

Figure 10.6: Evolution of five consecutive conductance resonances in a magnetic field. The peak positions fluctuate by about 20% of their spacing, while the amplitude varies by up to 100%. After [Luscher2001].

Question 10.1: Show that the same expression for Δ is obtained from the energy spectrum of a two-dimensional square well in the limit of large quantum numbers; estimate Δ and the single electron charging energy for a GaAs quantum dot with a diameter of 300 nm!

Apparently, in order to add an electron to the dot, the *addition energy* is required, which is the sum of the electrostatic and the kinetic part of the energy, as discussed already during our discussion of capacitance spectroscopy in section 6.1. One estimate for the addition energy of the jth electron would be to simply add the single electron charging energy and the single-particle separation of the jth energy level from its occupied neighbor. This approach seems perfectly okay at first sight, but makes in fact the crucial assumption that the kinetic energy of the dot states is independent of the number of electrons in the dot. Due to the electron-electron interactions, screening, as well as exchange and correlation effects, this is not strictly the case. It is well known that interactions strongly modify the energy spectra of atoms. Even in the case of the helium atom with just two electrons, the addition spectrum is tremendously complicated. The approach outlined above, known as the *constant interaction (CI) model*, disregards such difficulties.

The CI model is a valuable tool for analyzing quantum dot addition spectra, and it provides a good explanation of the data in several cases, as we will see in the following section. Deviations of the CI model are both interesting and numerous, however; some examples will be discussed as well.

It is straightforward to include the additional discrete energy levels in our single electron tunnelling model of chapter 9. Suppose state j with single particle energy ϵ_j is the highest occupied state in the quantum dot. An additional electron will occupy the empty state with the lowest energy, ϵ_{j+1}. This energy can be simply added to the energy difference ΔE in eq. (9.4). Likewise, for processes that reduce the electron number in the dot, we subtract ϵ_j, the energy of the highest occupied level, in that equation. In the previous chapter, we took it for granted that the number of electrons in the island is large compared to the number of additional charges we forced on the island with the gate voltage. This is no longer necessarily the case in quantum dots, since the electron densities are smaller by a factor of ≈ 1000. Typical occupation numbers range between 0 and 100. It is thus natural to define the number j of electrons as zero for the empty dot. In principle, the dot can be filled with holes, but this would require to tune the dot potential across the band gap of the semiconductor host. We do not consider this possibility.[1] For simplicity, let us assume that there is no background charge. The boundaries of the stable regions that form the diamonds (eqns. (9.13)) are modified and now read

$$V(V_G, j) > \frac{C_{1G}}{C_{11} - C_{1S}}(V_G - V_{dep}) - \frac{e(j+1/2)}{C_{11} - C_{1S}} - \epsilon_{j+1}\frac{C_{11}}{e(C_{11} - C_{1S})} \qquad \delta\vec{q} = e(-1, 1)$$

$$V(V_G, j) < \frac{C_{1G}}{C_{11} - C_{1S}}(V_G - V_{dep}) - \frac{e(j-1/2)}{C_{11} - C_{1S}} - \epsilon_j\frac{C_{11}}{e(C_{11} - C_{1S})} \qquad \delta\vec{q} = e(1, -1)$$

$$V(V_G, j) < -\frac{C_{1G}}{C_{1S}}(V_G - V_{dep}) + \frac{e(j+1/2)}{C_{1S}} + \epsilon_{j+1}\frac{C_{11}}{eC_{1S}} \qquad \delta\vec{q} = e(-1, 0)$$

$$V(V_G, j) > -\frac{C_{1G}}{C_{1S}}(V_G - V_{dep}) + \frac{e(j-1/2)}{C_{1S}} + \epsilon_j\frac{C_{11}}{eC_{1S}} \qquad \delta\vec{q} = e(1, 0) \quad (10.2)$$

Here, V_{dep} denotes the depletion voltage as indicated in Fig. 10.5, while V_G is the gate voltage used to tune the dot. For the special case of $j = 0$, of course, we cannot remove a further electron from the island, and thus the second and the fourth inequality do not apply. The stability diagram thus consists of a semi-infinite set of diamonds of equal shape. Their sizes vary due to the varying single particle level spacings. In addition, a "semi-diamond" is obtained for $j = 0$, see Fig. 10.5. In analogy to the diamonds in the pure electrostatic case, their maximum extension in V - direction equals

$$\Delta V = \frac{1}{e}\left(\frac{e^2}{C_{11}} + \epsilon_{j+1} - \epsilon_j\right)$$

The peak spacing in gate voltage at $V \approx 0$ is given by

$$\Delta V_G = \alpha\left(\frac{e^2}{C_{11}} + \epsilon_{j+1} - \epsilon_j\right) \qquad (10.3)$$

The ratio $\alpha = C_{11}/eC_G$ is a lever arm that translates the addition energies into gate voltages, while the dot's total energy changes by $\frac{e^2}{C_{11}} + \epsilon_{j+1} - \epsilon_j$ as one electron is added.

---

[1] Note that although we speak of quantum dots in electron gases, everything is analogous for hole gases.

10.2 The constant interaction model

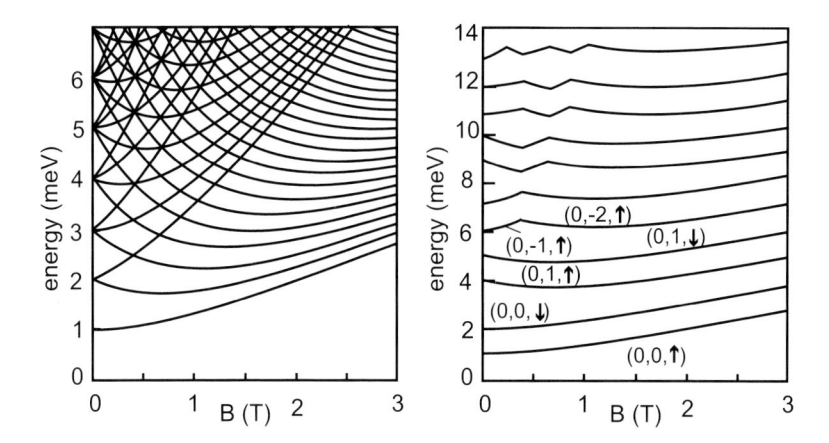

Figure 10.7: A section of the Fock-Darwin spectrum (left), calculated for $\hbar\omega_0 = 1$ meV, and the predicted evolution of the conductance resonances as a function of the gate voltage and the magnetic field (right), when the single electron charging energy equals 1 meV as well. The labelling is explained in the text.

Question 10.2: The quantum dots of Figs. 10.1 and 10.4 do have more than one gate. How does this fact enter in eq. 10.2?

Within the CI model, we can subtract E_C from the measured peak spacings according to eq. (10.3), as has been done to get Fig. 10.4(b) from Fig. 10.4(a). The remainder should correspond to the single-particle energy spectrum of the dot. In order to appreciate the effect of a magnetic field, we consider a model which has the advantage of being analytically solvable, namely the Fock-Darwin model [Fock1928],[Darwin1931] encountered already in chapter 7. Now, the electron motion is no longer free in the third direction, and the parabolic potential is orientated in the x-y plane. The energy spectrum is not changed by these modifications, due to the superposition principle. Hence, we assume that the quantum dot is circular in shape and has a parabolic confinement potential,

$$V(x,y) = \frac{1}{2}m^*\omega_0^2(x^2+y^2) = \frac{1}{2}m^*\omega_0^2 r^2$$

with r being the radius of the dot. The corresponding Schrödinger equation can be solved analytically, even with a magnetic field applied in z-direction [Jacak1998]. The energy spectrum is given by

$$E_{n,\ell}(B) = (2n+|\ell|+1)\hbar\sqrt{\omega_0^2 + \frac{1}{4}\omega_c^2} - \frac{\ell}{2}\hbar\omega_c \pm g^*\mu_B B \tag{10.4}$$

The radial quantum number is $n = 0, 1, 2, ...$, while ℓ is the angular momentum quantum number, i.e., $\ell = 0, \pm 1, \pm 2, ...$. At B=0, the energy levels are located at $j \cdot \hbar\omega_0$, with $j = 2n+|\ell|$), and with an orbital degeneracy of j. In addition, there is a twofold spin degeneracy.

In analogy to atomic energy spectra, we can speak of the jth Fock-Darwin shell. The orbital degeneracies in each shell get removed by a perpendicular magnetic field, since all states within one shell have different angular momenta and thus respond differently to the magnetic field. Let us for now assume that the effective g-factor of the dot g^* is negligible, such that the levels remain spin degenerate. This spin-degenerate Fock-Darwin spectrum is shown for $\omega_c \leq \omega_0$ to the left in Fig.10.7. We see that a sufficiently strong magnetic field induces level crossings. For the confining strength shown in this figure ($\hbar\omega_0 = 1$ meV), for example, a crossing of level $(n, \ell) = (0, 2)$ with $(n, \ell) = (0, -1)$ occurs at $B \approx 0.4$ T, and the ground state configuration of the dot changes. Similar level crossings occur more frequently at higher energies.

Sometimes, the behavior expected within the Fock-Darwin model agrees even quantitatively with the experimental observations, see [P 10.1]. In the typical experiment shown in Fig. 10.4, of course, one cannot expect a perfect agreement, since the dot's confining potential is neither circular nor strictly parabolic.[2] The Fock-Darwin model predicts filled shells for $N = 2, 6, 12, 20...$. Although these spacings are slightly enhanced in Fig. 10.4(b), there is certainly no 1:1 correspondence. Note further that there is also no spin pairing visible. The CI model is thus a reasonable first approximation for this regime, although it cannot explain all the details of the experimental observations.

Quantum dots in intermediate magnetic fields

We proceed by looking at the quasi-periodic cusps of the peak positions that can be seen in Fig. 10.4(b). To begin with, it is useful to transform the Fock-Darwin model to a different set of quantum numbers, which emphasizes the behavior of the energy levels in magnetic fields. Intuitively, we expect that the stronger B is, the less important should the electrostatic confinement be, and the Fock-Darwin levels should bunch together and form Landau levels. It is therefore appropriate to relabel the energy levels by the Landau level quantum number $m = 1, 2, 3, ...$, and a quantum number p that enumerates the energy levels within a Landau level. With the transformation (see, e.g., [Jacak1998])

$$p = n + \frac{|\ell| + \ell}{2}$$

and

$$m = n + \frac{|\ell| - \ell}{2} + 1$$

the energy levels of the Fock-Darwin spectrum read

$$E_{m,p} = \hbar(m+p)\sqrt{\omega_0^2 + \frac{1}{4}\omega_C^2} + \frac{1}{2}\hbar(m - p - 1)\omega_C \quad (10.5)$$

Here, we again neglect spin splitting. In the regime of filling factors $2 < \nu < 4$, this spectrum develops a very simple structure, see Fig. 10.8. States with $m = 1$ decrease in energy as

---

[2]Numerical simulations have revealed that in many samples, a circular dot shape and a parabolic confinement is actually a better approximations than might be expected from the gate geometry [Kumar1990].

10.2 The constant interaction model

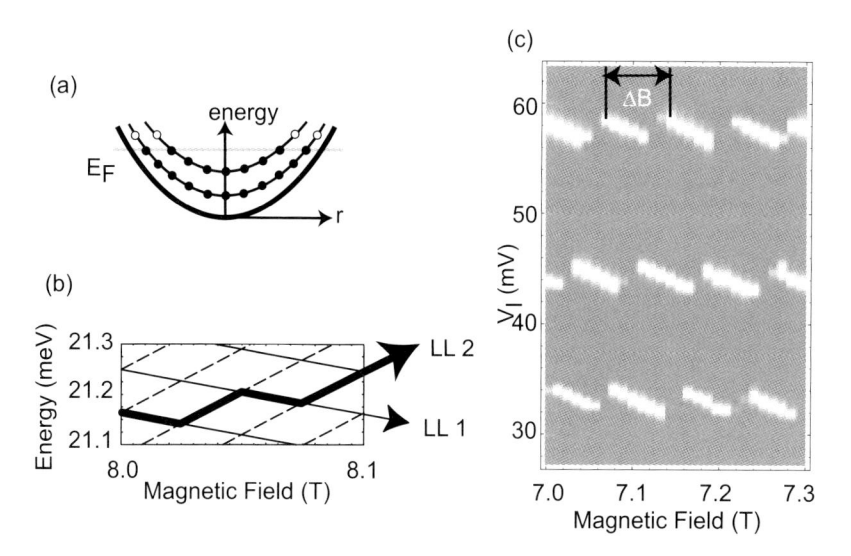

Figure 10.8: (a): two Landau levels in a parabolic quantum dot. Below the Fermi energy, the discrete states of each Landau level are occupied, as indicated by the full circles. Higher Landau levels are not shown, since all their states are empty. The filling factor is thus in the regime $2 < \nu < 4$. A section of the corresponding energy spectrum is shown in (b), calculated for a circular dot with a parabolic confinement of $\hbar\omega_0 = 1$ meV. States belonging to LL 1 (thin full lines) reduce their energy as B is increased, while those states belonging to LL 2 (dashed lines) are running upwards in energy. The bold lines represent the Fermi level as a function of the magnetic field B, for a constant number of electrons in the dot. In (c) the conductance of a quantum dot as a function of the gate voltage and the magnetic field in this regime is shown. Bright shades correspond to a high conductance.

B is increased, while Landau level 2 - states show a positive magneto-dispersion. This fact simply reflects the increasing degeneracy of LL 1 and the depopulation of LL 2 as B is increased. The energies as a function of B develop a quasi-periodic pattern of diamonds, see Fig. 10.8(b). At constant electron number, the electrochemical potential of the dot, and with it the energy of the transmission resonance, moves in zigzag lines as B is tuned. The period can be approximated by $\Delta B \approx (\frac{\omega_0}{\omega_C})^2 B$. Furthermore, the energy levels are approximately equally spaced in energy, with a spacing of $\Delta E = E_{m,p+1} - E_{m,p} \approx \hbar\frac{\omega_0^2}{\omega_C}$, which is independent of the energy level quantum numbers. These approximate expressions are part of [E 10.1]. The spatial location of these states is shown schematically in Fig. 10.8(a). At the Fermi level, the $m = 1$ states are close to the edge of the dot, while states with $m = 2$ are located towards the dot center. This means that states belonging to LL 1 couple much better to the leads than LL 2 states. Although we have developed this scenario within the Fock-Darwin model, most conclusions remain valid for other dot shapes and confining potentials. Independent of these details, the LL 2 states couple more weakly to the leads, since they are further inside the dot, and thus the tunnel barrier they form with the reservoirs is larger. These states will get depopulated as B is increased, independent of the confinement shape. In fact, the measurement shown in Fig. 10.8(c) agrees reasonably well with the Fock-Darwin model. Bright

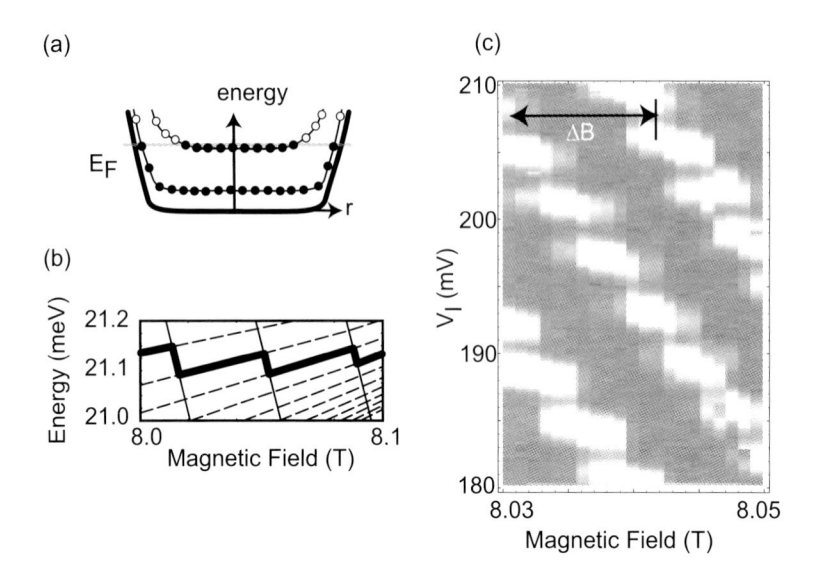

Figure 10.9: Landau levels (a) and the section of a calculated energy level diagram for $2 \leq \nu \leq 4$. In (c), a corresponding set of experimental data is shown.

lines of high conductance with a negative slope are observed. They measure the magnetic field dispersion of LL 1 - states. The LL 2 - states couple very weakly to the leads, due to the exponentially suppressed tunnelling. The conductance via those states is not detectable at the low source-drain bias voltages used in the experiment. For a further interpretation of these data, see [E 10.1].

Experimentally, different confinement potentials can be established by, e.g., varying the parameters of the sample and of the fabrication process accordingly. The corresponding data in a quantum dot with an approximate hard-wall confinement are discussed in Fig. 10.9. The energy spectrum of a circular disk with hard walls cannot be solved analytically. Rather, the spectrum is obtained by numerical calculation of the zeros of the hypergeometric function usually denoted as $_1F_1$ in the literature of special functions [Geerinckx1990]. This energy spectrum is shown schematically in Figs. 10.9 (a,b): most strikingly, the density of states at the Fermi level in LL 2 is higher than in LL 1, provided the Fermi level is not far above the energy of LL 2 at the center of the dot. Fig. 10.9 (c) shows a corresponding measurement [Fuhrer2001a].

The reconstruction of the energy spectrum of this dot within the CI model is performed in [E 10.1]. The result reveals that the spacing between LL 1 states is significantly larger as compared to that one for Ll 2 - states, which indicates a steep-wall confinement.

Quantum rings

As a third example, we have a look at the reconstructed energy spectrum of a quantum ring in moderate magnetic fields, Fig. 10.10 [Fuhrer2001b]. From the single particle spectrum of a one-dimensional quantum ring (the topic of [E 8.3]), we expect a pattern formed by a set of

10.3 Beyond the constant interaction model

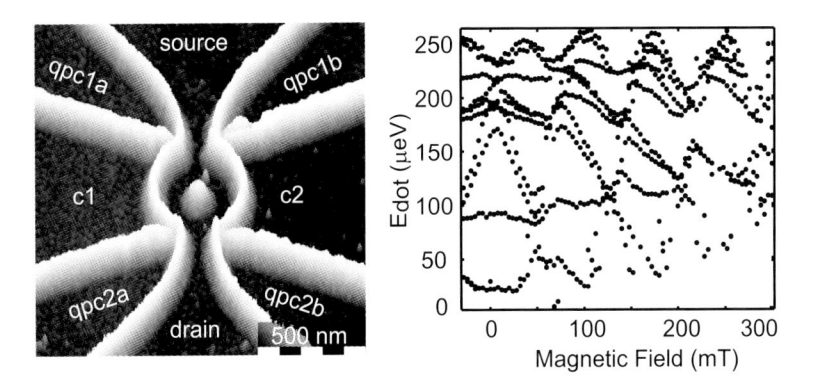

Figure 10.10: Left: sample geometry of a quantum ring, defined in the 2DEG of a Ga[Al]As - HEMT by local oxidation of the surface. As usual, the QPC gates tune the dots coupling to source and drain, while the ring can be tuned with the center gates. The ring contains about 100 electrons. The reconstruction of the energy spectrum is shown to the right.

parabolas

$$E_{\ell,n} = \frac{\hbar^2}{2m^* r^2}[\ell + n]^2 \qquad (10.6)$$

Here, $n = eBr^2/2\hbar$ is the number of magnetic flux quanta that penetrate the ring, and $\ell = 0, \pm 1, \pm 2...$ is the angular momentum quantum number. This is partly observed in the data of Fig. 10.10. The periodicity and the amplitude of the well-pronounced zigzag lines are in agreement with the expectations from the single-particle spectrum. In addition, quasi dispersionless states are observed, which most likely reflect the imperfections of the ring. Due to azimuthal thickness variations, the actual states may be an admixture of various eigenstates with different ℓ, which damp the amplitude of the zigzag lines. Although strong deviations from the single-particle spectrum of a perfect one-dimensional ring are observed, the CI model thus seems to be a good approximation.

10.3 Beyond the constant interaction model

The CI model is bound to fail if residual interactions come into play, like exchange and correlation effects, or like screening which depends on the number of electrons in the dot. Actually, the absence of spin pairing in 10.4 is a good example for exchange and correlation energies which can be dealt with only numerically. We now look at some effects in quantum dots which remain unexplained within the CI model, but can be interpreted within simple models for residual interactions.

Hund's rules in quantum dots

Hund's rules tell us in what sequence the states within an atomic shell are filled with electrons. Hund's first rule states that the total spin gets maximized, without violating the Pauli principle.

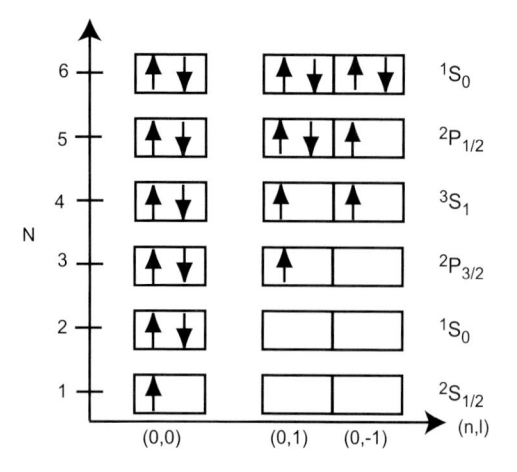

Figure 10.11: Schematic occupation of the Fock-Darwin energy levels (n, ℓ) as the dot is filled with N electrons, according to Hund's rules.

This rule originates in the exchange interaction, due to which electrons with parallel spin are kept spatially separated, which reduces their mutual Coulomb energy. Hund's second rule forces the electrons to maximize the total orbital angular momentum, under the constraint of Hund's first rule. How the first 6 electrons are filled in a Fock-Darwin potential at $B = 0$ according to Hund's rules is shown in Fig. 10.11. The third electron filled in this potential occupies the level $(n, \ell) = (0, 1)$.[3] The fourth electron occupies the $(0, -1)$ level, with the same spin direction as the third electron. In analogy to the nomenclature used in atomic physics, we denote the electronic configurations by $^{2S+1}L_J$, where S is the total spin, J the total angular momentum, and L gives the total orbital momentum, which is usually denoted by S for $L = 0$, P for $L=1$ and so on. Forcing the 4th electron in the $(0, 1)$ level as well requires to pay the exchange energy Δ_{xc}, and the configuration $^3D_3$ would result. The 5th electron goes again in level $(0, 1)$, with its spin in opposite direction, while the 6th one fills the last empty state in this shell. How Hund's rules can be broken in quantum dots by applying magnetic fields is the topic of [P 10.1].

Quantum dots in strong magnetic fields

A plain failure of the CI model occurs in quantum dots under strong magnetic fields, i.e., for filling factors $\nu < 2$. In Fig. 10.4, this regime is located above the upper magnetic field threshold for the cusps. The experimental findings are summarized in Fig. 10.12. In the previous section, we have seen that frequent and quasi-periodic level crossings are observed in quantum dots for filling factors just above 2, which is in agreement with a single particle spectrum. For filling factors below 2, there are no orbital level crossings. Only the spin splitting causes slightly different magnetodispersions for spin-down states as compared to spin-up states. Very infrequent level crossings are therefore expected within a single-particle picture

---

[3]The spin orientation is determined by Hund's third rule, which states how the total spin and the total angular momentum couple. This depends on the host material and is of minor relevance here.

10.3 Beyond the constant interaction model

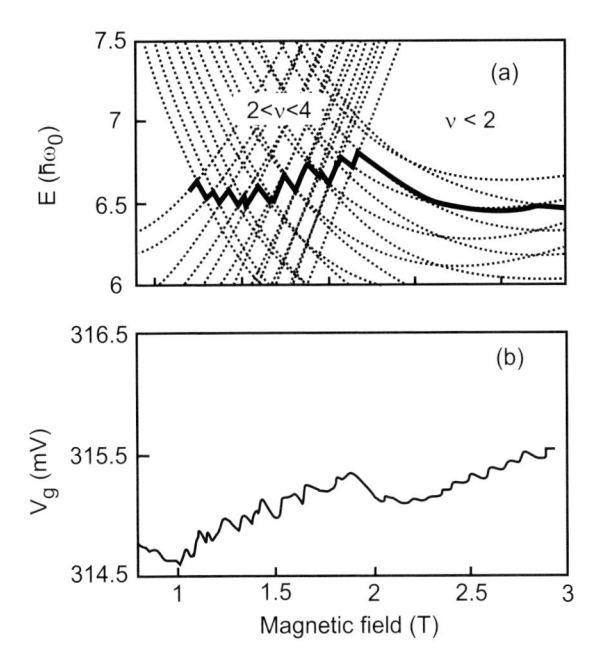

Figure 10.12: The dashed lines in (a) represent the magnetic field dispersion of the levels 30 to 50 of the Fock-Darwin spectrum, including spin splitting. The bold line follows the energy of the 39th state. Below filling factor 2, levels cross only because the two spin directions have different magnetodispersions. Hence, very rare level crossings are expected in this regime. Experimentally, however, rapid oscillations of the conductance peak positions are found, as exemplified by a resonance observed for about 39 electrons in the dot. After [McEuen1992].

and for reasonable effective g-factors. However, [McEuen1992] have observed *frequent* level crossings in this regime, which remain unexplained within the CI model. To understand this effect, we revisit the Chklovskii picture of edge channels mentioned in section 7.3, and adapt it to a quantum dot. Imagine the edge channel configuration of Fig. 7.20 is bent to form a circle. Qualitatively, we see right away that due to the spin-splitting and the modulated screening properties at the edge and the resulting electronic structure, a dot with only the spin-up and the spin-down sublevel of one Landau level occupied, segregates into a metallic ring around its edge and a metallic disk at its center, separated by an insulating stripe.[4] The resulting structure is sketched in Fig. 10.13. The location and width of the insulating stripe is determined by the effective g-factor inside the dot. From an electrostatic point of view, an additional capacitance is formed between the ring and the disk. The system can be thought of a variation of a double island system discussed in [E 9.3]: the ring corresponds to the first island coupled to the leads. The second island, i.e., the central disk, couples only indirectly to the leads, via the ring. The analogy is not complete, though: first of all, the intra-dot capacitance C_{12} is a function of the magnetic field and of the electron density. Second, tuning both islands independently

---
[4]This picture is, as in the case of straight edge channels, only physically meaningful if the insulating stripe is wider that the magnetic length.

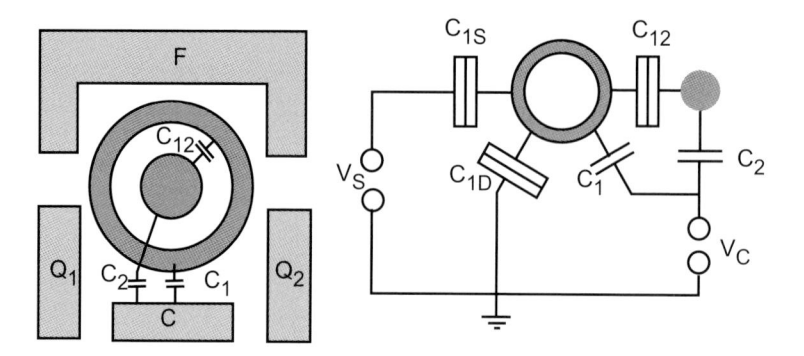

Figure 10.13: Electrostatic structure of a quantum dot in strong magnetic fields (left). The dark regions in the left figure denote metallic areas of the dot, while the white stripe is an insulating region. The metallic ring close to the dot edge is formed by the edge channel of the spin-up sublevel of Landau level 1, while the disk at the dot's center is formed by the spin-down sublevel. An intra-dot capacitance emerges, and the capacitance of the dot with respect to the gates is split up in two components stemming form the disk and the ring. The system is equivalent to the double island shown to the right.

with two gate voltages is impossible. Rather, the gate voltage couples differently to the two islands, with a ratio that changes with the gate voltage. As the magnetic field is increased, the spin-down sublevel gets depopulated, and the electrons get transferred to the spin-up sublevel. Such a transfer, however, requires to overcome the intra-dot charging energy, since the electron cannot be transferred to the spin-up sublevel within the disk, as all the states of this sublevel lie well below the Fermi level and are thus occupied.[5] The period of the zigzag lines of the transmission peaks as a function of B are now a measure for the intradot capacitance. Quantitative calculations within such a model agree well with the experimental data, see Fig. 10.14.

The distribution of nearest neighbor spacings

Above an occupation number of about 20 and for small magnetic fields, analytical single-particle energy spectra bear no resemblance to the observed energy level spacings of quantum dots. This experimental fact is usually explained in terms of *quantized chaos*. Due to the tremendous relevance of the underlying theory for all kinds of mesoscopic phenomena, we introduce the concept using some properties of quantum dots as an example.

A classical system is chaotic if its evolution in time depends exponentially on changes of the initial conditions. To make the connection, consider a point-like mass confined in a two-dimensional box. The point mass moves in a constant potential, i.e., along straight lines, and experiences specular reflection at the walls. The trajectory of the point mass can be parameterized in a suitable way, and its position at time t can be written as $p_1(t)$. Now, suppose we start the motion of the point mass with a slightly changed initial condition $p_2(0) = +p_1(0)\delta p$, and we ask how $\Delta p(t) = p_2(t) - p_1(t)$ evolves over time. It turns out that there are two fundamentally different kinds of evolutions, which depend on the shape of the box. If,

---

[5]The electrostatics of this system is the topic of paper [P 10.2].

10.3 Beyond the constant interaction model 265

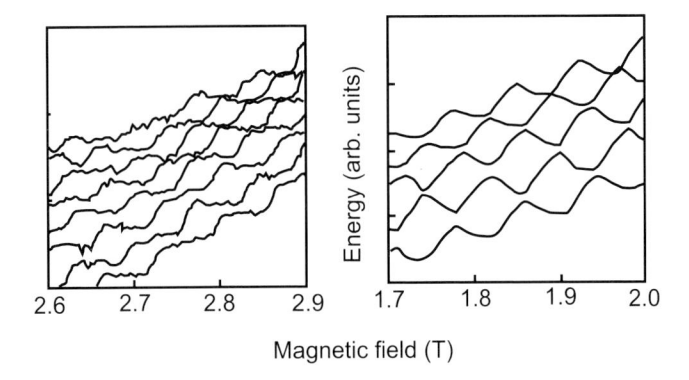

Figure 10.14: Measured (left) and calculated (right) evolution of consecutive conductance peak positions for quantum dot filling factors below two agree reasonably well, in particular with respect to the periodicity, if this electrostatic dot structure is taken into account. After [McEuen1992].

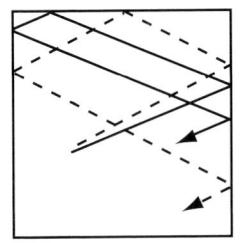

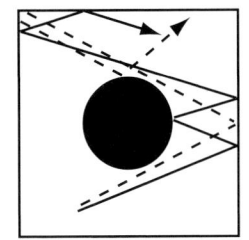

Figure 10.15: In a classically chaotic geometry like the Sinai billiard (right), trajectories diverge exponentially as the initial conditions change by arbitrary small amounts. This is not the case for regular structures, like the square to the right.

for long time scales ("long" meaning that the point mass has hit the wall many times), $\Delta p(t)$ diverges exponentially, the box is called chaotic. If the divergence is non-exponential, the box is regular. It turns out that only very few structures are regular, e.g., a square box, or a circle. Most shapes show a chaotic classical dynamics. A famous example widely discussed in the literature is the so-called Sinai billiard, which consists of a square box with a circular pillar at its center (see Fig. 10.15). Quantizing a classically chaotic system can be done by using the Gutzwiller trace formula [Gutzwiller1971]. The properties typical for classically chaotic systems, like the exponential sensitivity on initial conditions, get lost during the quantization process, but there are nevertheless remnants of classical chaos in the quantum regime. [6]

The two most widely investigated remnants of chaos in quantum dots are the probability distribution of nearest-neighbor peak separations (the NNS distribution), which we will discuss below, and the distribution of the transmission resonance amplitudes, which is the topic

---

[6]The subfield of chaos theory that investigates these remnants is referred to as quantized chaos. For further information, see [Gutzwiller1990] and [Haake1991].

of [P 10. 3].

Experimentally, the NNS distribution $p(s)$ is obtained in a straightforward way: determine the spacings s_j of adjacent energy levels E_j and E_{j+1}, plot them in a histogram, and normalize the distribution properly. For randomly placed energy levels, a Poisson distribution is obtained, i.e., $p(s) = exp(-s)$. For some quantum mechanical systems, we can say right away how the NNS distribution looks like. For the Fock-Darwin potential at B=0, for example, it consists of two δ-functions, one at $s = 0$, which contains the degenerate levels including spin degeneracies, and one at $s = \hbar\omega_0$. Each regular system has its characteristic $p(s)$, i.e., $p(s)$ is non-universal. This is not the case for classically chaotic systems. Naively, one is probably tempted to assume that in a chaotic system, the positions of the energy levels are completely random, and thus $p(s)$ would be a Poisson distribution. This, however, is not the case. Rather, $p(s)$ takes one of three universals forms, which depends solely on the symmetry properties of the Hamiltonian. These distributions can be calculated within the so-called *random matrix theory - RMT*, which has turned out to be highly successful in many branches of physics, including several aspects of mesoscopic transport [Beenakker1997].

We have a look at the concept of RMT by sketching the derivation of $p(s)$. Suppose we represent the Hamiltonian of our quantum dot in some basis, such that it can be written as a Hamiltonian matrix H. For a Hamiltonian that is invariant under time inversion (i.e., no magnetic field should be present), the matrix is hermitian, and $p(H)$ should be invariant under orthogonal basis transformations. If time reversal symmetry is broken, the Hamiltonian matrix is unitary, and $p(H)$ should not change under unitary transformations. These conditions clearly set some constraints for the matrix elements. It is assumed that within these constraints, the matrix elements are completely random for classically chaotic systems. The above conditions define the orthogonal and the unitary ensemble of random matrices, called *Gaussian orthogonal ensemble - GOE* and *Gaussian unitary ensemble - GUE*, respectively. For an arbitrary large number of levels, the Wigner-Dyson distributions for $p(s)$ result from the ensemble properties. The calculation is carried out in [Mehta1991]. These complicated distributions can be very well approximated by the Wigner surmises, which are the corresponding distributions for a two-level system with the same symmetry properties [Bohr1969]. One distinguishes between pure distributions, where a spin degeneracy is absent, and bimodal distributions, where a δ - function at $s = 0$ is introduced ad hoc, which takes a twofold spin degeneracy into account. Their most important features are summarized in Table 10.1.

The Wigner surmises for spin-degenerate systems are plotted in Fig. 10.16. Furthermore, calculating the Wigner surmise for the orthogonal ensemble is the topic of exercise [E 10.3]

There is up to now no strict proof that quantized chaos should obey the predictions of RMT. The crucial point is whether the Hamiltonian matrix elements of a chaotic system are really random. Empirically, however, the agreement between RMT and experimental results is overwhelming. For example, the measured NNS distributions of microwave cavities, the excitation spectra of nuclei, or of a hydrogen atom in strong magnetic fields, are indistinguishable to the Wigner surmise. Numerical simulations, show very good agreement as well, see Fig.10.18. Most strikingly, small level separations are suppressed in chaotic systems, an effect known as level repulsion. It can be traced back to more anticrossings in systems with reduced symmetry. In quantum dots, however, the experimentally obtained NNS distributions deviate from the bimodal Wigner-Dyson distribution expected within the constant-interaction model. Experimentally, one subtracts the single-electron charging energy from the measured

10.3 Beyond the constant interaction model

Table 10.1: Properties of the Wigner surmises of relevance. TRS means time-reversal symmetry, while $\bar{s}$ denotes the average spacing and σ its standard deviation.

Ensemble	GOE	GUE	bimodal GOE	bimodal GUE
$p(s)$	$\frac{\pi}{2}se^{-\frac{\pi}{4}s^2}$	$\frac{32}{\pi^2}s^2 e^{-\frac{4}{\pi}s^2}$	$\frac{1}{2}(\delta(s) + p(s))$	$\frac{1}{2}(\delta(s) + p(s))$
$\bar{s}$	1	1	0.5	0.5
σ	$\sqrt{\frac{4}{\pi} - 1}$	$\sqrt{\frac{3\pi}{8} - 1}$	$\sqrt{\frac{2}{\pi} - \frac{1}{4}}$	$\sqrt{\frac{35\pi}{16} - \frac{1}{4}}$

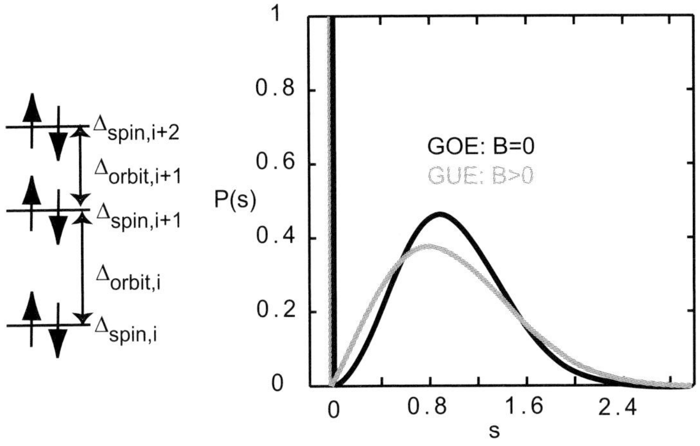

Figure 10.16: Bimodal Wigner surmises for the Gaussian orthogonal (GOE) and the Gaussian unitary (GUE) ensembles.

addition spectrum and plots a histogram of the remaining peak spacings. A typical example of such a measurement is shown in Fig. 10.19. $P(s)$ does not resemble the expected traces at all. There is no signature of a bimodal distribution, nor does the FWHM agree well with the RMT prediction, see [Sivan1996], [Simmel1997], for example. It is natural to suspect that residual electron electron interactions beyond the CI model cause the discrepancy. Clearly, the spin degeneracy is removed by such interactions, which broadens the spin peak and displaces it to a non-zero value. If these residual interactions are strong enough, they can deform the bimodal Wigner surmise distributions into singly peaked distributions of a different shape. In that respect, the experimental NNS distributions serve as a reference measurement, to optimize various models for residual interactions in small electronic systems [Ullmo2001].

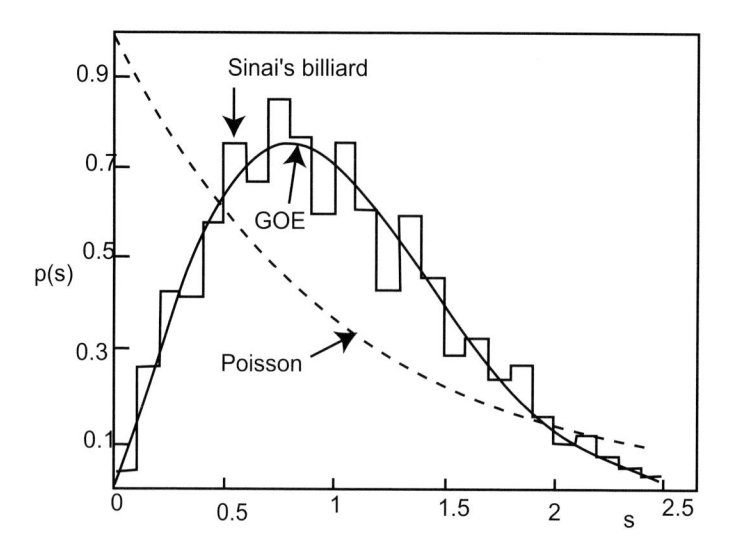

Figure 10.17: Numerically calculated NNS distribution of a Sinai billiard (histogram, about 1000 eigenvalues have been used) vs. the GOE Wigner surmise (full line). For comparison, the Poisson distribution (dashed line) is shown as well. After [Bohigas1984].

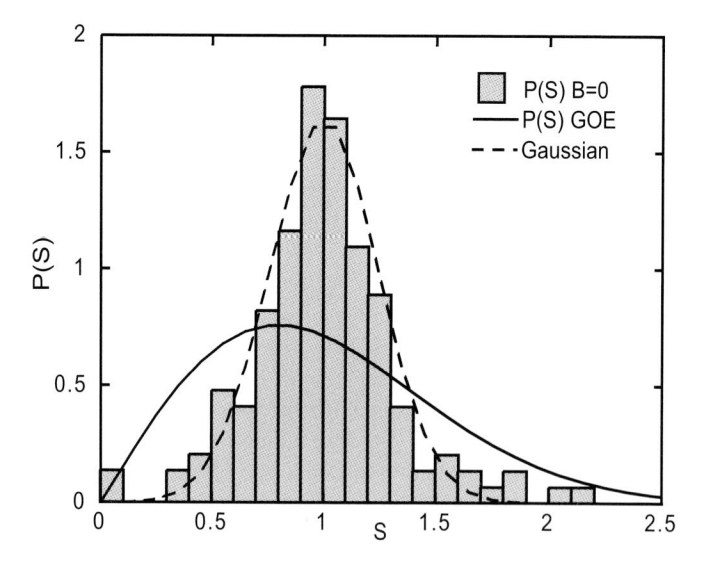

Figure 10.18: Measured NNS distribution of a quantum dot in GaAs (histogram), in comparison to the Wigner surmise. The distribution look more like a Gaussian, and the bimodal structure is absent. After [Simmel1997].

10.4 Shape of conductance resonances and current-voltage characteristics

As we have just seen, tuning a quantum dot with a gate voltage and looking at its conductance as a function of small source-drain bias voltages is an important experimental technique. So far, we have extracted the information from the peak positions. It is self-evident to ask what kind of information is contained in the line shapes and the amplitudes of the conductance resonances. Clearly, the fluctuating amplitude is a measure for the coupling of the current-carrying state in the dot to the leads, and hence of corresponding wave function amplitude close to the tunnel barriers. It varies from state to state and as a function of a magnetic field. This fact can be used to compare the statistical properties of the wave functions with RMT, see [P 10.3]. Note that this is in marked contrast to the single electron transistors discussed in chapter 9, where many states couple very weakly to the leads, which results in an approximately constant peak amplitude. First of all, these considerations allow us to interpret the results of Fig. 10.5(b) qualitatively, see Fig. 10.19. In (a), no current flows, since the source and drain electrochemical potentials μ_S, μ_D lie inside the Coulomb gap. As V_G is increased (b), level 2 gets aligned with μ_D, and an electron may tunnel into this level, a process which establishes the scenario depicted in (c). Now, one of the electrons in state 1 or 2 may tunnel into source, such that both level contribute to the coupling between the dot and source. As V_G is increased further, scenario (d) gets established at some point. Here, only level 2 contributes to the current, and the system oscillates between (d) and (e). Hence, the overall conductance will be smaller than in (b) and (c). Finally, consider the situation depicted in (f) and (g) at higher gate voltages. Here, the two empty states 2 and 3 lie in between μ_S and μ_D, which increases the coupling between the dot and drain. One of them will get occupied and re-emptied by the electron tunnelling into source. Hence, we expect a conductance resonance with a doublet shape, as observed in Fig. 10.5, trace B.

From these considerations, it becomes clear that the single particle level spacings can be determined by high-bias transport experiments. In fact, a gate is not even necessary. As we increase the source-drain voltage, additional quantum dot states become subsequently accessible for transport, which is reflected in current steps, or in peaks in the differential conductance dI/dV, as a function of V, see Fig. 10.20. This is an important experimental tool for investigating quantum dots where a gate electrode is impossible to define; some examples will be mentioned in the next section. The underlying theory has been developed by [Averin1991a].

We have already discussed the resonance line shape at negligible source-drain voltages in the metallic regime $h\Gamma \ll \Delta \ll k_B\Theta \ll E_C$. This situation is usually not encountered in quantum dots. In many experiments, $k_B\Theta$, Δ as well as $h\Gamma$ are actually of the same order of magnitude. There is no general expression for the line shape for arbitrary values of these quantities. Complications arise due to the Coulomb interactions which correlate the tunnelling events across the two barriers, and due to the fact that the distribution function inside the quantum dot is usually not a Fermi-Dirac function. For small couplings $h\Gamma \ll k_B\Theta, \Delta$, the corresponding theory has been worked out in [Beenakker1991a]. Even in this regime, the line shapes have to be calculated numerically. An analytical result is available, however, for one important limiting case, namely for $h\Gamma \ll k_B\Theta \ll \Delta \ll E_C$. In this regime, only a

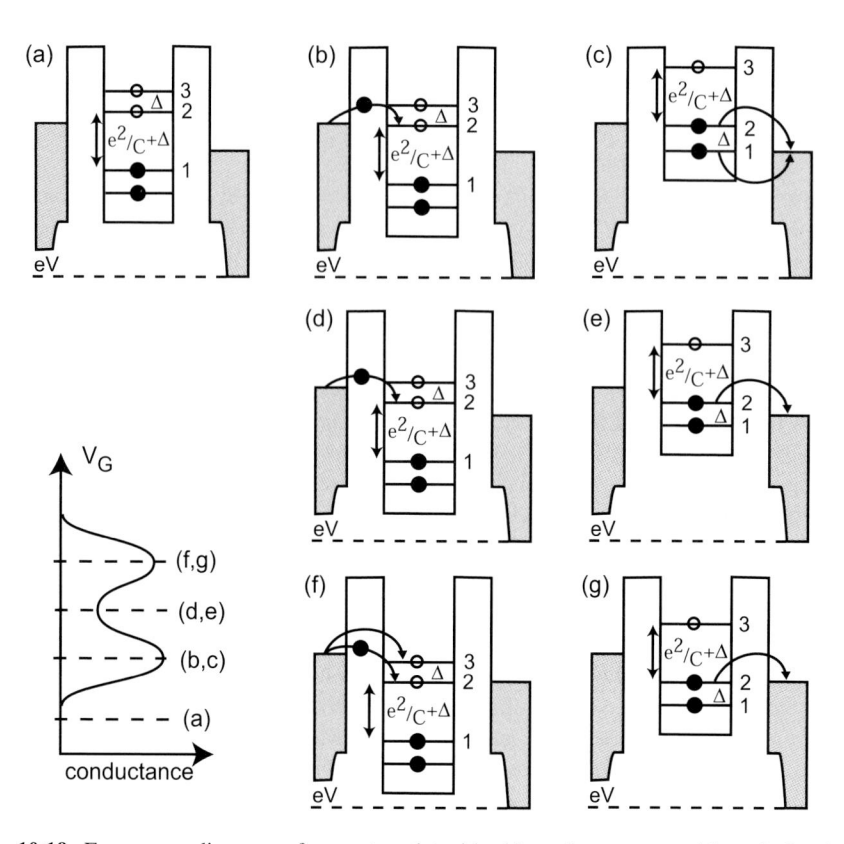

Figure 10.19: Free energy diagrams of a quantum dot with a bias voltage comparable to the level spacing Δ applied. Full circles denote occupied dot states, while empty circles indicate empty states. To the lower left, the resulting conductance resonance and the corresponding energies for each scenario (a)-(g) are sketched. It should be compared to the observed trace B in Fig. 10.5.

single level of negligible homogeneous broadening carries current. Now, the line shape equals

$$G(E) = \frac{e^2}{4k_B\Theta} \frac{\Gamma^S \Gamma^D}{\Gamma^S + \Gamma^D} \cdot \cosh^{-2}\left(\frac{E - E_{max}}{2k_B\Theta}\right) \tag{10.7}$$

which is a generalization of the line shape discussed in [E 8.4]. Note that the peak conductance now increases as $1/\Theta$, as long as $h\Gamma \ll k_B\Theta$, in contrast to the peaks in the metallic regime.

10.5 Other types of quantum dots

A laterally confined region in a semiconductor heterostructure is just one variation of a quantum dot. In this section, some other types are presented. It should be noted that large ensemble of quantum dots have been investigated for a long time, see for example the experiment by Giaever and Zeller in Fig. 9.6. Later on, capacitance measurements on self-assembled

10.5 Other types of quantum dots

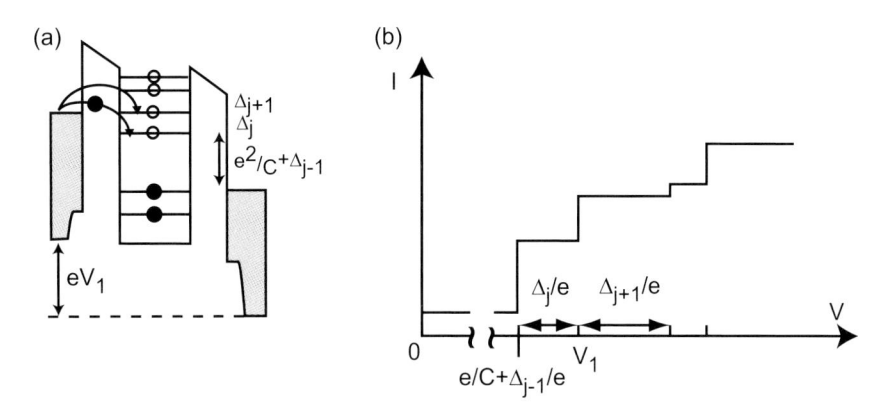

Figure 10.20: Spectroscopy on quantum dots by $I-V$ measurements. At the voltage V_1, two energy levels can carry current. Note that the voltage is large enough to overcome the Coulomb blockade. Tuning the voltage changes the number of current-carrying states, which is observed as steps in the $I-V$ characteristics (b).

quantum InAs dots embedded in GaAs have demonstrated for the first time experimentally that the single-particle level spacing Δ can be larger than the single electron charging energy [Drexler1994a]. Also, single-electron charging of individual atoms which are part of metal-organic molecules has been demonstrated by measurements on arrays of such molecules [Fischer1994]. Below, we will have a glance at experiments in which individual quantum dots are probed.

As pointed out in the previous chapter, the very first experiments related to single electron tunnelling have been performed on granular films of extremely small metal grains, which were embedded in an insulating matrix. The size of the Sn grains fabricated by Giaever and Zeller (Fig. 9.6), for example, are in principle small enough to observe discrete energy levels. As a rule of thumb, a grain radius of 10 nm suffices to observe size quantization in metals at a temperature of 100 mK. The challenge consists in contacting individual grains in order to avoid ensemble averaging. The length scale is clearly below the resolution limit of conventional lithographic techniques. In principle, one way to access individual grains is by contacting it with a scanning tunnelling microscope, like in the experiment of Fig. 9.9. Discrete energy levels in individual InAs nanocrystals have in fact been observed with this approach, see [Banin1999]. In [Ralph1995], the first transport experiment on single metallic quantum dots is reported. The authors used an ingenious fabrication technique, which combines self-assembly of nanometer-sized grains with conventional electron beam lithography, see Fig. 10.21. In a first step, a Si_3N_4 layer is etched. A patterned resist on top of the layer serves as etch mask. The etch is stopped at a point where a tiny hole of a few nanometers in diameter only has been formed. Subsequently, the bowl-shaped hole is filled with Al by thermal evaporation, and the Al is oxidized. Now a granular film of the metallic quantum dots to be investigated is evaporated on the bottom. This layer is subsequently covered by another oxide layer and a homogeneous Al electrode. By chance, one obtains devices this way where one grain sits right below the hole. A combination of self-assembly with the angle evaporation technique (Fig. 4.15) also produced working samples [Davidovic1999]. In none of the schemes, a gate electrode could be defined, such that up to now, all the information has been collected by current-voltage characteristics or by differential conductance measurements. Typical grain diameters range between 5 and 20 nm. The single electron charging energies can be estimated from the self-capacitance of a sphere, $C = 4\pi\epsilon\epsilon_0 r$, while measurements give values of $E_C \approx 5$ meV $-$ 50 meV. The single particle level spacings (they depend on the energy in three dimensions) and scale with the radius and the Fermi wave vector according to $\Delta \approx 2\pi^2\hbar^2/(m^* k_F r^3)$. Typical values of $\Delta \approx 0.1$ meV $-$ 1 meV have been measured. Gold particles, as well as CdSe particles, have been attached to leads also by a hybrid assembly method, which is based on a combination of angle evaporation with organic layer deposition by wet chemistry. In this scheme, two metal electrodes with a gap of a few nanometers are patterned by electron beam lithography and angle evaporation. Next, an organic molecule is deposited on the electrodes. The molecule 1,6 - hexanedithiol can be used for this purpose. It has the property to bind with one end to the electrodes, while the molecules are oriented perpendicular with respect to the surface. Hence, the electrodes are covered with a molecular monolayer this way. In a subsequent step, the nanoparticles are deposited on this substrate from a solution. They bind to the second, dangling end group of the monolayer molecules, and there is a good chance that one grain gets deposited in between the two electrodes. In this case, the organic molecules bound to the grain serve as tunnel barriers. This approach has been used in [Klein1996] and [Klein1997].

What can we learn from such experiments? First of all, it is no doubt of fundamental interest to investigate size quantization in metals. Up to now, Al [Ralph1995], Au [Davidovic1999], Cu and Ag [Petta2001], as well as Co [Gueron1999] grains have been investigated. Second, these experiments offer a variety of novel options not available in semiconductor quantum dots. The leads can be made superconducting [Ralph1995],for example, or energy levels of ferromagnetic grains like Co can be studied. In contrast to semiconducting quantum dots, the energy levels in metallic dots have been found to be spin-degenerate at $B = 0$, which suggests that exchange interactions are less important. Energy levels in Al grains show a remarkable clustering, while the effective g-factors in the grains are found to be reduced as compared to the bulk metal.

In contrast to metallic quantum dots, single electron transistors have been built both carbon nanotubes [Tans1997, Bockrath1997], as well as from the C_{60} fullerene [Park2000]. Since carbon nanotubes (CNs) are usually several microns long, it is possible to contact them by wires fabricated by conventional electron beam lithography. In several experiments, a suitable array of gold electrodes was defined on an insulating substrate, and the carbon nanotubes were deposited from a suspension on the surface. By chance, a CN makes contact with two electrodes, while a third metal finger can be used as a gate [Tans1997]. Since CNs are quasi-one dimensional structures, both E_C and Δ scale with $1/L$, where L denotes the length of the CN. Elementary considerations give $\Delta = h v_F/(4L)$, while $E_C \approx 5\Delta$, [Cobden2002]. A unique picture of the electronic properties of CNs has not yet emerged. While some experiments indicate deviations from a Fermi-gas behavior [Bockrath1999], others show excellent agreement with the constant interaction picture, including spin degeneracy and shell filling [Cobden2002].

10.5 Other types of quantum dots

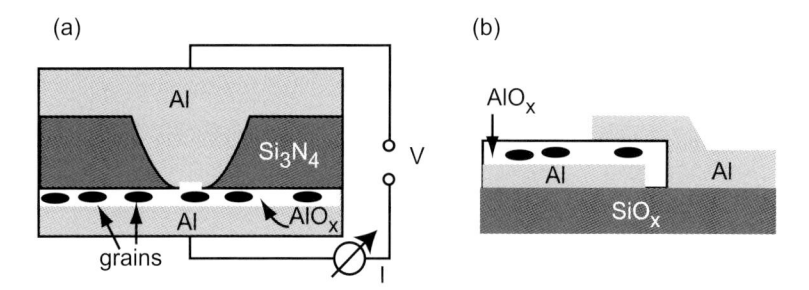

Figure 10.21: Experimental setups for measuring individual metallic quantum dots. Scheme (a) has been demonstrated by [Ralph1995]), while scheme (b) has been successful in [Davidovic1999].

The C_{60} quantum dot measured by [Park2000] not only shows an extremely large single electron charging energy of $0.27eV$, but also a vibrational excitation of the C_{60} molecule as its charge is altered. These samples were made by depositing a C_{60} in between two electrodes with a separation as small as 1 nm. This electrode geometry has been fabricated by passing a large current through a thin gold wire made be an angle evaporation technique. The current induces migration of the gold atoms in the wire, which finally breaks, and a gap in the regime of 1 nm opens up. This phenomenon is known as electromigration. The same kind of technique has also been used in [Park1999] to contact individual CdSe clusters. Both types of carbon-based quantum dot devices are the subject of ongoing research, and a lot of further fruitful scientific results can be expected here.

The $I - V$ characteristics of an individual molecule, namely of benzene-1,4-dithiol, has been measured by [Reed1997]. This molecule consists of a benzene ring with sulfur atoms at opposite corners. The sulfur atoms bind to two gold electrodes, which have been made by breaking a gold wire in a setup similar to that one shown in Fig. 7.21. Here, a Coulomb gap can be observed even at room temperature. However, the energy spectrum of the combined system, including the gold atoms in the leads the sulfur binds to, must be considered for an interpretation of such data.

This brief survey was supposed to show that scientists are about to establish novel fabrication techniques and assembly schemes to overcome the limitations posed by conventional lithographic techniques, and they learn how to attach wires and gates to particles as small as a nanometer. These are important steps towards routine operation of quantum dots at room temperature in electronic circuits, and towards molecular electronics.

Papers and Exercises

P 10.1: In [Tarucha1996], the density of states is reconstructed from the transmission resonances of a quantum dot. Describe the sample design and the reconstruction of the dot's energy spectrum. How do Hund's rules influence the data?

P 10.2: Evans et al. [Evans1993] have developed an electrostatic model for a quantum dot with a filling factor below 2. Explain the dot's "phase diagram" within this model!

P 10.3: The statistical properties of the transmission resonance *amplitudes* of quantum dots are discussed in [Folk1996]. What are the results?

P 10.4: Describe the experiment carried out by [Berman1999]!

P 10.5: In [Porath2000] a DNA quantum dot has been measured. Summarize the sample fabrication and the results.

E 10.1:

(i) Show that eqns. 10.4 and 10.5 are equivalent. What happens for $B \to \infty$?

(ii) Consider the Fock-Darwin spectrum for $\omega_0 \ll \omega_c/2$. Show that adjacent states with identical m have an energy spacing of $\Delta E = E_{m,p+1} - E_{m,p} \approx \hbar \frac{\omega_0^2}{\omega_c}$, and a spacing of $\Delta B \approx B \frac{\omega_0^2}{\omega_c}$.

(iii) Use these approximate expressions to analyze the data of Fig. 10.8!

E 10.2: In this exercise, the Wigner surmise shall be derived for a 2x2 Hamilton matrix in the orthogonal case.

(i) Consider the Hamiltionian of a chaotic system which, in some basis, can be written as

$$(H) = \begin{pmatrix} H_{11} & H_{12} \\ H_{12} & H_{22} \end{pmatrix}$$

Assume that the matrix has eigenvalues λ_+ and λ_-. Express these in terms of the matrix elements H_{ij}.

(ii) Each matrix element is supposed to be random, which -per definition- means that the probability for a certain matrix to occur can be written as a product of the probabilities for the matrix elements, i.e.,

$$p(H) = p_{11}(H_{11}) p_{12}(H_{12}) p_{22}(H_{22})$$

$p(H)$ further has to be invariant under orthogonal basis transformations, which means that

$$p(O^T H O) = p(H)$$

10.5 Other types of quantum dots

with

$$(O) = \begin{pmatrix} \cos\alpha & \sin\alpha \\ -\sin\alpha & \cos\alpha \end{pmatrix}$$

For our purposes, it suffices to consider only very small transformation angles $\alpha \ll 2\pi$.

Approximate O to first order in α, and derive a system of differential equations for $p_{ij}(H_{ij})$ (use the requirement that $O^T H O$ must be equal to H, independent of α). Solve the differential equations.

(iii) Show that by a suitable choice of the zero point of the energy, one can write

$$p(H) = c_1 exp(-Tr[H^2])$$

(iv) Substitute the variables in $p(H)$, such that it becomes a function of $\Delta = \lambda_+ - \lambda_-$, plus a second variable. Recall that the transformation law for probability densities, $p(y) = p(x)|\partial y/\partial x|$ generalizes to

$$p(y_1,\ldots y_n) = p(x_1,\ldots x_n)|det\left(\frac{\partial(y_1,\ldots y_n)}{\partial(x_1,\ldots x_n)}\right)|$$

where $\left(\frac{\partial(y_1,\ldots y_n)}{\partial(x_1,\ldots x_n)}\right)$ denotes the Jacobian matrix of the transformation. To determine it, consider a suitable transformation, that maps H onto its diagonal form.

(v) Finally, Integrate over the second variable that comes into play.

Further Reading

A good introduction to all aspects of quantum dot physics is given in the book by [Jacak1998]. A must-read related to the transport properties of quantum dots is [Kouwenhoven1997].

11 Mesoscopic Superlattices

In the previous chapters, we have been mostly studying individual quantum films, - wires, or - dots, respectively. A rich phenomenology emerges from packing these structures into periodic arrays. In section 6.2.3, for example, a stack of quantum films was used to investigate the behavior of the quantum Hall effect as the dimension is gradually changed from 2 to 3. Multilayers of epitaxially grown quantum films are fascinating objects, but beyond our scope here. In fact, the elementary Kronig-Penney potential, i.e., a periodic array of rectangular barriers, can be easily fabricated by molecular beam epitaxy. Such samples have been used to demonstrate fundamental effects, like the Wannier-Stark localization, or Bloch oscillations. The reader is referred to textbooks on solid state physics for further information, like, e.g. [Grosso2000].

In this short chapter, we focus on the most elementary mesoscopic phenomena that occur in lateral periodic structures, *lateral superlattices*, which are imposed onto a 2DEG by lithographic techniques. Here, the superlattice period is much larger than the lattice constant of the crystal, but comparable to mesoscopic length scales, in particular to the Fermi wavelength and the elastic mean free path.

11.1 One-dimensional superlattices

Lateral superlattices can patterned by holographic schemes, see Fig. 11.1, or by electron beam lithography. While the interference pattern of a laser generates extremely accurate periodicity, the superlattice constants a are subject to the usual limitations, which means $a \gtrsim 200$ nm. Smaller periods can be generated by electron beam lithography or scanning probe lithography, but the deviations from perfect periodicity become larger. If the one-dimensional superlattice on the sample surface is used to impose a *weak* density modulation in the 2DEG, novel magneto-resistivity oscillations are observed in ρ_{xx}, i.e., perpendicular to the modulated (y) direction [Weiss1989], [Winkler1989]. At first sight, they resemble Shubnikov - de Haas oscillations, see Fig. 11.2. They occur only at small magnetic fields and vanish for cyclotron radii above the superlattice period. Note that in Fig. 11.2, the Shubnikov - de Haas oscillations set in at $B = 0.4$ T. Like Shubnikov - de Haas oscillations, these additional oscillations are periodic in $1/B$. Essentially no effect is observed in y-direction. These oscillations are known as *Weiss oscillations*.

It seems strange that a density modulation of a couple of percent only is able to produce such a strong modification of ρ_{xx}. Theoretical considerations revealed that the Weiss oscillations can be understood in terms of a resonant drift of the cyclotron orbits induced by

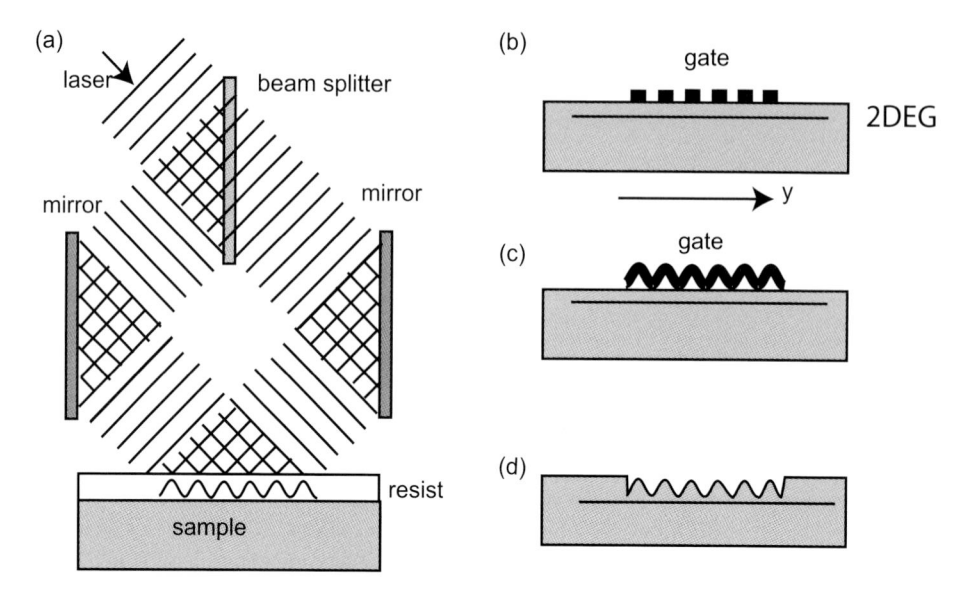

Figure 11.1: Sketch of various techniques to define a periodic lateral superlattice. A photoresist is illuminated by the periodically modulated intensity of the pattern that emerges from the interference of two partial laser beams (a). The resist can be used as a mask for a lift-off process (b), to modulate the distance between the 2DEG and a homogeneous top gate (c), or simply as an etch mask (d).

the electric field of the superlattice. The drift can be calculated by treating the superlattice electric field as a perturbation to the Hamiltonian for the cyclotron motion of free electrons [Winkler1989], [Gerhardts1989]. The effect can be also understood in a semiclassical picture, by considering the total $\vec{E} \times \vec{B}$-drift an electron experiences during one complete cyclotron orbit [Beenakker1989]. We assume $r_C \gg a$. At certain cyclotron radii, this drift will average to zero. At slightly different magnetic fields, however, the drift will average out along those parts of the cyclotron trajectory, where the electron moves in y-direction and thus crosses many modulation periods. A large integrated $\vec{E} \times \vec{B}$-drift is collected for the motion in x-direction, though. If this drift has the same sign for the electron motion in positive and in negative x-direction, a net drift will remain for a complete cyclotron motion. The result from the calculation in the limit of $r_C \gg a$ reads

$$\rho_{xx} \propto \frac{V_{sl}^2 B}{a E_F^{5/2}} \cos^2\left[\frac{2\pi r_C}{a} - \frac{\pi}{4}\right] \tag{11.1}$$

The expression predicts minima in ρ_{xx} for $r_C/a = (4j+3)/8$, and maxima at $r_C/a = (4j+1)/8$, where j is an integer including zero. The corresponding cyclotron orbits are sketched for $j=2$ in Fig. 11.2. Furthermore, it predicts that the oscillations are periodic in $1/B$, that the oscillation amplitude increases as B is increased, and that the oscillations vanish for $r_C < a$, in good agreement with the experiment. This result holds only for $r_C \gg a$, however. It emerges from approximating the Bessel function $J_0(x) \approx \sqrt{2/(\pi x)} \cos[x - \pi/4]$, which is a good approximation for large arguments only. In Fig. 11.2, the correct theoret-

11.2 Two-dimensional superlattices 279

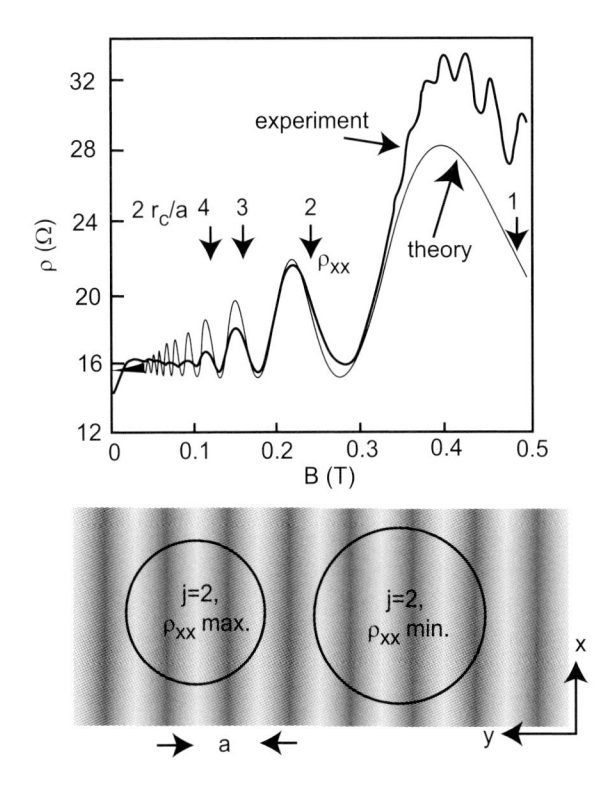

Figure 11.2: Top: measured and calculated longitudinal magnetoresistivities ρ_{xx} of a 2DEG with a density modulated in y - direction. Oscillations are observed in ρ_{xx} at small magnetic fields. The Shubnikov - de Haas oscillations are visible above $B = 0.4$ T (adapted from [Beenakker1989]). The resistivity in y-direction remains essentially unaffected by the superlattice. Bottom: cyclotron orbits in a maximum (left) and in a minimum (right) of ρ_{xx}. The gray scale indicates the electric field strength in y-direction.

ical expression containing the Bessel functions is compared to the experimental data. The agreement is excellent. It should be noted that the theories assumes a sinusoidal superlattice potential. Other potential shapes will give phases which differ form $\pi/4$ in (11.1), without changing the overall appearance.

11.2 Two-dimensional superlattices

After the previous discussion, it is self-evident to ask what happens to 2DEGs with potential modulations in both directions. Corresponding gate structures can be easily made by adopting the schemes mentioned in relation to Fig. 11.1, for example by performing a second illumination with the interfering laser beam, after rotating the sample by 90 degrees. We first focus on *antidot lattices*, which are two-dimensional arrays of holes in a 2DEG. In a simple semiclassical picture, one would expect that there exists a set of discrete magnetic fields at

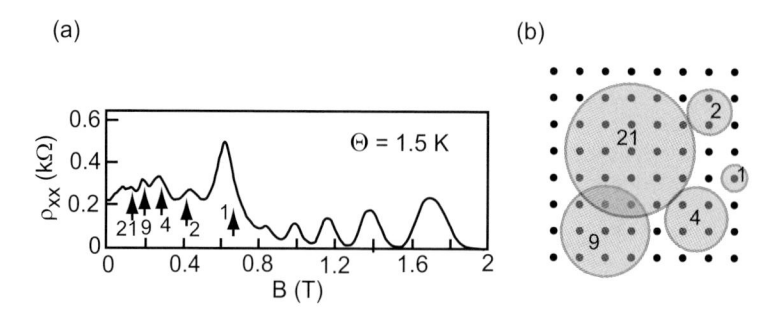

Figure 11.3: Longitudinal magnetoresistivity (a) of a square antidot lattice (full circles in (b)). Peaks in ρ_{xx} are observed if the cyclotron orbit fits in between the antidots. The arrows and numbers in (a) denote the magnetic fields at which the cyclotron orbit is commensurate with the antidot lattice, and the number of antidots enclosed in the cyclotron orbit, respectively. The lattice period in this experiment was 300 nm. Adapted from [Weiss1991].

which the cyclotron orbits fit in the lattice without hitting a scatterer. These orbits are called *commensurate* with the lattice. For such magnetic fields, the antidot lattice localizes the electrons, and we expect an increased resistivity. As can be easily verified, for a square lattice, this is possible for cyclotron orbits that enclose 1,2,4,9,21,... antidots. This behavior has been in fact observed in the longitudinal resistivity, see Fig. 11.3. A closer look, however, reveals that this is not the end of the story. First of all, the resistance maxima are not exactly at the expected positions. Also, a *negative* Hall resistance is observed [Weiss1991], which cannot be understood in this simple picture. We take these issues as an example to introduce a semiclassical tool for numerical conductance simulations. It is based on the Einstein relation and the Kubo formula, see section 5, and has turned out to be particularly successful in explaining the behavior of antidot lattices [Fleischmann1992].

In a generalized version of eq. (5.10), the Kubo formula tells us that the components of the conductivity tensor can be calculated by averaging the velocity correlation functions according to

$$\sigma_{ij} = \frac{ne^2}{k_B \Theta} \int_0^\infty e^{-t/\tau} C_{v_i v_j}(t) dt \tag{11.2}$$

Here, the term $e^{-t/\tau}$ is included, which models the momentum relaxation after the scattering time τ within the relaxation time approximation. In a numerical simulation, electrons are randomly placed within a unit cell of the antidot lattice. Their velocity vector has the amplitude of the Fermi velocity, and points in an arbitrary direction within the xy-plane. Of course, points where the antidot potential is above the Fermi level are no starting points. The motion of an ensemble of electrons in the antidot potential and the magnetic field is calculated. The antidot potential can be approximated by

$$V(x,y) = V_0 \left[\cos(\pi x/a) \right]^{2\beta} \left[\cos(\pi y/b) \right]^{2\beta\gamma} \tag{11.3}$$

11.2 Two-dimensional superlattices

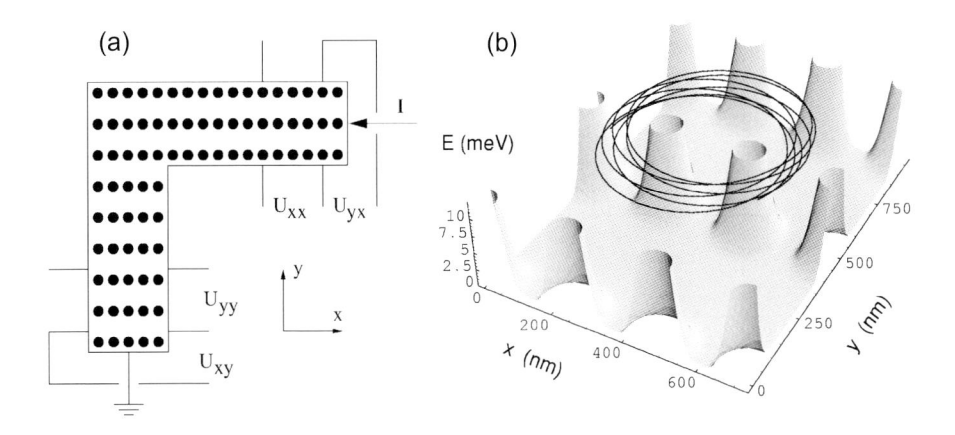

Figure 11.4: (a): Hall bar geometry for measuring the longitudinal resistivities of an antidot lattice with a lattice constant of $a = 240$ nm in x-direction, and $b = 480$ nm in y-direction. In (b), an electron trajectory for $r_C = a/2$ is shown. The model potential is given by a steepness parameter of $\beta = 2$, while V_0 and γ are chosen such that the antidot potential peaks out of the Fermi sea in circles with a radius of 43 nm. After [Rychen1998].

where β tunes the steepness of the antidot walls, while γ is a correction parameter that is adjusted such that the potential peaks are approximately circular. After the time τ, the direction of the velocity can randomized numerically, in order to take residual scattering at random positions into account. In Fig. 11.4, a sample layout, its model potential and a sample trajectory are shown.

After a time which is large compared to τ and after the trajectories of a sufficient number of electrons has been calculated, the average velocity correlation function, and hence the components of the conductivity tensor, can be determined. In Fig. 11.5 (a), the result of a simulation of σ_{xx} and σ_{yy}, for the sample of Fig. 11.4, is compared to these conductivity components as determined from the measured components of the resistivity tensor. While pronounced peaks in σ_{xx} are observed, σ_{yy} essentially drops smoothly towards zero as the magnetic field is increased. At magnetic fields indicated by A and C, the conductivity is approximately isotropic, while at B and D, they differ up to a factor of ≈ 25. For simulation parameters chosen as stated in the figure, the numerical results are in good agreement with the experiments. The structures in the magnetoconductivity can be traced back to certain typical trajectories that occur at the corresponding magnetic fields, see Fig. 11.5 (b). On the peaks of σ_{xx}, the antidot lattice tends to channel the electrons along the x-direction (B and D). At C, where the conductance is almost isotropic, the antidots to not favor a certain direction of motion. At large magnetic fields (D), The electrons get caught at the antidots, and the overall conductance is very low. Nevertheless, it is easier for them to diffuse along the x-direction, the direction with a *higher* density of scatterers.[1] This trend can also be seen in

---

[1] Likewise, it can be shown that for small magnetic fields, the antidots deflect the electrons predominantly in the direction opposite to the direction of the $\vec{E} \times \vec{B}$ drift, which explains the mentioned negative Hall effect,

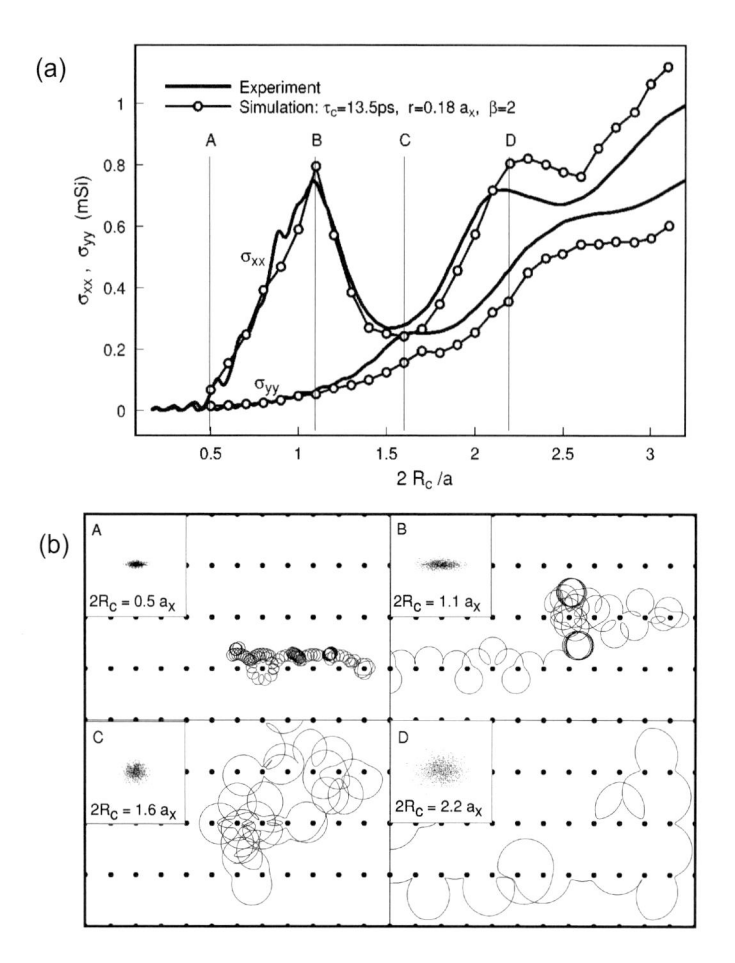

Figure 11.5: (a): comparison of σ_{xx} and σ_{yy} as measured on the sample sketched in Fig. 11.4(a), compared to numerical simulations with the potential shown in Fig. 11.4(b). Letters $A - D$ indicate magnetic fields for which tyical trajectories are shown in (b). Here, the insets show the positions of 1000 electrons after a time of $10\tau = 135$ ps in an area of 24×24 lattice periods. After [Rychen1998].

plots of the diffusing cloud of electrons (insets in Fig. 11.5 (b)). Since the electron density was $n = 3.6 \cdot 10^{15}$ m^{-2} and the scattering time is $\tau = 13.5$ ps, the electrons would diffuse (see chapter 5) ≈ 4.5 μm in the simulation time of 135 ps, which roughly corresponds to the areas of the insets in Fig. 11.5 (b). The antidots thus hamper the diffusion significantly, although they represent a periodic potential. The elementary cell of a square antidot lattice represents a Sinai billiard (see chapter 10), such that quantized chaos can be expected to be present. Are there signatures of quantized chaos in antidot lattices? The answer is yes, and it can be proved by such semiclassical simulations as well. It turns out that some of the trajectories behave

[Fleischmann1994].

11.2 Two-dimensional superlattices

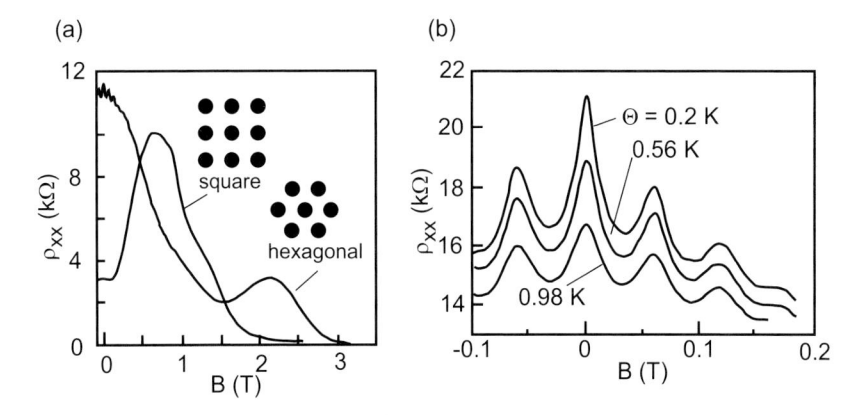

Figure 11.6: (a): longitudinal magnetoresistivities in square and hexagonal antidot lattices (the lattice geometries are shown in the inset). (b): enlargement of ρ_{xx} for the hexagonal lattice around $B = 0$, which oscillates with a period of $\Delta B = h/2eA$, where A is the average area of one antidot. The strong temperature dependence indicates a phase-coherent origin. After [Nihey1995].

chaotic, while others show regular behavior, in the sense of section 10.3. The character of the trajectory is determined by its initial conditions, i.e. the position within the unit cell and the magnitude and direction of its velocity. Since this represents a position in phase space, antidot lattices are said to have a mixed phase space, since it segregates into chaotic seas and stable islands, see, e.g. [Fleischmann1992].

In hexagonal antidot lattices, additional oscillations can be observed, which are periodic in B, see Fig. 11.6 [Nihey1995]. The strong temperature dependence of these oscillations is in contrast to the very weak temperature dependence of the commensurability peaks and indicates a phase coherent origin. This observation is thus explained in terms of an enhanced classical backscattering probability in hexagonal lattices as compared to that one in square, or rectangular lattices. The resulting enhanced backscattering due to phase coherence generates Altshuler-Aronov-Spivak oscillations (see chapter 8) of significant amplitude.

So far, we have treated antidot lattices semiclassically. We conclude our excursion to lateral superlattices by a glance at the interesting quantum mechanics of a weak, two-dimensional superlattice with a square geometry. In the limit of a free electron gas in a magnetic field, highly degenerate Landau levels are obtained, see chapter 6. The electrons move in cyclotron orbits, the wave functions are thus rotational invariant. In a periodic potential modulation and in the absence of a magnetic field, the wave functions are Bloch functions and have the corresponding discrete translational invariance. The combination of magnetic fields and periodic potentials leads to a very rich phenomenology,[2] which up to now, however, has not been observed experimentally, although first indications have been seen, see below. The effect of a weak periodic potential on the Landau levels can be studied both theoretically by a perturbation calculation, as well as experimentally in modulated 2DEGs. Turning on the (weak) periodic potential will lift the degeneracy of the Landau levels and generate Landau bands, with a width that should be proportional to the amplitude of the potential. In Fig. 11.7, such

---
[2] See [Pfannkuche1992] for an introduction to this problem.

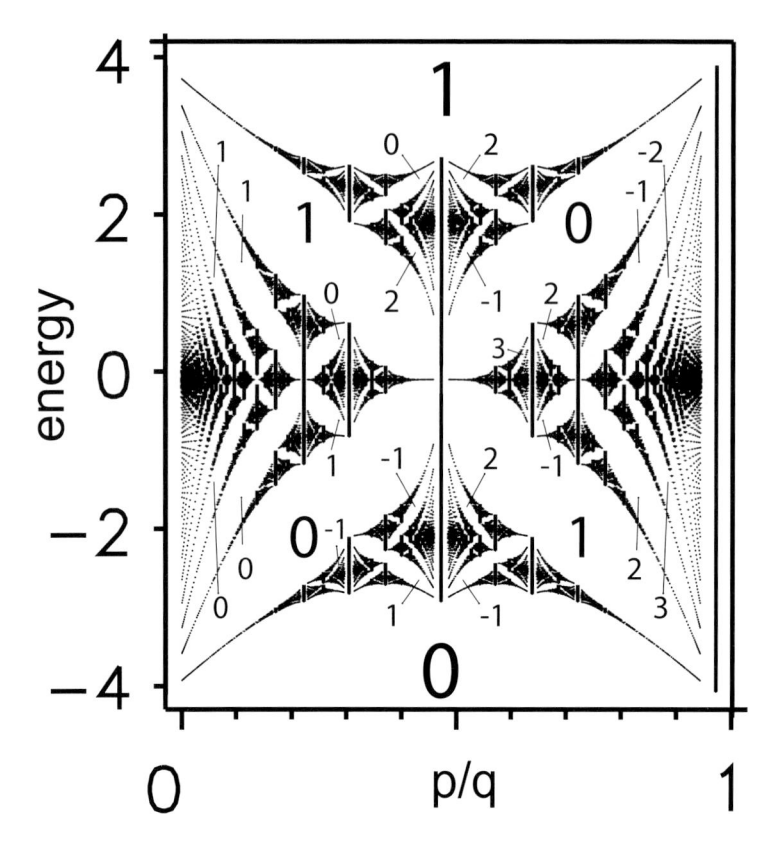

Figure 11.7: Calculated energy spectrum for one Landau band in a square antidot lattice. The ratio $p/q \propto 1/B$ measures the number of flux quanta h/e in units of $B \cdot A$, where A denotes the area of the unit cell. The numbers inside the figure indicate the value of the Hall resistance in the corresponding minigaps, in units of e^2/h. After [Springsguth1997].

a Landau band emerging from the lowest Landau level is shown for a modulation potential of

$$V(x,y) = \frac{V_0}{4}\Big[\cos(2\pi x/a) + \cos(2\pi y/a)\Big]$$

In this figure, the coordinate p/q gives the number p of flux quanta that penetrate q unit cells of the lattice, i.e., $p/q = h/(a^2 eB)$, while the energy is plotted in units of V_0.

This energy spectrum has a fractal structure, which means it is self-similar no matter at what scale we look at it. It is symmetric around $p/q = 0.5$ and repeats itself periodically at smaller magnetic fields. Higher Landau levels have the same appearance, although their energies are scaled. Due to its appearance, this spectrum is known as the *Hofstadter butterfly* [Hofstadter1976].

In an experiment, we would thus expect to observe additional structures in the longitudinal resistivity, namely in the peaks of the Shubnikov-de Haas oscillations. Furthermore, it has been shown theoretically that the Hall conductance jumps irregularly as a function of the magnetic field within one Landau band, while the height of the jumps is quantized in steps of e^2/h [Thouless1982]. The heights of these jumps are indicated in Fig. 11.7. Investigating the Hofstadter butterfly experimentally has turned out to be very challenging, because the required ranges for the parameters are quite small and hard to meet. First of all, residual disorder as well as disorder deviations from perfect periodicity in the superlattice will average out the fine structure. Second, in order to increase the widths of the butterfly's energy gaps, a larger modulation V_0 is desirable. This, however, couples the different Landau bands. As has been shown by [Springsguth1997], even a weak inter-Landau band coupling already changes the spectrum significantly. Screening can cause further modifications. Nevertheless, some signatures of the butterfly have been observed, both in the Shubnikov-de Haas oscillations[Schlosser1996, Albrecht2001], as well as in the Hall conductance [Albrecht2001].

Papers

P 11.1: In [Schuster1997], rectangular antidot lattices with $a \ll b$ have been investigated. Discuss the observations and relate them to the experiments on quantum wires with non-specular walls (section 7.1.2)! Compare the results also to those of section 11.1.

P 11.2: Arrays of antennas in microwave fields are very similar to electrons in periodic superlattices. Discuss this analogy; use [Kuhl1998] as an example.

P 11.3: [Weiss1993] found magneto-oscillations in antidot lattices which are *periodic in B*. Explain their origin!

Appendices

A SI and cgs Units

Some people use the cgs (Gaussian) unit system, while others prefer the SI system (also known as MKSA system). This results in both a constant source of irritation for students as well as an inconvenience for researchers. The following remarks should allow the reader to switch between them with confidence.

The SI and cgs systems originate in a different choice of units in equations containing electrodynamic quantities. Table A.1 lists how the prefactors must be replaced as the unit system is changed.

Table A.1: Prefactors in cgs and SI units

Quantity	cgs	SI
speed of light	c	$1/\sqrt{\epsilon_0\mu_0}$
electric field	$\vec{E}$	$\sqrt{4\pi\epsilon_0}\,\vec{E}$
dielectric shift	$\vec{D}$	$\sqrt{4\pi/\epsilon_0}\,\vec{D}$
polarization	$\vec{P}$	$(1/\sqrt{4\pi\epsilon_0})\vec{P}$
magnetic field	$\vec{B}$	$\sqrt{4\pi/\mu_0}\,\vec{B}$
magnetizing field	$\vec{H}$	$\sqrt{4\pi\mu_0}\,\vec{H}$
magnetization	$\vec{M}$	$\sqrt{\mu_0/4\pi}\,\vec{M}$
dielectric constant	ϵ	ϵ/ϵ_0
permeability	μ	μ/μ_0
current	I	$(1/\sqrt{4\pi\epsilon_0})I$
resistance	R	$4\pi\epsilon_0 R$
inductance	L	$4\pi\epsilon_0 L$
capacitance	C	$1/(4\pi\epsilon_0)C$

Examples:

- In cgs units, the generalized momentum is given by $\vec{p} + \frac{e}{c}\vec{A} = \vec{p} - \frac{e}{2c}\vec{r} \times \vec{H}$. In order to switch to SI units, we replace $e \to e/\sqrt{4\pi\epsilon_0}$, $\vec{H} \to \sqrt{4\pi\mu_0}\vec{H}$, and $c \to 1/\sqrt{\epsilon_0\mu_0}$, such that $\vec{p} - \frac{e}{2}\vec{r} \times \mu_0\vec{H}$.

- In cgs, the Bohr magneton reads $\mu_B = \frac{e\hbar}{2mc}$. This changes to $\mu_B = \frac{e\hbar}{2m}$, by replacing the magnetization, the charge and the speed of light.

Occasionally, quantities have to be transformed as well. Table A.2 lists the most important transformation factors.

Table A.2: Numerical factors in cgs and SI units

cgs	SI
1 cm	0.01 m
1 g	10^{-3} kg
1 dyn	10^{-5} N
1 erg	10^{-7} J
1 esE	$(1/3) \cdot 10^{-9}$ Cb
1 statvolt	300 V
1 cm	$(1/9) \cdot 10^{-11}$ F
1 Mx	10^{-8} Wb
1 G	10^{-4} T
1 Oe	$(1/4\pi) \cdot 10^{3}$ A/m

Q A.1: Sodium has an electric polarizability of $0.4 \cdot 10^{-24}$ cm^3. Express this quantity in SI units!

B Correlation and Convolution

B.1 Fourier transformation

The Fourier transform of a function $f(x)$ is the continuous version of its expansion into a Fourier series, namely

$$F(X) = \frac{1}{\sqrt{2\pi}} \int_{-\infty}^{\infty} f(x) e^{i2\pi Xx} dx \tag{B.1}$$

and, respectively,

$$f(x) = \frac{1}{\sqrt{2\pi}} \int_{-\infty}^{\infty} F(X) e^{-i2\pi Xx} dX \tag{B.2}$$

The units of the variables are inverse to each other. Fourier transformations are frequently used, due to their efficiency and versatility in performing certain analytical tasks. Examples can be seen in chapters 2, 8, 14, as well as below.

B.2 Convolutions

The convolution of two functions $f(x)$ and $g(x)$ is defined as

$$h(x) = f * g(x) = \int_{-\infty}^{\infty} f(\xi) g(x - \xi) d\xi \tag{B.3}$$

The effect of the convolution is to "smear out" $f(x)$ with $g(x)$. This is illustrated in Fig. B.1 Here, we convolute

$$f(x) = \theta(x - 1) - \theta(x - 2)$$

with

$$g(x) = \theta(x) - \theta(x - 1)$$

Hence, $g(x - \xi) = \theta(x - \xi) - \theta(x - \xi - 1)$. One finds for the convolution

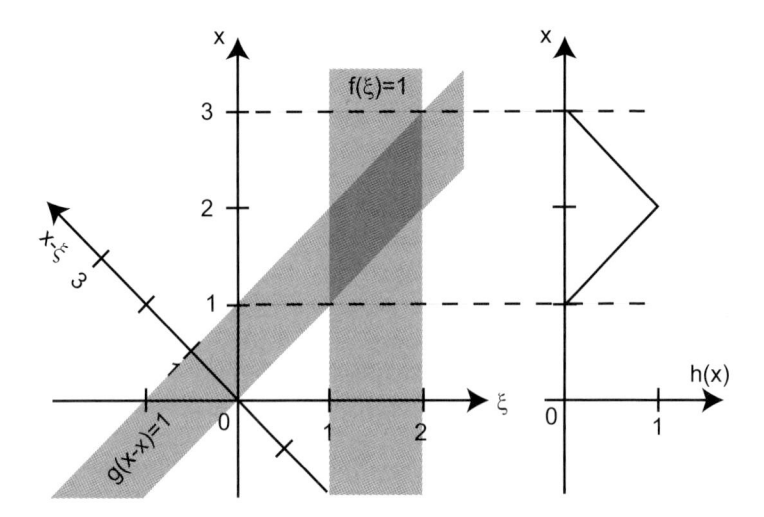

Figure B.1: Graphical representation of a convolution. The functions $f(x) = \theta(x-1) - \theta(x-2)$ and $g(x - \xi) = \theta(x - \xi) - \theta(x - \xi - 1)$ are drawn in the (x, ξ) plane in gray scale. Note that the $x - \xi$-axis is rotated with respect to the ξ axis by 135 degrees. For these functions, the convolution $h(x)$ is given by the extension of the overlapping area of $f(\xi)$ and $g(x - \xi)$ parallel to the x-axis, as sketched to the right.

$$h(x) = \begin{cases} 0 & x < 1 \\ x - 1 & 1 \leq x < 2 \\ 3 - x & 2 \leq x < 3 \\ 0 & x \geq 3 \end{cases}$$

The convolution theorem states that

$$H(X) = F(X)G(X) \tag{B.4}$$

i.e., the Fourier transform of the convoluted functions is the product of the Fourier transforms of the individual functions. This is useful for a process called deconvolution. Suppose we know that a signal, e.g., a QPC characteristics, is thermally smeared. We can then obtain the characteristics at $\Theta = 0$ by numerically Fourier transforming the measured data, divide it by the Fourier transform of the derivative of the Fermi function, and transform back the result.

B.3 Correlation functions

In mathematical terms, the correlation function of the two functions $f(x)$ and $g(x)$ is defined as

$$C_{fg}(x) = \int_{-\infty}^{\infty} f(\xi)g(x + \xi)d\xi \tag{B.5}$$

B.3 Correlation functions

For $f = g$, we speak of the autocorrelation function $C_f(x)$.

In mesoscopics, this notation is frequently used for the correlation function of the fluctuations around an average, i.e.,

$$C_{fg}(x) = \langle \delta f(\xi) \delta g(\xi + x) \rangle \tag{B.6}$$

with $\delta f(\xi) = f(\xi) - \langle f(\xi) \rangle$. The brackets denote the ensemble average. This means that

$$C_{fg}(x) = \lim_{N \to \infty} \frac{1}{N} \sum_{j=1}^{N} \delta f_j(\xi) \delta g_j(\xi + x)$$

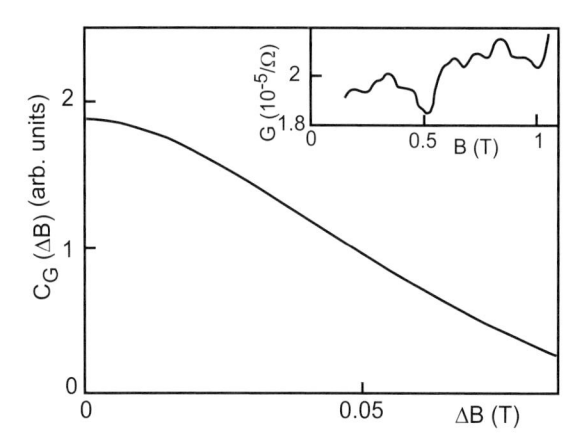

Figure B.2: The autocorrelation function (labelled here as F) of the conductance $G(B)$ of a quantum wire as a function of a magnetic field (the raw data are shown in the inset). The shape is typical for autocorrelation functions of experimental parametric fluctuations. The autocorrelation field is $B_c \approx 50mT$. After [Beenakker1987].

with j enumerating the N ensembles. Since, however, we assume that the system is ergodic, we can also average over the parameter. Assuming that the variable is continuous, the correlation function is then obtained by

$$C_{fg}(x) = \lim_{\xi_0 \to \infty} \frac{1}{\xi_0} \int_0^{\xi_0} \delta f(\xi) \delta g_j(\xi + x) d\xi \tag{B.7}$$

This, by the way, also indicates how the correlation function is obtained for finite measured intervals of the parameter ξ, which can be a magnetic field, a gate voltage, or the time, for example. $C_{fg}(x)$ effectively compares f with g and measures the degree of similarity. g is shifted with respect to f along the x-axis, and the product function is integrated. Therefore, the autocorrelation function of a fluctuating function has a characteristic structure. For very small shifts, the original and the shifted function have approximately identical values at each x. Almost everywhere, both functions have the same sign. For large shifts, however, the signs of the two curves are no longer correlated, and the average area under the product function averages to zero. Consequently, $C_f(x)$ will drop to zero within the generalized correlation length x_c. Usually, it is defined as the value of x where $C_f(x)$ has dropped to $1/e$ (e=2.71828) of $C_f(0)$, although sometimes, different definitions are used, which, however, do not change the order of magnitude. Note that this continuous drop occurs only for random fluctuations. An oscillatory function, for example, also has an oscillatory autocorrelation function. Generally speaking, x_c becomes smaller as the bandwidth of the fluctuations increases. For $x = 0$, the autocorrelation function is simply the variance:

$$C_f(0) = \langle (\delta f(\xi))^2 \rangle \tag{B.8}$$

The Wiener-Khintchine theorem states that the spectral power $S_f(X)$ is just twice the Fourier transform of the autocorrelation function of $f(x)$:

$$S_f(X) = 2C_f(X) \tag{B.9}$$

Furthermore, the variance must be the spectral power, integrated over all X:

$$\langle (\delta f(\xi))^2 \rangle = \int_{X=0}^{\infty} S_f(X) dx \tag{B.10}$$

C Capacitance Matrix and Electrostatic Energy

In this appendix, the electrostatic energy of a system of conductors is calculated. Consider a system of $n+m$ conductors, with n islands (floating conductors) and m electrodes (connected to voltage sources). As in chapter 9, the charge distribution is given by the charges at the

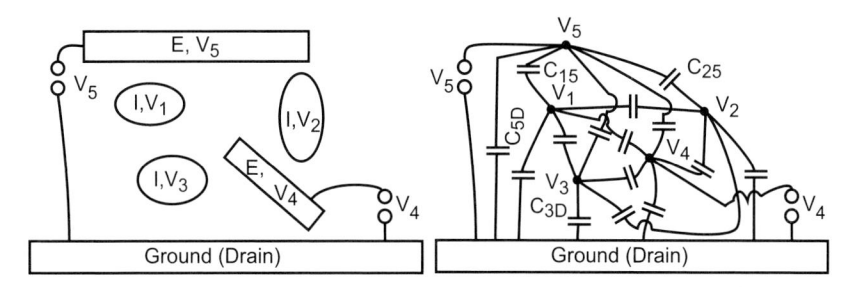

Figure C.1: Left: a system of islands (floating) and electrodes (connected to voltage sources). Right: the corresponding equivalent circuit composed of potential nodes and mutual capacitances.

electrodes, which can be written as a charge vector $\vec{q} = (\vec{q_I}, \vec{q_E})$ composed of the island charge vector $\vec{q_I}$ and the electrode charge vector $\vec{q_E}$. Similarly, a potential vector can be constructed: $\vec{V} = (\vec{V_I}, \vec{V_E})$, with V_i being the potential of conductor i with respect to drain (ground). The charge at each conductor is given by the potentials of all conductors. This can be expressed by

$$q_i = \sum_{j=1}^{n+m} d_{ij} V_j \tag{C.1}$$

with coefficients d_{ij} determined by the electric field distribution [Simonyi1963]. We would like to express this relation in terms of capacitance coefficients, which express the effect of conductor j on the charge at conductor i as a function of the voltage between the two conductors:

$$q_i = \sum_{j=1}^{n+m} C_{ij}(V_i - V_j) + C_{iD}(V_i - V_D) \tag{C.2}$$

Since $V_D = 0$, we can rearrange the sum in this equation. In order to express all d_{ij} in terms of C_{ij}, we write (C.1) as

$$q_i = \sum_{\substack{j \neq i}}^{n+m} -C_{ij} V_j + \sum_{\substack{j \neq i}}^{n+m} (C_{ij} + C_{iD}) V_i$$

Thus, it is immediately obvious that $d_{ij} = -C_{ij}$ for $i \neq j$, and that $d_{ii} = C_{iD} + \sum_{k=1, k \neq i}^{n+m} C_{ik}$.

This allows us defines the capacitance matrix $\underline{C}$ by

$$\vec{q} = \underline{C} \vec{V} \tag{C.3}$$

The coefficients of the capacitance matrix are given by

$$(\underline{C})_{ij} = \begin{cases} -C_{ij} & i \neq j \\ C_{iD} + \sum_{k=1, k \neq i}^{n+m} C_{ik} & i = j \end{cases}$$

So far, there has been no distinction between electrodes and islands. The definition of an ideal voltage source requires that the potential of electrodes is constant, no matter what. If, for example, an electron tunnels from electrode k to an island, the potential of the electrode must not change. The voltage source has to do some work to replace the electron. For the islands, this looks as if the electrode has an infinitely large capacitance with drain, $C_{kD} \to \infty \Rightarrow C_{kk} \to \infty$: no matter how the island potentials change, this will not modify the electrode potential.

We will proceed by applying this formalism to a single electron tunneling circuit. We are interested in studying how the electrostatic energy changes as $\vec{V}_E$ changes, which may induce charge transfers across "leaky" capacitors, i.e., those capacitors that allow tunneling. Since the voltage sources represent electron reservoirs, the electrostatic energy is in fact a free energy, given by the total energy stored in the system, minus the work W done by the voltage sources. Typically, it is convenient to specify the initial state by $\vec{q}_I$ and $\vec{V}_E$. A transition to a different state can be characterized by a change of the charge vector, $\Delta \vec{q}$. We therefore write

$$\Delta E[\vec{V}_E, \vec{q}, \Delta \vec{q}] = \frac{1}{2} (\vec{q} + \Delta \vec{q})(\underline{C})^{-1}(\vec{q} + \Delta \vec{q}) - \frac{1}{2} \vec{q}(\underline{C})^{-1} \vec{q} - \Delta W =$$

$$\Delta \vec{q} \underline{C}^{-1} \vec{q} + \frac{1}{2} \Delta \vec{q} \underline{C}^{-1} \Delta \vec{q} - \Delta W \tag{C.4}$$

Inverting $\underline{C}$ and approximating $C_{kD} = \infty$, $1/C_{kk} = \infty$ results in

$$\underline{C}^{-1} = \begin{matrix} 1 \\ : \\ n \\ n+1 \\ : \\ n+m \end{matrix} \begin{pmatrix} C_{II}^{-1} & 0 \\ & \\ 0 & 0 \end{pmatrix}$$

Hence, only the $n \times n$- submatrix of $\underline{C}^{-1}$ that describes inter-island coupling contains non-vanishing elements. Note that $(\underline{C}^{-1})_{II} = \underline{C}_{II}^{-1}$. The energy difference in eq. (C.4) is thus independent of $\vec{q_E}$. This equation now reads

$$\Delta E[\vec{V_E}, \vec{q_I}, \Delta \vec{q}] = \Delta \vec{q_I} \underline{C}_{II}^{-1} \vec{q_I} + \frac{1}{2} \Delta \vec{q_I} \underline{C}_{II}^{-1} \Delta \vec{q_I} - \Delta W \tag{C.5}$$

It remains to calculate the work done by the voltage sources as the charge vector is changed. This work is composed of two components:

1. As the charge of one island i changes, all islands that couple to island i will change their potentials accordingly, and their potential difference to an electrode k will change as well. In order to keep V_k constant, the voltage source connected to it has to take care of the charge changes influenced at electrode k. The work done is given by $\Delta W_{k,1} = \sum_{j=1}^{n} \Delta q_k V_k = \sum_{j=1}^{n} \Delta V_j C_{jk} V_k$, where $\Delta V_j = \sum_{i=1}^{n} (\underline{C}_{II}^{-1})_{ij} \Delta q_i$. The work done by all voltage sources can therefore be written as

$$\Delta W_1 = \Delta \vec{q_I} \underline{C}_{II}^{-1} \underline{C}_{IE} \vec{V_E} \tag{C.6}$$

The capacitance coefficients between islands and electrodes form the matrix $\underline{C}_{IE}$.

2. Some electrodes may be connected to some islands via tunnel junctions. In case electrons tunnel between electrode k and an island, the voltage source has to neutralize this charge change Δq_k, which requires the work $\Delta W_{k,2} = -\Delta q_k V_k$. Note that such a process will also change the charge configuration at the islands, and the voltage source will therefore have to perform the corresponding work $\Delta W_{k,1}$ in addition. The contribution to the work done by all voltage sources due to such tunnel processes is given by

$$\Delta W_2 = -\vec{q_E} \vec{V_E} \tag{C.7}$$

Therefore, the total work done by the voltages sources in response to a change in the island charge configuration is given by

$$\Delta W = \Delta W_1 + \Delta W_2 \tag{C.8}$$

We therefore obtain the final result (eq. 9.4)

$$\Delta E[\vec{V_E}, \vec{q_I}, \Delta \vec{q}] = \Delta \vec{q_I} \underline{C}_{II}^{-1} [\vec{q_I} + \frac{1}{2} \Delta \vec{q_I} - \underline{C}_{IE} \vec{V_E}] + \Delta \vec{q_E} \vec{V_E}$$

D The Transfer Hamiltonian

Occasionally, the assumption that the energies before and after the tunnel event are identical, cannot be made. This may be due to tunnelling induced by Photons or phonons, or due to electron electron interactions, as in chapter 9. In such cases, the transfer Hamiltonian model is useful, which is based upon time-dependent perturbation theory.

The problem can be elegantly dealt with within the so-called transfer Hamiltonian model. We start from two electron gases separated by an impenetrable barrier. Now, a time-dependent perturbation Hamiltonian is considered, which allows transfer of electrons across the barrier. Time-dependent perturbation theory shows that the transfer rate can be described with Fermi's golden rule, which we consider in the static limes:

$$\Gamma_{i \to f} = \frac{2\pi}{\hbar} |\langle i|H_t|f\rangle|^2 \delta(E_f - E_i) \tag{D.1}$$

This is just the transmission probability per unit time for a single electron in state $|i\rangle$, with energy E_i, to be transferred into state $|f\rangle$, with energy E_i, to the other side of the barrier. The δ-function ensures an elastic event. In order to relate the transfer rate to a current at a voltage drop V across the barrier, we have to consider:

1. the electron density in $[E_i, E_i + dE_i]$, given by the density of states $D_i(E_i)$ times the Fermi-Dirac distribution $f_i(E_i)$. Here the index i denotes the side of the barrier that hosts state i.

2. Since we are dealing with fermions, the electron can tunnel only into an empty state $|f\rangle$. The transfer rate for an electron in $|i\rangle$ will thus be proportional to $D_f(E_f) \cdot [1 - f_i(E_f)]$.

3. Electrons can tunnel in both directions. The current is the sum of the two partial currents in opposite directions.

Let us assume the voltage drop is from left to right. The spectral current at energy E is given by

$$I(E) = e\frac{2\pi}{\hbar}|\langle i|H_t|f\rangle|^2 (D_l(E)f(E - E_F)D_r(E + eV)[1 - f(E - E_F - eV)] - $$
$$D_r(E + eV)f(E - E_F - eV)D_l(E)[1 - f(E - E_F)])$$

For large energy barriers, the matrix elements of the perturbation Hamiltonian will be independent of energy. Second, we assume that the density of states does not depend on energy, either since the electron gas is two-dimensional, or since the voltage drop is sufficiently small. In this approximation, the total current is obtained by integration over all relevant energies:

$$I(V) = e\frac{2\pi}{\hbar}|\langle i|H_t|f\rangle|^2 D^2 \int_{E_cb,l}^{\infty} (f(E - E_F) - f(E - E_F - eV))dE$$

If the thermal energy and eV are small compared to the E_F, the Fermi functions can both be approximated by step functions, and the integral simply gives eV, resulting in

$$I(V) = \frac{2\pi e^2}{\hbar}|\langle i|H_t|f\rangle|^2 D^2 \cdot V$$

Hence, large tunnel barriers show a linear I-V characteristics for small voltages, with a resistance given by $R = \frac{\hbar}{2\pi e^2 |\langle i|H_t|f\rangle|^2 D^2}$, and consequently, we can speak of a voltage-independent conductance, which is directly related transmission $T = 4\pi|\langle i|H_t|f\rangle|^2 D^2$.

Let us now study a tunnel event in a SET device. Here, the electrostatic energy may change, and E_f can differ from E_f. In that case, we have to change Fermi's golden rule accordingly:

$$\Gamma_{i \to f} = \frac{2\pi}{\hbar}|\langle i|H_t|f\rangle|^2 \delta(E_f - E_i - \Delta E) \tag{D.2}$$

The tunnelling rates as a function of the voltage applied now reads

$$\Gamma^{pm}(V) = \frac{1}{e^2 R}\int_{E_cb,l}^{\infty} f(E)[1 - f(E + \Delta E^{\pm})]dE \quad = \frac{1}{R_i e^2}\frac{\Delta E}{1 - exp(\Delta E/k_B \Theta)}$$

E Solutions to Selected Exercises

Chapter 2

E 2.1: A zincblend lattice hosts 4 atoms of each type per unit cell. Therefore, $m_{GaAs} = 12.12$ g. The unit cell of the diamond lattice contains 8 atoms, hence $m_{Si} = 5.58$ g.

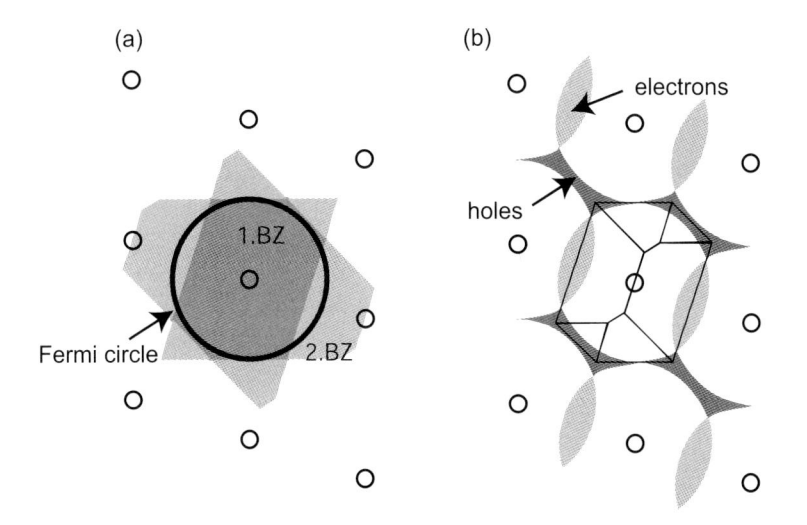

Figure E.1: Reciprocal lattice, first and second Brillouin zone, and the Fermi circle of [E 2.3] is shown in (a). The repeated zone scheme in (b) reveals the regions filled with holes (dark gray) and those filled with electrons (light gray). Also shown is how the elements of the second Brillouin zone combine to the reduced zone scheme.

E 2.2: (a)

(i): The reciprocal lattice vectors are $\vec{b_1} = \frac{\pi}{6\,\text{nm}}(3,-1)$ und $\vec{b_2} = \frac{\pi}{6\,\text{nm}}(0,4)$

(ii): The electron density is $n = \frac{2\pi}{3 \cdot 12 nm^2} \rightarrow |k_F| = \sqrt{2\pi n} = \frac{\pi}{3}$ nm^{-1}. The Fermi circle just touches the edge of the first Brillouin zone in $\vec{b_2}$-direction (see Fig. E.1(a)). Both the first and the second Brillouin zones are partly filled. We do have a metal here, as always for materials where the number of conduction electrons per unit cell is not an even integer.

(ii): The repeated zone scheme of Fig. E.1(b) reveals that there is one hole-type de Haas - van Alphen (dHvA) oscillation. From the enclosed area, a dHvA period of $\Delta(1/B) \approx \frac{2\pi e}{\hbar A} \approx 0.035$ T^{-1} is expected. In addition, the Fermi surface in the second Brillouin zone gives an electron-type orbit with $\Delta(1/B) \approx 0.024$ T^{-1}.

E 2.3: For $\vec{q} \to 0$, the function $F(s)$ (see eq. (2.50)) approaches 1, and the dielectric function can be approximated by

$$\epsilon(\vec{q}) \approx 1 + \frac{k_{TF}^2}{q^2}$$

The potential of the point charge is

$$V_{ext}(\vec{r}) = -\frac{Ze^2}{r}$$

where Z is the number of protons. Fourier transformation yields

$$V_{ext}(\vec{r}) = -\frac{Ze^2}{(2\pi)^3} \int_{-\infty}^{\infty} \frac{4\pi}{q^2} e^{i\vec{q}\vec{r}} d\vec{q}$$

The Fourier components are thus given by

$$V_{ext}(\vec{q}) = -\frac{4\pi Ze^2}{q^2}$$

The Fourier components of the screened potential are given by

$$V_{eff}(\vec{q}) = -\frac{4\pi Ze^2}{q^2 \epsilon(\vec{q})}$$

The Fourier transform of the screened potential thus reads

$$V_{eff}(\vec{r}) = -\frac{Ze^2}{(2\pi)^3} \int_{-\infty}^{\infty} \frac{4\pi}{q^2 + k_{TF}^2} e^{i\vec{q}\vec{r}} d\vec{q}$$

An evaluation of this integral gives the Yukawa potential

$$V_{eff}(\vec{r}) = -\frac{Ze^2}{r} e^{-ik_{TF}r}$$

E 2.4: (i): The Schrödinger equation reads

$$-\frac{\hbar^2}{2m} \frac{d^2\Phi(x)}{dx^2} - V_0 \delta(x) \Phi(x) = E\Phi(x)$$

$\Phi(x)$ must be continuous at $x = 0$; this requirement can be written as $\lim_{\eta \to 0} \Phi(\eta) = \lim_{\eta \to 0} \Phi(-\eta)$. The second necessary condition is obtained from integrating the Schrödinger equation, with the integration limits approaching $x = 0$:

$$\lim_{\eta \to 0} [-\frac{\hbar^2}{2m} \int_{-\eta}^{\eta} \frac{d^2\Phi(x)}{dx^2} dx - V_0 \int_{-\eta}^{\eta} \delta(x)\Phi(x) dx = E \int_{-\eta}^{\eta} \Phi(x) dx]$$

$$-\frac{\hbar^2}{2m}[\Phi'(+0) - \Phi'(-0)] - V_0 \Phi(0) = 0$$

Since the wave function is evanescent everywhere except for $x = 0$, the ansatz $\Phi(x) = A e^{-\kappa|x|}$ makes sense. By inserting this expression in the conditions above, we obtain $\kappa = \frac{mV_0}{\hbar^2}$. From the normalization condition $\int |A|^2 e^{-2\kappa|x|} = 1$, the amplitude $A = \sqrt{\kappa}$ is found. Hence,

$$\Phi(x) = \sqrt{\kappa} e^{-\kappa|x|}$$

The eigenvalue is obtained from

$$E_0 = \frac{(\hbar \cdot i\kappa)^2}{2m} = -\frac{mV_0^2}{2\hbar^2}$$

(ii): $\Psi_k(x)$ satisfies the Bloch theorem if $\Psi_k(x+na) = e^{ikna}\Psi_k(x)$. This is in fact the case:

$$\Psi_k(x+na) = \sum_{j=-\infty}^{\infty} \Phi_0(x+na-ja)e^{ikja} =$$

$$e^{ikna} \sum_{j=-\infty}^{\infty} \Phi_0[x-(j-n)a]e^{ik(j-n)a} = e^{ikna}\Psi_k(x)$$

(iii): We carry out the integration as suggested in the exercise:

$$\langle \Phi_0 | H_0 + \Delta V | \Psi_k \rangle = E(k)\langle \Phi_0 | \Psi_k \rangle$$

Here ist H_0 denotes the Hamiltonian of a single δ-function at x=0, and $\Delta V(x)$ is the residual crystal potential without the δ potential at the origin. The last equation can be rewritten as

$$E_0 I_0(k) + I_1(k) = E(k) I_0(k) \Rightarrow E(k) = E_0 + \frac{I_1(k)}{I_0(k)}$$

We proceed by calculating I_0:

$$I_0(k) = \langle \Phi_0 | \Psi_k \rangle = 1 + \sum_{j=1}^{\infty} \left[e^{ikja}\langle \Phi_0(x)|\Phi_0(x-ja)\rangle + e^{-ikja}\langle \Phi_0(x)|\Phi_0(x+ja)\rangle \right]$$

The first term stems from the contribution of $j = 0$. The two integrals entering here are identical, therefore

$$I_0(k) = 1 + \sum_{j=1}^{\infty} 2\cos(jka)\langle\Phi_0(x)|\Phi_0(x+ja)\rangle$$

where the overlap integral $\langle\Phi_0(x)|\Phi_0(x+ja)\rangle = \alpha_j$ is given by

$$\alpha_j = \kappa\langle e^{-\kappa|x|}|e^{-\kappa|x+ja|}\rangle = \kappa\left[e^{\kappa ja}\int_{-\infty}^{-ja} dx e^{2\kappa x} + \right.$$

$$\left. e^{-\kappa ja}\int_{-ja}^{0} dx + e^{-\kappa ja}\int_{0}^{ja} dx e^{-2\kappa x}\right] = (1+j\kappa a)e^{-j\kappa a}$$

We finally obtain

$$I_0(k) = 1 + 2\sum_{j=1}^{\infty}(1+j\kappa a)e^{-j\kappa a}\cos(jka)$$

For I_1, the expression reads

$$I_1(k) = \langle\Phi_0|\Delta V|\Psi_k\rangle = \beta + 2\sum_{j=1}^{\infty}\gamma_j\cos(jka)$$

with $\beta = \langle\Phi_0|\Delta V|\Phi_0\rangle$, and $\gamma_j = \langle\Phi_0(x)|\Delta V|\Phi_0(x+ja)\rangle$. This is a transfer integral. Inserting the wave function leads, via a geometric series, to

$$\beta = \frac{E_0}{1 - e^{2\kappa a}}$$

and after some algebra, one finds

$$\gamma_j = E_0\left[-je^{-j\kappa a} + \frac{2\cosh[\kappa a]}{1 - e^{2\kappa a}}\right]$$

We have determined $E(k)$ for this model potential exactly. For an interpretation, we make two approximations. First of all, only nearest-neighbor transfer integrals are assumed to be non-vanishing. Second, terms of the order $e^{-2\kappa a}$ are neglected, i.e., $1/\kappa \ll a$ is assumed. we obtain

$$\alpha_1 = (1+\kappa a)e^{-\kappa a} \approx e^{-\kappa a}; \qquad \beta = 0; \qquad \gamma_1 = E_0 e^{-\kappa a}$$

und from this

$$E(k) = E_0 + \frac{2\gamma_1\cos(ka)}{1 + 2\alpha_1\cos(ka)}$$

(iv): Since $1/\kappa \ll a$, $\alpha_1 \ll 1$, such that the denominator in the dispersion relation can be set to 1. A Taylor expansion to second order gives $E(k) = E_0 + \gamma_1(1 - \frac{1}{2}k^2 a^2)$. Hence,

$$m^* = -\frac{\hbar^2}{\gamma_1 a^2} = \frac{2\hbar^4}{mV_0^2 a^2}$$

Intuitively, this is a very reassuring result: the effective mass depends exponentially on $\kappa \cdot a$. As the nearest neighbor overlap increases, it becomes easier for the electron to move from site to site, and its effective mass decreases.

E 2.5: (a) The effective densities of states are

$$N_c(T) = g\frac{1}{4}[\frac{2m_c^* k_B T}{\pi \hbar^2}]^{\frac{3}{2}} \quad \text{und} \quad P_v(T) = g\frac{1}{4}[\frac{2m_v^* k_B T}{\pi \hbar^2}]^{\frac{3}{2}},$$

where $m_{c,v}^*$ is the geometric mean of the eigenvalues of the effective-mass tensor, i.e., $m_{c,v}^* = (\Pi_{i=1}^{3} m_{c,v,i}^*)^{1/3}$. This can be seen as follows: we intend to substitute the Fermi ellipsoid given by

$$\frac{\hbar^2}{2E_F}(\frac{k_x^2}{m_x} + \frac{k_y^2}{m_y} + \frac{k_z^2}{m_z}) = 1$$

by a Fermi sphere with isotropic effective mass m_{eff}, such that the volume is maintained:

$$(\frac{2E_F}{\hbar^2})^{3/2} \cdot \frac{4\pi}{3} \sqrt{m_x m_y m_z} = (\frac{2E_F}{\hbar^2})^{3/2} \cdot \frac{4\pi}{3} (m_{eff})^{3/2} \rightarrow m_{eff} = \sqrt{m_x m_y m_z}$$

Furthermore, g is the degeneracy of the band in addition to spin degeneracy. Thus, $g = 6$ for electrons in Si, and $g = 1$ else. For Si, this gives $m_c^* = (m_l^*(m_t^*)^2)^{1/3} = 0.321 m_e$ and thus $N_c = 2.76 \cdot 10^{25}$ m^{-3}.

In case of the holes, we simply have to add up the two effective densities of states: $P_v = P_{v,lh} + P_{v,hh} = 1.14 \cdot 10^{25}$ m^{-3}, such that an intrinsic carrier concentration of $n_{i,Si}(300 \text{ K}) = \sqrt{N_c P_v} e^{-E_g/2k_B \cdot 300 \text{ K}} = 7 \cdot 10^{15}$ m^{-3} is found.

There is no valley degeneracy in GaAs, and one obtains $N_c = 4.35 \cdot 10^{23}$ m^{-3}, $P_v(300 \text{ K}) = 9.72 \cdot 10^{24}$ m^{-3}, and $n_{i,GaAs}(300 \text{ K}) = 2.42 \cdot 10^{12}$ m^{-3}.

At $\Theta = 77$ K, $n_{i,Si}(77 \text{ K}) \approx 10^{-15}$ m^{-3}, and $n_{i,GaAs}(77 \text{ K}) \approx 10^{-26}$ m^{-3}, which is irrelevant in both cases.

Chapter 3

E 3.1: The expectation value for the electron position in z-direction is given by

$$\langle z \rangle = \int_0^\infty \Phi^*(z) z \Phi(z) dz = \int_0^\infty \frac{b^3}{2} z^3 e^{-bz} dz = \frac{b^3}{2} \frac{\Gamma(4)}{b^4} = \frac{3}{b}$$

Here, we have used

$$\int_0^\infty x^n e^{-ax} dx = \frac{\Gamma(n+1)}{a^{n+1}}$$

with $\Gamma(n+1) = n!$ for integer n. This means that the size quantization removes the electrons from the O - S interface, which increases the mobility.

Chapter 4

E 4.1: Consider a gas at low pressure in a vacuum chamber. The number of molecules N_c that hit an area A of the wall within a time interval t is given by

$$N_c = n \bar{v}_x t A$$

here, n denotes the density of the gas, and $\bar{v}_x$ is their average velocity in x-direction, which is perpendicular to the wall. The quantity $\bar{v}_x t A$ is the volume that contains all molecules which hit the area A within t. On the other hand, the pressure is given by $p = n k_B \Theta$, and thus

$$N_c = \frac{p}{k_B T} \bar{v}_x t A \Rightarrow n_c = \frac{p}{k_B T} \bar{v}_x$$

n_c is the scattering rate at the wall (i.e., the number of hits per area and time). We obtain $\bar{v}_x$ from the Maxwell velocity distribution

$$f(\vec{v}) d\vec{v} = \left(\frac{m}{2\pi k_B T}\right)^{3/2} e^{-\frac{mv^2}{2k_B T}} d\vec{v}$$

Since we are only interested in the x-component, we integrate over dv_y and dv_z and find

$$g(v_x) = \left(\frac{m}{2\pi k_B T}\right)^{1/2} e^{-\frac{mv_x^2}{2k_B T}}$$

In order to get $\bar{v}_x$ from $g(v_x)$, we have to calculate the expectation value of v_x under the constraint $v_x > 0$:

$$\bar{v}_x = \int_0^\infty v_x g(v_x) dv_x = \ldots = \sqrt{\frac{k_B T}{2\pi m}} \Rightarrow n_c = \frac{p}{\sqrt{2\pi m k_B T}}.$$

We estimate the time it takes until a monolayer of oxygen has formed at the wall. A sticking coefficient of 1 is assumed, which means that all molecules that hit the wall remain there. We denote the area density of molecules within a monolayer by N_s. The time required to form a monolayer is $t_m = N_s/N_c$. As a simple guess, assume that an oxygen molecule has an effective diameter of $d = 0.36$ nm. Suppose further that the molecules form a hexagonal lattice, which means the area $A_m = \frac{d^2 \sqrt{3}}{2}$ contains one O_2 molecule. Then, $N_s = 1/A_m = 8.7 \cdot 10^{18}$ m^{-2}, which means that at a pressure of $p = 10^{-10}$ mbar, a monolayer forms within $t_m = 6.5$ h: at a pressure of $p = 10^{-6}$ mbar, this only takes 2.4 s! This simple estimate shows that really high vacuum is needed for molecular beam epitaxy!

E 4.2: (i) Inserting gives $D \leq 0.03$ cm^{-2}.
  (ii) $N \leq 9$.
  (iii) $y \geq 0.9 \Rightarrow D \leq 4.4 \cdot 10^{-3}$. Within (8 in.)^3, we must have less that $A \cdot D/6$ = 0.23 particles. Since 1 ft. =12 in. and the class of a clean room is given by the number of particles per cubic foot with sizes larger than 500 nm, $R \approx 1$ is necessary. This is a realistic number for an industrial clean room.

E 4.3: The dosage is distributed among 2^{26} points, such that a dosage per point of $d = 3 \cdot 10^{-16}$ Cb is required. The dwell time is therefore $t_{dwell} = d/I_{beam} = 30$ μs. The spots form a square lattice with a lattice constant of 12.2 nm. Hence, each spot must have an illuminating diameter of about 12.2 nm $\cdot \sqrt{2} = 17$ nm for complete coverage.

Finally, increasing the current means shorter dwell times, which is limited by the speed of the beam control. This can be circumvented by reducing the bit resolution, but then the spatial resolution is lost as well.

E 4.4: Applying the rules for operational amplifiers gives the condition

$$V_{out}(t) = -\left[\frac{R_2}{R_1}V_{in}(t) + \frac{1}{R_1 C}\int_0^t V_{in}(\tau)d\tau\right]$$

For $V_{in} = 0$, the output voltage is constant. Suppose the input voltage changes as indicated in the exercise. The response is

$$V_{out}(t) = -V_0\left[\frac{R_2}{R_1}\theta(t-t_0) + \frac{t-t_0}{R_1 C}\right]$$

This circuit as known as proportional-integral (PI) - controller. Suppose some parameter, like the temperature in a gas flow cryostat, has to be held constant. The difference between the measured temperature and the required temperature is translated by some circuit into a voltage, which is applied at the input. The output voltage is then used to adjust the temperature to the desired value by some control function, like, in our example, the He gas flow through a needle valve of a gas floe cryostat. Suppose the temperature is too high. The gas flow must be increased, and the output voltage is used to open the needle valve with a step motor. This opening increases with time, until the input voltage has reached zero again, which means the temperature is back at its required value. PI controllers are widely used for such kind of tasks.

Chapter 5

E 5.1: With the dielectric constants $\epsilon_{GaAs} \approx 13$, $\epsilon_{Si} \approx 11$, one finds the results as listed in Table E.1 Note that in Si-MOSFETs, there is an additional valley degeneracy of 2! Apparently, GaAs-HEMTs are perfect for investigating ballistic and phase coherent effects, while Si-MOSFETs are particularly interesting for studying interaction effects.

Table E.1: Results for E 5.1

	GaAs (T=4.2 K)	Si (T=4.2 K)
Drude scattering time [10^{-12} s]	38	4.3
Fermi velocity [10^4 m/s]	27	1.3
diffusion constant [m^2/s]	1.43	0.00035
Fermi wavelength [nm]	40	95
phase coherence length [nm]	6500	59
inelastic scattering length [nm]	8200	127
thermal length [nm]	1610	25
interaction parameter	0.87	13.8

Chapter 6

E 6.1: The filling factor $\nu = 4$ is at B=6.4 T, as can be seen directly from the position of the Hall plateau. Analyzing the Hall slope at small magnetic fields gives $d\rho_{xy}/dB \approx 1060\ \Omega/T = -1/n_{2D}e$ which corresponds to $n_{2D} = 5.8 \cdot 10^{15}$ m^{-2}. This is close to the upper limit of electron densities possible in Ga[Al]As-HEMTs, if the second subband must remain empty. The figure tells us $\rho_{xx}(B=0) \approx 8\ \Omega$. Since $\rho_{xx}(B=0) = (n_{2D}e\mu)^{-1}$, we find an electron mobility of $\mu = 134$ m^2/Vs. Because of $\mu = e\tau/m^*$ and $\ell_e = v_F\tau$, the elastic mean free path $\ell_e = \frac{\hbar}{e}\mu\sqrt{2\pi n_{2D}} = 16.8\ \mu$m is obtained. This corresponds to the elastic scattering time of $\tau = 51$ ps.

Spin splitting states in at $B = 1.6$ T. This allows us to estimate the effective g-factor via

$$2g^*\mu_B B_{split-start} \approx \hbar/\tau_q \Rightarrow g^* \approx 5.5.$$

Here, it is assumed that the peaks in the density of states have the width $\hbar/\tau_q$, which is a reasonable approximation.

E 6.2: (i) The effective mass is obtained from the slope of $ln(A/\Theta)$ vs. Θ: $m^* = 0.032 m_e$. The Dingle plot gives a quantum scattering time of $\tau_q = 0.18$ ps. Once we know m^*, a Drude scattering time $\tau = \mu m^*/e = 14$ ps is calculated.

(ii) The ratio $\tau/\tau_q = 78$ is extremely large. This means the dominant source of scattering are remote scattering centers. In fact, the sample is a $InAs$ quantum well $30 nm$ below the surface, embedded in a $AlSb$ barrier, and capped with a GaSb layer, see Fig. 3.25. It is known that in this material system, the charge neutrality level of the $GaSb$ surface states lie above the conduction band bottom of $InAs$, and thus electrons are transferred from the surface into the quantum well. The remaining space charge region close to the surface represents the scattering potential.

E 6.3: The potential is given by $U(z) = -e\epsilon z$, where z denotes the growth direction, and ϵ

is the electric field. The Schrödinger equation thus reads.

$$\left(-\frac{\hbar^2}{2m}\frac{d^2}{dz^2}+\frac{1}{2}m\omega_0^2 z^2 - e\epsilon z\right)\Psi(z) = E\Psi(z)$$

which, by completing the square and by substituting $z = u + \frac{e\epsilon}{m\omega_0^2}$, can be rewritten as

$$\left(-\frac{\hbar^2}{2m}\frac{d^2}{du^2}+\frac{1}{2}m\omega_0^2 u^2\right)\Psi(u) = E^*\Psi(u)$$

with $E^* = E + \frac{q^2\epsilon^2}{2m\omega_0^2}$. The energy eigenvalues are

$$E_n = (n+\frac{1}{2})\hbar\omega_0 - \frac{e^2\epsilon^2}{2m\omega_0^2}$$

The electric field thus displaces the parabolic potential without modifying its shape. The minimum is given by

$$z_{min}(\epsilon) = \frac{e\epsilon}{m\omega_0^2}$$

$$E_{min} = \frac{1}{2}\hbar\omega_0 - \frac{q^2\epsilon^2}{2m\omega_0^2}$$

Chapter 7

E 7.1: (i) The problem is very similar to the discussion of the effects of a parallel magnetic field on a 2DEG given in chapter 6. Using the gauge given in the exercise, we obtain

$$\left[-\frac{\hbar^2}{2m^*}\frac{\partial^2}{\partial x^2}+i\frac{eB\hbar y}{m^*}\frac{\partial}{\partial x}+\frac{e^2B^2}{2m^*}y^2-\frac{\hbar^2}{2m^*}\frac{\partial^2}{\partial y^2}+\frac{1}{2}m^*\omega_0^2 y^2\right]\psi(y)e^{ik_x x} = E\psi(y)e^{ik_x x}$$

$$\left[\frac{\hbar^2}{2m^*}k_x^2-\frac{eB\hbar y}{m^*}k_x+\frac{e^2B^2}{2m^*}y^2-\frac{\hbar^2}{2m^*}\frac{\partial^2}{\partial y^2}+\frac{1}{2}m^*\omega_0^2 y^2\right]\psi(y) = E\psi(y)$$

$$\left[-\frac{\hbar^2}{2m^*}\frac{\partial^2}{\partial y^2}+\frac{1}{2}m^*(\omega_0+\omega_c)^2 y^2-\frac{eB\hbar y}{m^*}k_x+\frac{\hbar^2}{2m^*}k_x^2\right]\psi(y) = E\psi(y)$$

Completing the square by adding and subtracting $\bar{y}_0^2$, with $\bar{y}_0 = \frac{\hbar k_x}{m^*}\frac{\omega_c}{\omega^2} = y_0(\frac{\omega_c}{\omega})^2$ (this relation holds since $y_0 = \frac{\hbar k_x}{m^*\omega_c}$), it follows

$$\left[-\frac{\hbar^2}{2m^*}\frac{\partial^2}{\partial y^2}+\frac{1}{2}m^*\omega^2(y-\bar{y}_0)^2+\frac{\hbar^2}{2m^*}k_x^2(\frac{\omega_0}{\omega})^2\right]\psi(y) = E\psi(y)$$

(ii) The third term on the right hand side in the above equation represents the energy dispersion in x-direction. The electron mass now depends on B and is known as the magnetic mass $m^*(B) = m^*(\frac{\omega}{\omega_0})^2$. For large B, the solution approaches the Landau quantization, with the magnetic mass going towards infinity.

(iii) With the above solution, the electron density of a quantum wire in a magnetic field can be written as

$$n_{QWR} = \frac{2}{\pi\hbar} \sum_{j=0}^{\infty} \sqrt{2m^*(E_F - E_j)}(1 - \Theta(E_F - E_j))$$

For integer filling factors N, $E_F = \hbar\omega(N + \frac{1}{2})$, such that

$$n_{QWR} = \frac{2}{\pi\hbar} \sum_{j=0}^{N} \sqrt{2m^*(\frac{\omega}{\omega_0})^2 \hbar\omega(N-j)} \approx \frac{2}{\pi}\sqrt{\frac{2m^*}{\hbar}}(\frac{\omega^{3/2}}{\omega_0}) \int_{\ell=0}^{N} \ell^{1/2} d\ell$$

This gives the relation

$$n_{QWR} = \frac{4}{3\pi}\sqrt{\frac{2m^*}{\hbar}}(\frac{\omega^{3/2}}{\omega_0}) N^{3/2}$$

The fit parameters are ω_0 and $n_{QWR} (\Rightarrow E_F)$. Since $E_F = \frac{1}{2}m^*\omega_0^2(\frac{w}{2})^2$, we can determine the electronic wire width w.

E 7.2: We write down the current that flows at a bias voltage V:

$$I = \int \left(D_1^{\rightarrow}(E)v(E)f(E-\mu)[1 - f(E-\mu+eV)] - \ldots \right.$$

$$\left. D_1^{\leftarrow}(E+eV)v(E+eV)f(E-\mu+eV)[1 - f(E-\mu+eV)] \right) \Theta(E - E_1) dE$$

$$= \frac{2e}{h} \int \Theta(E - E_1)[f(E-\mu) - f(E-\mu+eV)] dE$$

With

$$f(e-\mu+eV) \approx f(E-\mu) + \frac{\partial f}{\partial eV}(eV = 0) \cdot eV = f(E-\mu) + \frac{\partial f}{\partial E}(eV = 0) \cdot eV$$

we find

$$I = \frac{2e^2 V}{h} \int -\frac{\partial f}{\partial E}(eV = 0)\Theta(E - E_1) dE$$

Partial integration gives

$$I = \frac{2e^2}{h} f(E_1 - \mu) V \Rightarrow G = I/V = \frac{2e^2}{h} f(E_1 - \mu)$$

The steps are thermally smeared as soon as the full widths at half maximum of $\frac{\partial f}{\partial E}$ equals Δ. This is the case for $\Delta = 2k_B\Theta \ln(3 + 2\sqrt{2}) \approx 3.52 k_B T$.

E 7.3: (i) From the Landauer-Büttiker formalism, the system

$$\begin{pmatrix} I_S \\ I_D \\ I_1 \\ I_2 \\ I_3 \\ I_4 \end{pmatrix} = \frac{e^2}{h} \begin{pmatrix} N & 0 & -N & 0 & 0 & 0 \\ 0 & N & 0 & 0 & 0 & -N \\ 0 & 0 & N & -M & M-N & 0 \\ 0 & -N & 0 & N & 0 & 0 \\ -N & 0 & 0 & 0 & N & 0 \\ 0 & 0 & 0 & M-N & -M & N \end{pmatrix} \cdot \begin{pmatrix} V_S \\ V_D \\ V_1 \\ V_2 \\ V_3 \\ V_4 \end{pmatrix}$$

is obtained. We set the drain potential to zero and use the fact that $I_S + I_D = 0$. the remaining 5×5 matrix equation has the solution

$$\begin{pmatrix} V_S \\ V_1 \\ V_2 \\ V_3 \\ V_4 \end{pmatrix} = \frac{hI_S}{e^2} \begin{pmatrix} 1/M \\ 1/M - 1/N \\ 0 \\ 1/M \\ 1/N \end{pmatrix}$$

(ii) The resistances are obtained from (i) as $R_{ij} = (\mu_i - \mu_j)/eI_S$:

$$R_{12} = R_{34} = \frac{h}{e^2} \left(\frac{1}{M} - \frac{1}{N} \right)$$

$$R_{13} = R_{24} = \frac{h}{e^2} \frac{1}{N}$$

$$R_{14} = \frac{h}{e^2} \left(\frac{1}{M} - \frac{2}{N} \right)$$

$$R_{23} = \frac{h}{e^2} \frac{1}{M}$$

By a proper choice of our setup, we can measure just the barrier!

E 7.4: (i) The Landauer-Büttiker matrix is

$$\begin{pmatrix} I_S \\ I_D \\ I_1 \\ I_2 \\ I_3 \\ I_4 \end{pmatrix} = \frac{e^2}{h} \begin{pmatrix} 2 & 0 & -2 & 0 & 0 & 0 \\ 0 & 2 & 0 & 0 & 0 & -2 \\ 0 & 0 & 2 & 0 & -1 & 0 \\ 0 & -2 & 0 & 2 & 0 & 0 \\ -2 & 0 & 0 & 0 & 2 & 0 \\ 0 & 0 & 0 & -1 & 0 & 2 \end{pmatrix} \cdot \begin{pmatrix} V_S \\ V_D \\ V_1 \\ V_2 \\ V_3 \\ V_4 \end{pmatrix} + \frac{e^2}{h} \begin{pmatrix} 0 \\ 0 \\ -V_2^* \\ 0 \\ 0 \\ -V_3^* \end{pmatrix}$$

Next, we have to find V_i^*. They depend on p, V_2 and V_3. Current conservation and using the definition of p lead to

$$V_3 + V_c^* = V_3^* + V_c$$

$$V_2^* + V_c^* = V_2 + V_c$$

$$p = 1 - \frac{V_2^* - V_c^*}{V_2 - V_c}$$

$$p = 1 - \frac{V_3^* - V_c}{V_3 - V_c^*}$$

And we obtain the system of equations

$$\begin{pmatrix} V_2 \\ V_3 \\ V_2 \\ V_3 \end{pmatrix} = \begin{pmatrix} 1 & 0 & -1 & 1 \\ 0 & 1 & 1 & -1 \\ \frac{1}{1-p} & 0 & 1 & \frac{1}{p-1} \\ 0 & \frac{1}{1-p} & \frac{1}{p-1} & 1 \end{pmatrix} \cdot \begin{pmatrix} V_2^* \\ V_3^* \\ V_c \\ V_c^* \end{pmatrix}$$

with the solution

$$\begin{pmatrix} V_2^* \\ V_3^* \\ V_c \\ V_c^* \end{pmatrix} = \frac{1}{p-4} \begin{pmatrix} (2p-4)V_2 - pV_3 \\ -pV_2 + (2p-4)V_3 \\ (p-2)V_2 - 2V_3 \\ -2V_2 + (p-2)V_3 \end{pmatrix}$$

Inserting this in the above linear system gives

$$\begin{pmatrix} I_S \\ I_D \\ I_1 \\ I_2 \\ I_3 \\ I_4 \end{pmatrix} = \frac{e^2}{h} \begin{pmatrix} 2 & 0 & -2 & 0 & 0 & 0 \\ 0 & 2 & 0 & 0 & 0 & -2 \\ 0 & 0 & 2 & -\frac{2p-4}{p-4} & -1+\frac{p}{p-4} & 0 \\ 0 & -2 & 0 & 2 & 0 & 0 \\ -2 & 0 & 0 & 0 & 2 & 0 \\ 0 & 0 & 0 & -1+\frac{p}{p-4} & -\frac{2p-4}{p-4} & 2 \end{pmatrix} \cdot \begin{pmatrix} V_S \\ V_D \\ V_1 \\ V_2 \\ V_3 \\ V_4 \end{pmatrix}$$

Again, we reduce the matrix by the drain column and row (as in the previous exercise) to

$$\begin{pmatrix} V_S \\ V_1 \\ V_2 \\ V_3 \\ V_4 \end{pmatrix} = \frac{hI_S}{2e^2} \begin{pmatrix} \frac{p-4}{p-2} \\ \frac{2}{2-p} \\ 0 \\ \frac{p-4}{p-2} \\ 1 \end{pmatrix}$$

(ii) From (i), we find immediately $R_{xx} = \frac{V_1 - V_2}{I_S} = \frac{h}{e^2} \frac{1}{2-p}$. Other resistances are not of interest here.

(iii) Inserting the numerical values given in the exercise, one gets $R_{12} = 0.53 h/e^2 \Rightarrow p = 0.113$. To define the equilibration length L_{eq}, we require that along the distance $L = L_{eq}$ the potential difference between the edge states has been reduced to $1/e = 0.368$ of its initial value:

$$\frac{\Delta \mu^*}{\Delta \mu} = \frac{1}{e}; \quad \frac{\Delta \mu - \Delta \mu^*}{\Delta \mu} = p \Rightarrow p = 1 - \frac{1}{e}$$

or, more generally

$$p = 1 - e^{-L/L_{eq}}$$

For the numbers given, we thus obtain $L_{eq} = 0.42$ mm! Equilibration between spin polarized edge states takes place on macroscopic length scales, see e.g., G. Müller et al., Phys. Rev. B 45, p.3932 (1992).

E 7.5: (i) $\vec{C} = n_1\vec{a}_1 + n_2\vec{a}_2$, with $(n_1, n_2) = (4, 1)$. For symmetry reasons, it suffices to consider $n_1 \geq 0$ and $0 \leq n_2 \leq n_1$. For zigzag tubes, $(n_1, n_2) = (n, 0)$, for armchair tubes, $(n_1, n_2) = (n, n)$.

(ii) We write $\vec{A} = m_1\vec{a}_1 + m_2\vec{a}_2$. From $\vec{A}\vec{C} = 0$, $m_1 = N/(2n_1 + n_2)$ and $m_2 = N/(n_1 + 2n_2)$ is found, where N is the smallest common multiple of $(2n_1 + n_2)$ and $(n_1 + 2n_2)$.

(iii) Since $\vec{A}\vec{B} = 2\pi$, we obtain $\vec{B} = \frac{2\pi}{m_1^2 + m_1 m_2 + m_2^2}(m_1\vec{a}_1 + m_2\vec{a}_2)$. For an illustration, we express $\vec{B}$ in terms of the reciprocal lattice vectors of the graphite sheet, $\vec{b}_1 = \frac{4\pi}{3a^2}(2\vec{a}_1 - \vec{a}_2)$, $\vec{b}_2 = \frac{4\pi}{3a^2}(-\vec{a}_1 + 2\vec{a}_2)$, where a denotes the lattice constant of the graphite sheet, and find $\vec{B} = \frac{a^2}{2(m_1^2 + m_1 m_2 + m_2^2)}\Big((2m_1 + m_2)\vec{b}_1 + (m_1 + 2m_2)\vec{b}_2\Big)$. The mode spacing in k_y - direction is found to be $\Delta k_y = \frac{a^2}{2(n_1^2 + n_1 n_2 + n_2^2)}\Big((2n_1 + n_2)\vec{b}_1 + (n_1 + 2n_2)\vec{b}_2\Big)$.

(iv) This condition is derived in the literature on CNs given at the end of chapter 7. The CN under study here is therefore metallic.

(v) For these zigzag CNs, we can estimate the distance $\Gamma - X$ to about π/a. Assuming a parabolic dispersion around Γ, one estimates $m^* \approx 0.8m$. For the metallic tube, the concept of effective mass is not good, since a parabolic dispersion is a very poor approximation. To calculate the density of states, we set $E(k) = \alpha \cdot k$; $\Delta k = \pi/L \Rightarrow D(k) = 2L/\pi \Rightarrow d(k) = 2/\pi$. This is translated in energy via $d(E) = d(k)dk/dE = \frac{2}{\pi}\frac{1}{\alpha}$. From the figure, we estimate $\alpha \approx 4eV/(\pi/a) \Rightarrow d(E_F) \approx 1.25 \cdot 10^{28}$ J^{-1}m^{-1}. Hence, in 1D, a linear energy dispersion gives a constant density of states. Consequently, the chemical potential does not depend on temperature.

E 7.6: Let us add a real Ohmic contact to the middle region and connect it to ground, while we connect the collector to a voltmeter. The current is now flowing via m, and the Landauer-Büttiker equations now read

$$\frac{h}{2e^2}I_i = NV_i - T_{ci}V_C$$

$$\frac{h}{2e^2}I_m = T_{cm}V_c - T_{im}V_i$$

$$0 = NV_c - T_{ic}V_C$$

Since we consider a situation where $T_{ic} \ll N$, we can approximate $N \pm T_{ic} \approx N$ and find for $T_{cm} = T_{im}$

$$\frac{V_c}{I_i} = \frac{h}{2e^2}\frac{T_{ic}}{N^2}$$

The quantity of interest is thus no longer a small signal on a large background. This setup is frequently used for measurements on ballistic samples.

Chapter 8

E 8.1: For the upper branch, one obtains

$$\theta_{upper} = -\frac{e}{\hbar}\int_\Gamma \vec{A}d\vec{\Gamma} = -\frac{e}{\hbar}\int_{\alpha=0}^{\pi} R\begin{pmatrix}\sin\alpha\\ \cos\alpha\\ 0\end{pmatrix}\cdot\begin{pmatrix}0\\ -BR\cos\alpha\\ 0\end{pmatrix}d\alpha =$$

$$\pi R^2 B\frac{e}{\hbar}\int_{\alpha=0}^{\pi}\cos^2\alpha\,d\alpha = \pi R^2\frac{e}{2\hbar}B = \pi\frac{\Phi}{\Phi_0}$$

Correspondingly, an electron collects a phase of $\theta_{lower} = \pi\frac{\Phi}{\Phi_0}$ as it traverses the lower branch. The interference between these two waves generates the Aharonov-Bohm effect:

$$t = \sqrt{\epsilon}(e^{i\phi}+e^{-i\phi})\sqrt{\epsilon} \Rightarrow T = t^*t = 4\epsilon\cos^2\phi$$

E 8.2: (i) We divide the time t into N intervals of equal length. N is so large that none of the intervals hosts two scattering events. An individual interval is occupied with a probability of $p = \gamma t/N$ The probability for j scattering events follows from the probability that j of the intervals are occupied, times the number of possible arrangements of the occupied intervals among all intervals. Hence,

$$P(j) = p^n(1-p)^n \times \frac{N!}{j!(N-j)!}$$

In the limit $N\to\infty$, this probability becomes

$$P(j) = \frac{\gamma^j t^j}{j!}e^{-\gamma t}$$

This is the Poisson distribution of random processes.

(ii) Clearly, ℓ_{e-e} should be the average distance an electron travels before it hits one of its colleagues. Therefore, we require $j = 0$ in the Poisson distribution, which then reads $P(0) = e^{-\gamma t}$. Mapping γ and t on length scales is easy: $t = L/v_F$, and $\gamma = 1/\tau_{e-e} = v_F/\ell_{e-e}$. Here, v_F is the Fermi velocity and L the flight distance under consideration (remember that we are in the ballistic regime). This gives

$$P(0) = e^{-\frac{L}{\ell_{e-e}}}$$

Since the amplitude as defined in the text equals $P(0)$, and the assumption has been made that complete dephasing occurs in individual e-e scattering events, eq. (8.9) follows.

E 8.3: (i) The Schrödinger equation of the system reads

$$\frac{1}{2m^*}(\vec{p}+e\vec{A})^2\Psi(\phi) = E\Psi(\phi)$$

In cylindrical coordinates,

$$\vec{\nabla} \times \vec{A} = \left(\frac{1}{r}\frac{\partial A_z}{\partial \phi} - \frac{\partial A_\phi}{\partial z}, \frac{\partial A_r}{\partial z} - \frac{\partial A_z}{\partial r}, \frac{1}{r}\frac{\partial(r \cdot A_\phi)}{\partial r} - \frac{1}{r}\frac{\partial A_r}{\partial z}\right)$$

and

$$\vec{\nabla}\Psi = \left(\frac{\partial}{\partial r}, \frac{1}{r}\frac{\partial}{\partial \phi}, \frac{\partial}{\partial z}\right)\Psi$$

The vector potential given in the exercise gives $\vec{\nabla} \times \vec{A} = (0, 0, B)$. With the ansatz for the wave function

$$\Psi(\phi) = \frac{1}{\sqrt{2\pi r}} e^{i\ell\phi}$$

the Schrödinger equation becomes

$$\frac{1}{2m^*}\left(-i\hbar\frac{1}{r}\frac{\partial}{\partial \phi} + \frac{erB}{2}\right)^2 e^{i\ell\phi} = E e^{i\ell\phi}$$

leading to the energy eigenvalues

$$E_\ell(B) = \frac{\hbar^2}{2m^* r^2}\left[\ell + \frac{eBr^2}{2\hbar}\right]^2$$

Using the magnetic flux quantum $\Phi_0 = h/e$, we can rewrite this as

$$E_{\ell,n} = \frac{\hbar^2}{2m^* r^2}[\ell + n]^2$$

with n being the number of magnetic flux quanta penetrating the ring. ℓ is, of course, the angular momentum quantum number. Note that the probability density is independent of ℓ, ϕ and B!

(ii) The current is obtained from

$$I_\ell = -\frac{i\hbar e}{2m^*}(\Psi^* \vec{\nabla}\Psi - \Psi \vec{\nabla}\Psi^*)$$

Inserting a wave function gives

$$I_\ell = \frac{\hbar e}{4\pi m^* r^2}\ell$$

Since $\vec{L} = \vec{r} \times \vec{p} = \hbar\sqrt{\ell(\ell+1)} \approx \hbar\ell$, we can write it as

$$I_\ell = \frac{ev_F}{2\pi r} = e\nu$$

Here, ν denotes the circulation frequency of the electron in the ring. States with ℓ and $-\ell$ are degenerate, such that the corresponding currents cancel each other. If, however, the number of electrons in the ring is odd, an *equilibrium current* flows in the ring. This current is known as *persistent current*.

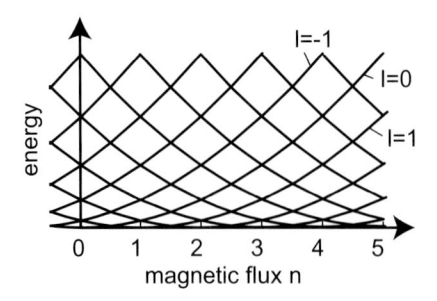

Figure E.2: Energy spectrum of a one-dimensional quantum ring.

(iii) Suppose there are about 100 electrons ($\ell = 50$) in the ring with radius $r = 300$ nm. A persistent current of 24 nA is found. This is a large current! It can be, and has been, measured by different techniques. One way is to detect the magnetic field generated by the current loop, using a superconducting quantum interference device (SQUID)[Mailly1993]. Another way becomes apparent as soon as one realizes that

$$I_\ell = \frac{1}{2A_{ring}} \frac{\partial E_\ell(B=0)}{\partial B}$$

Hence, the magnetic field dispersion of E_ℓ measures directly the persistent current. This can be done in a resonant tunneling experiment, see chapter 10.

E 8.4: (i) In analogy to chapter 7, one finds

$$I = \frac{2e}{h} \int -\frac{\partial f}{\partial E}(eV = 0)\delta(E - E_r)dE$$

$$\Rightarrow G = I/V = -\frac{2e^2}{h}\frac{\partial f}{\partial E}(E_r - \mu)$$

The peak transmission at $E_r = \mu$ equals $T(E_r) = 1/4k_B\Theta$. The FWHM is obtained from

$$\frac{1}{8k_B\Theta} = \frac{1}{k_B\Theta}\frac{e^{(E_{1/2}-\mu)/k_B\Theta}}{(1+e^{(E_{1/2}-\mu)/k_B\Theta})^2}$$

$$\Rightarrow E_{1/2} = k_B\Theta \ln(3 \pm 2\sqrt{2}) = \pm 1.7627 k_B T \Rightarrow \text{FWHM} \approx 3.52 k_B\Theta$$

(Note the remarkable relation $a^2 - b^2 = 1 \Rightarrow -\ln(a-b) = \ln(a+b)$).

(ii) The general line shape is a convolution of a Lorentzian with the derivative of the Fermi function:

$$G(\mu, E_r) = \frac{2e^2}{h}\frac{\Gamma_a\Gamma_b}{\Gamma}\int -\frac{\partial f}{\partial E}(E-\mu) \cdot L(\Gamma, E)dE$$

$L(\Gamma, E)$ is the Lorentzian. Experimentally, one can either fit the data to the above expression, using both Θ and Γ as fit parameters. Alternatively, one could vary the temperature and plot the FWHM as a function of Θ. The saturation temperature should give a good estimate for Γ.

Chapter 9

E 9.1: The single electron box consists of one electrode and one island. The capacitance matrix of the circuit in Fig. 9.20 reads

$$\underline{C} = \begin{pmatrix} C_{11} & -C_{1G} \\ -C_{1G} & C_{GG} \end{pmatrix}$$

with $C_{11} = C_{1G} + C_{1D}$ and $C_{GG} = C_{1G}$. The island charge equals $q = q_0 - ne$, such that the charge vector is $\vec{q} = (q_0 - ne, q_G)$. The voltage vector is just V_G. Here, n is the excess number of electrons at the island.

Two charge transfers are possible via the leaky capacitor between island and drain, $\Delta \vec{q} = e(\pm 1, 0)$, which means the energy relation

$$\Delta E[V, q_0 - ne, e(\pm 1, 0)] \geq 0$$

has to hold. Therefore, n excess electrons are on the island for

$$\frac{1}{C_G}\left(n(e - \frac{1}{2}) - q_0\right) < V_G < \frac{1}{C_G}\left(n(e + \frac{1}{2}) - q_0\right)$$

Fig. E.3 shows $n(V)$ for $q_0 = 0$.

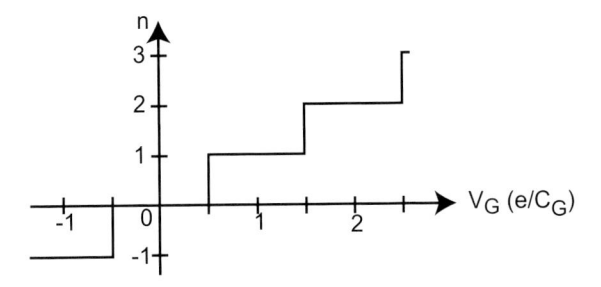

Figure E.3: Number of excess electrons in a single electron box (circuit of Fig. 9.20).

E 9.2: In that case,

$$\Gamma_{i \to f} = \frac{2\pi}{\hbar}|\langle i|H_t|f\rangle|^2 \delta(E_f - E_i)$$

which means that

$$\Gamma_{1 \to 2}(\Delta E = 0) = \frac{2\pi}{\hbar} \int_{E_{cb,max}}^{\infty} |\langle i|H_t|f\rangle|^2 D_i(E) D_f(E) f(E)[1 - f(E - eV)]dE$$

For low temperatures, we approximate the Fermi functions by step functions. Furthermore, in the limit of small voltages, the densities of states on both sides of the barrier are identical (in $d = 2$, this is the case anyway), and we obtain

$$I = e\Gamma_{1 \to 2} = \frac{2\pi}{\hbar}|\langle i|H_t|f\rangle|^2 D^2(E)V$$

E 9.3: Although the circuit resembles somewhat the double dot in series, it behaves quite different. The current through island 1 can be tuned by both gate voltages, although V_B couples to it only via island 2. Since all capacitances are supposed to be equal, we have

$$\underline{C}_{II} = \begin{pmatrix} 4C & -C \\ -C & 2C \end{pmatrix}$$

$$\underline{C}_{IE} = \begin{pmatrix} -C & 0 & -C \\ 0 & -C & 0 \end{pmatrix}$$

The charge transfers to be considered are

$$\Delta \vec{q_I} = e(\pm 1, 0)$$

$$\Delta \vec{q_I} = e(\pm 1, \mp 1)$$

In addition, the processes

$$\Delta \vec{q_I} = e(0, \pm 1)$$

is taken into account, as will be discussed below. The conditions for a stable configuration (n_1, n_2) to be established, we evaluate the system using eq. (9.4), and one obtains the stability conditions

$$V_B \leq -2V_A + \frac{e}{C}(2n_1 + n_2 + 1)$$

$$V_B \geq -2V_A + \frac{e}{C}(2n_1 + n_2 - 1)$$

$$V_B \leq \frac{1}{3}V_A + \frac{e}{C}\left(n_2 - \frac{1}{3}n_1 + \frac{2}{3}\right)$$

$$V_B \geq \frac{1}{3}V_A + \frac{e}{C}\left(n_2 - \frac{1}{3}n_1 - \frac{2}{3}\right)$$

and

$$V_B \leq -\frac{1}{4}V_A + \frac{e}{C}\left(n_2 + \frac{1}{4}n_1 + \frac{1}{2}\right)$$

$$V_B \geq -\frac{1}{4}V_A + \frac{e}{C}\left(n_2 + \frac{1}{4}n_1 - \frac{1}{2}\right)$$

for the above charge transfers. The stability diagram in the $V_A - V_B$ plane is shown in Fig. E.4. Current flows for the boundaries of the stable regions that correspond to condition, as indicated by the bold lines in the figure. As for the relevance of the direct charge transfer between island 2 and the reservoirs, we follow the arrow in Fig. E.4, and increase V_B at constant $V_A \approx 1$ (in units of e/C). As we cross $V_B = 1/2$ (point i), the free energy of the configuration $(1, 1)$ becomes smaller than that one of the initial $(1, 0)$ configuration. But the charge transfer should not be possible! It will nevertheless take place via tunnelling with a large time constant, in order to relax

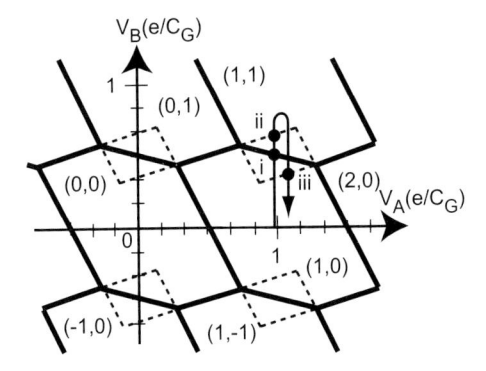

Figure E.4: Stability diagram of the circuit considered in [E 9.3]

the system into the ground state. If it can be neglected, the system will remain in a metastable state within the hatched region, until the energies of the configurations $(1,0)$ and $(0,1)$ are equal. This is the case at point ii and finally allows island 2 to obtain an electron form the reservoirs via island 1. As we go back, the system has to wait until the intermediate state $(2,0)$ becomes accessible, which happens at point iii. Via this state, the electron is transferred back into the reservoir. Hence, if the direct charge transfer between island 2 and the reservoirs is not possible, the system shows hysteresis effects within the diamonds formed by the dashed lines.

Chapter 10

E 10.1:
(iii) Only states with $m = 1$ couple sufficiently strongly to the leads, such that a current can be detected. In Fig. 10.8, we thus see the fraction of zigzag lines that corresponds to Landau level 1 states. Removing the charging energy gives the discrete spectrum of the island. Suppose it is approximately a Fock-Darwin spectrum. The beginning and the end of each bright line corresponds to level crossings of a Landau level 1 state witha Landau level 2 state. One finds $(\Delta B)_{measured} = 75$ mT at $B \approx 7$ T. So far, we have neglected the spin, though. The spin splitting of both Landau levels reduces the average period in B by a factor of 2. Hence, we find $\omega_0 = \omega_c \sqrt{\frac{(\Delta B)_{measured}}{B}} = 1.9 \cdot 10^{12} s^{-1}$. The reconstructed energy spectrum is shown in Fig. E.5.

E 10.2: (i) From $det(H - \lambda 1) = 0$, one finds

$$\lambda_\pm = \frac{1}{2}(H_{11} + H_{22}) \pm \frac{1}{2}\sqrt{(H_{11} - H_{22})2 + 4H_{12}^2}$$

(ii) For small transformation angles, the transformation matrix becomes

$$(O) = \begin{pmatrix} 1 & \alpha \\ -\alpha & 1 \end{pmatrix}$$

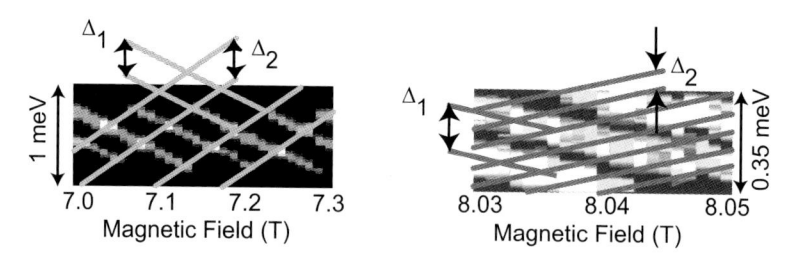

Figure E.5: The data of Fig. 10.8 (c) with the single electron charging energy removed, and the measured level spacings (left). To the right, the corresponding reconstruction of the data in Fig. 10.9 are shown for comparison.

From $p(O^T H O) = p(H)$, the condition

$$p(H) = p(H)\left[1 - \alpha\left(2H_{12}\frac{d\ln p_{11}}{dH_{11}} - 2H_{12}\frac{d\ln p_{22}}{dH_{22}} - (H_{11} - H_{22})\frac{d\ln p_{12}}{dH_{12}}\right)\right]$$

is obtained. Since α is arbitrary, it requires that the coefficient in front of α vanishes. The set of differential equations has the solution

$$p_{11}(H_{11}) = c_1 e^{-c_2 H_{11}^2 - c_3 H_{11}};$$

$$p_{12}(H_{12}) = c_1 e^{-2c_2 H_{12}^2};$$

$$p_{22}(H_{22}) = c_1 e^{-c_2 H_{22}^2 - c_3 H_{22}};$$

where the c_i are integration constants.

(iii) $p(H)$ now reads

$$p(H) = c_1 e^{-c_2(H_{11}^2 + 2H_{12}^2 + H_{22}^2) - c_3(H_{11} + H_{22})}$$

Choosing the energy reference such that $H_{11} + H_{22} = 0$, we see right away that this can be written as

$$p(H) = c_1 e^{-c_2 Tr(H^2)}$$

(iv) Let β be the transformation angle that maps H onto a diagonal matrix via an orthogonal transformation, namely

$$ODO^T = H$$

with

$$(D) = \begin{pmatrix} \lambda_+ & 0 \\ 0 & \lambda_- \end{pmatrix}$$

and

$$(O) = \begin{pmatrix} \cos\beta & \sin\beta \\ -\sin\beta & \cos\beta \end{pmatrix}$$

Note that here, we cannot assume that a transformation by a small angle will do the job. This transformation gives the functional dependence of H_{ij} on λ_+, λ_-, and α. Hence, the determinant of the Jacobian transformation matrix

$$J = \frac{\partial(H_{11}, H_{12}, H_{22})}{\partial(\lambda_+, \lambda_-, \beta)}$$

can be calculated, which gives

$$\det(J) = \lambda_+ - \lambda_-$$

In terms of the eigenvalues $\lambda_\pm$ of H, $Tr(H^2)$ equals $\lambda_+^2 + \lambda_-^2$. We can thus write $p(H) = p(H_{11}, H_{12}, H_{22}) = p(\lambda_+, \lambda_-, \alpha) \cdot \det[J(H_{11}, H_{12}, H_{22}, \lambda_+, \lambda_-, \alpha)]$ and get

$$p(\lambda_+, \lambda_-) = c_1(\lambda_+ - \lambda_-)e^{-c_2(\lambda_+^2 + \lambda_-^2)}$$

We transform variables here and write

$$\Delta = \lambda_+ - \lambda_-$$

$$\Sigma = \lambda_+ + \lambda_-$$

such that

$$p(\Delta, \Sigma) = c_1 \Delta e^{-\frac{c_2}{2}(\Delta^2 + \Sigma^2)}$$

Integration over Σ gives

$$p(\Delta) = \int_{-\infty}^{\infty} p(\Delta, \Sigma) d\Sigma = \sqrt{\frac{2\pi}{c_2}} c_1 \Delta e^{-\frac{c_2}{2}\Delta^2}$$

(v) The two normalization conditions read

$$\int_0^\infty p(\Delta) d\Delta = 1; \quad \int_0^\infty \Delta p(\Delta) d\Delta = 1$$

Hence, $c_1 = \pi/4$ and $c_2 = \pi/2$, which gives the corresponding expression in Table 10.1 for normalized peak spacings s.

References

[Aharonov1959] Y. Aharonov and D. Bohm, Phys. Rev. 115, 485 (1959)

[Albrecht2001] C. Albrecht, J. H. Smet, K. v. Klitzing, D. Weiss, V. Umansky, and H. Schweizer, Phys. Rev. Lett. 86, 147 (2001)

[Altshuler1981] B. L. Altshuler, A. G. Aronov, and B. Z. Spivak, JETP Lett. 33, 94 (1981)

[Altshuler1982] B. L. Altshuler, A. G. Aronov, and D. E. Khmelnitsky, J. Phys. C 15, 7367 (1982)

[Amlani1997] I. Amlani, A. O. Orlov, G. L. Snider, C. S. Lent, and G. H. Bernstein, Appl. Phys. Lett. 71, 1730 (1997)

[Amman1991] M. Amman, R. Wilkins, E. BenJacob, P. D. Maker, and R. C. Jaklevic, Phys. Rev. B 43, 1146 (1991)

[Ancona1996] M. G. Ancona, J. Appl. Phys. 79, 526 (1996)

[Ando1974] T. Ando, J. Phys. Soc. Japan 36, 959 (1974); ibid. 36, 1521 (1974); ibid. 37, 1044 (1974); Jpn. J. Appl. Phys. 2, 329 (1974))

[Ando1976] T. Ando, Phys. Rev. B 13, 3468 (1976)

[Ando1982] T. Ando, A.B. Fowler, and F. Stern, Rev. Mod. Phys. 54, 437 (1982)

[Andronikashvili1946] A. Andronikashvili, E.J. Phys. 10, 201 (1946). This article in a Russian journal has been translated in *Helium 4*, ed. A. Galasiewicz, Pergamon Press (1971)

[Ashcroft1985] N. W. Ashcroft and N. D. Mermin, *Solid State Physics*, Saunders College Publishing (1975)

[Averin1989] D. V. Averin and A. A. Odintsov, Phys. Lett. A 140, 251 (1989)

[Averin1990] D. V. Averin and Y. V Nazarov, Phys. Rev. Lett. 65, 2446 (1990)

[Averin1991] D.V. Averin and K.K. Likharev, in *Mesoscopic Phenomena in Solids*, eds. B.L. Altshuler, P.A. Lee, and R.A. Webb, Elsevier, Oxford, (1991)

[Averin1991a] D. V. Averin, A. N. Korotkov, and K. K. Likharev, Phys. Rev. B 44, 6199 (1991)

[Bachtold1998] A. Bachtold, M. Henny, C. Terrier, C. Strunk, C. Schnenberger, J.-P. Salvetat, J.-M. Bonard, and L. Forro, Appl. Phys. Lett. 73, 274 (1998)

[Banin1999] U. Banin, Y. Cao, D. Katz, and O. Millo, Nature 400, 542 (1999)

[Bardeen1948] J. Bardeen and W. Brattain, Phys. Rev. 74, 230 (1948)

[Bastard1989] G. Bastard, *Wave Mechanics Applied to Semiconductor Heterostructures*, Les Ulis, France (1989)

[Beenakker1987] C. W. J. Beenakker and H. van Houten, Phys. Rev. B 37, 6544 (1987)

[Beenakker1988] C.W.J. Beenakker and H. van Houten, Phys. Rev. B 37, 6544 (1988)

[Beenakker1989] C. W. J. Beenakker, Phys. Rev. Lett. 62, 2020 (1989)

[Beenakker1991] C. W. J. Beenakker and H. Van Houten, in *Solid State Physics* Vol. 44, pp. 1-228, Eds. H. Ehrenreich and D. Turnbull, Academic Press (1991)

[Beenakker1991a] C. W. J. Beenakker, Phys. Rev. B 44, 1646 (1991)

[Beenakker1997] C. W. J. Beenakker, Rev. Mod. Phys. 69, 732 (1997)

[Bennett1998] B. R. Bennett, M. J. Yang, B. V. Shanabrook, J. B. Boos, and D. Park, Appl. Phys. Lett. 72, 1193 (1998)

[Berggren1988] K.F. Berggren, G. Roos, and H. van Houten, Phys. Rev. B 37, 10118 (1988)

[Berman1999] D. Berman, N. B. Zhitenev, R. C. Ashoori, and M. Shayegan, Phys. Rev. Lett. 82, 161 (1999)

[Bernstein1987] G. Bernstein and D. K. Ferry, J. Vac. Sci. Technol. B 5, 964 (1987)

[Berthod1998] C. Berthod, N. Binggeli, and A. Baldereschi, Phys. Rev. B 57, 9757 (1998)

[Bhattacharya1982] S.K. Bhattacharya, Phys. Rev. B 25, 3756 (1982)

[Binnig1982] G. Binnig and H. Rohrer, Helv. Phys. Acta 55, 726 (1982); G. Binnig and H. Rohrer, Rev. Mod. Phys. 59, 615 (1987)

[Binnig1983] G. Binnig, H. Rohrer, Ch. Gerber, and E. Weibel, Phys. Rev. Lett. 50, 120 (1983)

[Bockrath1997] M. Bockrath, D. H. Cobden, P. L. McEuen, N. G. Chopra, A. Zettl, A. Thess, and R. E. Smalley, Science 275, 1922 (1997)

[Bockrath1999] M. Bockrath, D. H. Cobden, J. Lu, A. G. Rinzler, R. E. Smalley, L. Balents, and P. L. McEuen, Nature 397, 598 (1999)

[Boguslawski1989] P. Boguslawski and A. Baldereschi, Phys. Rev. B 39, 8055 (1989)

[Bohigas1984] O. Bohigas, M. J. Giannoni, and C. Schmit, Phys. Rev. Lett. 52, 1 (1984)

[Bohr1969] A. Bohr and B. R. Mottelson, in *Nuclear Structure*, W.A. Benjamin (1969)

[Bolognesi1996] C. R. Bolognesi, J. E. Bryce, and D. H. Chow, Appl. Phys. Lett. 69, 3531 (1996)

[Burt1992] M. G. Burt, J. Phys. Cond. Mat. 4, 6651 (1992)

[Buttiker1985] M. Büttiker, Y. Imry, R. Landauer, and S. Pinhas, Phys. Rev. B 31, 6207 (1985)

[Buttiker1986] M. Büttiker, Phys. Rev. Lett. 57, 1761 (1986)

[Buttiker1986a] M. Büttiker, Phys. Rev. B 33, 3020 (1986)

[Buttiker1988] M. Büttiker, IBM J. Res. Dev. 32, 317 (1988)

[Buttiker1990] M. Büttiker, Phys. Rev. B 41, 7906 (1990)

[Cahay1988] M. Cahay, M. McLennan, and S. Datta, Phys. Rev. B 37, 10125 (1988)

[Chadi1979] D. J. Chadi, Phys. Rev. Lett. 43, 43 (1979)

[Choi1986] K. K. Choi, D. C. Tsui, and S. C. Palmateer, Phys. Rev. B 33, 8216 (1986)

[Choi1987] K. K. Choi, D. C. Tsui, and K. Alavi, Phys. Rev. B 36, 7751 (1987)

[Chklovskii1992] D. B. Chklovskii, B. I. Shklovskii, and L. I. Glazman, Phys. Rev. B 46, 4026 (1992)

[Ciorga2000] M. Ciorga, A. S. Sachrajda, P. Hawrylak, C. Gould, P. Zawadzki, S. Jullian, Y. Feng, and Z. Wasilewski, Phys. Rev. B 61, R16315 (2000)

References

[Cleland1990] A. N. Cleland, J. M. Schmidt, and John Clarke, Phys. Rev. Lett. 64, 1565 (1990)

[Cobden2002] D. H. Cobden and J. Nygard, Phys. Rev. Lett. 89, 046803 (2002)

[CostaKramer1996] J. L. Costa-Kramer, Physics Today 49, 9 (1996)

[CostaKramer1997] J. L. Costa-Kramer, Phys. Rev. B 55, 4875 (1997)

[Cronenwett2002] S. M. Cronenwett, H. J. Lynch, D. Goldhaber-Gordon, L. P. Kouwenhoven, C. M. Marcus, K. Hirose, N. S. Wingreen, and V. Umansky, Phys. Rev. Lett. 88, 226805 (2002)

[Dagata1990] J.A. Dagata, Science 270, 1625 (1990), and references therein

[Darwin1931] C. G. Darwin, Proc. Cambridge Philos. Soc. 27, 86 (1931)

[Datta1997] S. Datta, *Electronic Transport in Mesoscopic Systems*, Cambridge University Press (1997)

[Davidovic1999] D. Davidovic and M. Tinkham, Phys. Rev. Lett. 83, 1644 (1999)

[Davidson1992] S. G. Davidson and M. Steslicka, *Basic Theory of Surface States*, Oxford University Press (1992)

[Desjonqueres1998] M.C. Desjonqueres and D. Spanjaard, *Concepts in Surface Physics* 2nd edition, Springer (1998)

[Devoret1990] M. H. Devoret, D. Esteve, H. Grabert, G.L. Ingold, H. Pothier, and C. Urbina, Phys. Rev. Lett. 64, 1824 (1990)

[Dingle1978] R. Dingle, H. L. Strmer, A. C. Gossard, and W. Wiegmann, Appl. Phys. Lett. 33, 665 (1978)

[Dolan1977] G. Dolan, Appl. Phys. Lett. 31, 337 (1977)

[Dresselhaus1992] P. D. Dresselhaus, C. M. A. Papavassiliou, R. G. Wheeler, and R. N. Sacks, Phys. Rev. Lett. 68, 106 (1992)

[Drexler1994] H. Drexler, Phd thesis, LMU Munich (1994)

[Drexler1994a] H. Drexler, D. Leonhard, W. Hansen, J. P. Kotthaus, and P. M. Petroff, Phys. Rev. Lett. 73, 2252 (1994)

[Eaglesham1990] D. J. Eaglesham and M. Cerullo, Phys. Rev. Lett. 64, 1943 (1990)

[Eigler1990] D. M. Eigler and E.K. Schweizer, Nature 344, 524 (1990)

[Einevoll1994] G. T. Einevoll and L. J. Sham, Phys. Rev. B 49, 10533 (1994)

[Esaki1974] L. Esaki and L. L. Chang, Phys. Rev. Lett. 33, 495 (1974)

[Evans1993] A. K. Evans, L. I. Glazman, and B. I. Shklovskii, Phys. Rev. B 48, 11120 (1993)

[Ezawa2000] Z.F. Ezawa, *Quantum Hall Effects: Field Theoretical Approach and related Topics*, World Scientific Press (2000)

[Facer1997] G. R. Facer, B. E. Kane, R. G. Clark, L. N. Pfeiffer, and K. W. West, Physical Review B 56, 10036 (1997)

[Fang1966] F. F. Fang and W. E. Howard, Phys. Rev. Lett. 16, 797 (1966)

[Ferry1997] D.K. Ferry and S.M. Goodnick, *Transport in Nanostructures*, Cambridge University Press (1997)

[Fischer1994] C.M. Fischer, M. Burghard, S. Roth, and K. v. Klitzing, Europhys. Lett. 28, 129 (1994)

[Fisher1981] D. S. Fisher and P. A. Lee, Phys. Rev. B 23, 6851 (1981)

[Fitzgerald1998] R. J. Fitzgerald, S. L. Pohlen, and M. Tinkham, Phys. Rev. B 57, 11073 (1998)

[Fleischmann1992] R. Fleischmann, T. Geisel, and R. Ketzmerick, Phys. Rev. Lett. 68, 1367 (1992)

[Fleischmann1994] R. Fleischmann, T. Geisel, and R. Ketzmerick, Europhys. Lett. 25, 219 (1994)

[Fock1928] V. Fock, Z. Phys. 47, 446 (1928)

[Folk1996] J. A. Folk, S. R. Patel, S. F. Godijn, A. G. Huibers, S. M. Cronenwett, C. M. Marcus, K. Campman, and A. C. Gossard , Phys. Rev. Lett. 76, 1699 (1996); A. M. Chang, H. U. Baranger, L. N. Pfeiffer, K. W. West, and T. Y. Chang, Phys. Rev. Lett. 76, 1695 (1996)

[Franco1997] S. Franco, *Design with Operational Amplifiers and Analog Integrated Circuits*, McGraw-Hill (1997)

[Frank1949] F. C. Frank and J. H. van der Merve, Proc. Royal Soc. London A 198, 205 (1949)

[Frank1998] S. Frank, P. Poncharal, Z. L. Wang, and W. de Heer, Science 280, 1744 (1998)

[Frensley1977] W. R. Frensley and H. Kroemer, Phys. Rev. B 16, 2642 (1977)

[Fromhold1995] T. M. Fromhold, P. B. Wilkinson, F. W. Sheard, L. Eaves, J. Miao, and G. Edwards, Phys. Rev. Lett. 75, 1142 (1995)

[Fuhrer2001a] A. Fuhrer, S. Lscher, T. Heinzel, K. Ensslin, W. Wegscheider, and M. Bichler, Phys. Rev. B 63, 125309 (2001)

[Fuhrer2001b] A. Fuhrer, S. Lüscher, T. Ihn, T. Heinzel, K. Ensslin, W. Wegscheider, and M. Bichler, Nature 413, 822 (2001)

[Fulton1983] T. A. Fulton and G. J. Dolan, Appl. Phys. Lett. 42, 752 (1983)

[Fulton1987] T. A. Fulton and G. J. Dolan, Phys. Rev. Lett. 59, 109 (1987)

[Furlan2000] M. Furlan, T. Heinzel, B. Jeanneret, S. V. Lothkov, and K. Ensslin, Europhys. Lett. 49, 369 (2000)

[Geerinckx1990] F. Geerinckx, F. M. Peeters, and J. T. Devreese, J. Appl. Phys. 68, 3435 (1990)

[Geerligs1990] L. J. Geerligs, V. F. Anderegg, P. A. M. Holweg, J. E. Mooij, H. Pothier, D. Esteve, C. Urbina, and M. H. Devoret , Phys. Rev. Lett. 64, 2691 (1990)

[Gerhardts1989] R. R. Gerhardts, D. Weiss, and K. v. Klitzing, Phys. Rev. Lett. 62, 1173 (1989)

[Ghandhi1994] S. K. Ghandhi, *VLSI Fabrication Principles: Silicon and Gallium Arsenide*, Wiley-Interscience (1994)

[Giaever1968] I. Giaever and H.R. Zeller, Phys. Rev. Lett. 20, 1504 (1968)

[Giannozzi1991] P. Giannozzi, S. de Gironcoli, P. Pavone, and S. Baroni , Phys. Rev. B 43, 7231 (1991)

[Giuliani1982] G. F. Giuliani and J. J. Quinn, Phys. Rev. B 26, 4421 (1982)

[Glazman1988] L. I. Glazman, G. B. Lesovik, D. E. Khmelnitsky, and R. I. Shekter, JETP Lett. 48, 238 (1988)

[Goodwin1939a] E. T. Goodwin, Proc. Cambridge Philos. Soc. 35, 205 (1939)

[Goodwin1939b] E. T. Goodwin, Proc. Cambridge Philos. Soc. 35, 221 (1939)

References

[Gorter1951] C. J. Gorter, Physica 17, 777 (1951)

[Grabert1991] H. Grabert, G.-L. Ingold, M. H. Devoret, D. Esteve, H. Pothier, and C. Urbina, Z. Phys. Cond. Mat. 84, 143 (1991)

[Grabert1992] H. Grabert and M. Devoret, *Single Charge Tunneling*, NATO ASI Series B 294, Plenum Press (1992)

[Grahn1999] H. Grahn, *Introduction to Semiconductor Physics*, World Scientific Press (1999)

[Grosso2000] G. Grosso and G. Parravicini, *Solid State Physics*, Academic Press (2000)

[Gruner1977] G. Grüner and M. Minier, Adv. Phys. 26, 231 (1977)

[Gueron1999] S. Gueron, M. M. Desmukh, E. B. Myers, and D. C. Ralph, Phys. Bev. Lett. 83, 4148 (1999)

[Guilani1982] G. F. Giuliani and J. J. Quinn, Phys. Bev. B 26, 4421 (1982)

[Gurevich2000] L. Gurevich, L. Canali, and L. P. Kouwenhoven, Appl. Phys. Lett. 76, 384 (2000)

[Gutzwiller1971] M.C. Gutzwiller, J. Math. Phys. 12, 343 (1971); see also W. H. Miller, J. Chem Phys. 63, 996 (1975)

[Gutzwiller1990] M. C. Gutzwiller, *Chaos in Quantum and Classical mechanics*, Springer (1990);

[Haake1991] F. Haake, *Quantum Signatures of Chaos*, Springer (1991)

[Hajdu1994] J. Hajdu (ed.), *Introduction to the theory of the integer quantum Hall effect* Wiley-VCH (1994)

[Halperin1982] B. I. Halperin, Phys. Rev. B 25, 2185 (1982)

[Hamada1992] N. Hamada, S. Sawada, and A. Oshiyama, Phys. Rev. Lett. 68, 1579 (1992).

[Hanna1991] A. E. Hanna and M. Tinkham, Phys. Rev. B 44, 5919 (1991).

[Harris1999] P. Harris, *Carbon Nanotubes and Related Structures*, Cambridge University Press (1999)

[Harrison1999] W. A. Harrison, *Elementary Electronic Stucture*, World Scientific Publishing Corp., Singapore (1999).

[Hattori1998] T. Hattori (ed.), *Ultraclean surface processing of silicon wafers*, Springer (1998)

[Haug1987] R. J. Haug, R. R. Gerhardts, K. v. Klitzing, and K. Ploog, Phys. Rev. Lett. 59, 1349 (1987)

[Haug1989] R. J. Haug, J. Kucera, P. Streda, and K. von Klitzing, Phys. Rev. B 39, 10892 (1989)

[Heine1965] V. Heine, Phys. Rev. 138, A1689 (1965)

[Heinzel2000] T. Heinzel, G. Salis, R. Held, S. Lscher, K. Ensslin, W. Wegscheider, and M. Bichler, Phys. Rev. B 61, 13353 (2000).

[Held1998] R. Held, T. Vancura, T. Heinzel, K. Ensslin, M. Holland, and W. Wegscheider, Appl. Phys. Lett. 73, 262 (1998)

[Hofstadter1976] D. Hofstadter, Phys. Rev. B 14, 2239 (1976)

[Horowitz1989] P. Horowitz and W. Hill, *The Art of Electronics*, Cambridge University Press (1989)

[Horowitz1999] G. Horowitz, R. Hajlaoui, D. Fichou, and A. El Kassmi, J. Appl. Phys. 85, 3202 (1999)

[Houten1988] H. van Houten, C. W. J. Beenakker, P. H. M. van Loosdrecht, T. J. Thornton, H. Ahmed, M. Pepper, C. T. Foxon, and J. J. Harris, Phys. Rev. B 37, 8534 (1988)

[Houten1988a] H. van Houten, J. G. Williamson, M. E. I. Broekaart, C. T. Foxon, and J. J. Harris, Phys. Rev. B 37, 2756 (1988)

[Houten1989] H. van Houten and C. W. J. Beenakker, Phys. Rev. Lett. 63, 1893 (1989)

[Houten1989a] H. van Houten, C. W. J. Beenakker, J. G. Williamson, M. E. I. Broekaart, P. H. M. van Loosdrecht, B. J. van Wees, J. E. Mooij, C. T. Foxon, and J. J. Harris, Phys. Rev. B 39, 8556 (1989)

[Houten1992] H. van Houten, C. W. J. Beenakker, and B. J. van Wees, in *Semiconductors and Semimetals* 35, pp. 9-112, Eds. H. Ehrenreich and H. Turnbull, Academic Press Inc. (1992)

[Iijima1991] S. Iijima, Nature 354, 56 (1991)

[Ilani2001] S. Ilani, A. Yacoby, D. Mahalu, and H. Shtrikman, Science 292, 1354 (2001)

[Irmer1998] B. Irmer, R. H. Blick, F. Simmel, W. Gödel, H. Lorenz, and J. P. Kotthaus, Appl. Phys. Lett. 73, 2051 (1998)

[Ismail1995] K. Ismail, M. Arafa, K. L. Saenger, J. O. Chu, and B. S. Meyerson, Appl. Phys. Lett. 66, 1077 (1995)

[Itterbeek1947] A. van Itterbeek and L. de Greeve, Experimentia 3, No.7 (1947)

[Jacak1998] L. Jacak, P. Hawrylak, and A. Wojs: *Quantum Dots*, Springer (1998)

[Jacoboni1985] C. Jacoboni and L. Reggiani Rev. Mod. Phys. 55, 645-705 (1983)

[Jalabert1992] R. A. Jalabert, A. D. Stone, and Y. Alhassid, Phys. Rev. Lett. 68, 3468 (1992)

[Jeckelmann1997] B. Jeckelmann, B. Jeanneret, and D. Inglis, Phys. Rev. B 55, 13124 (1997)

[Johnson1992] A. T. Johnson, L. P. Kouwenhoven, W. de Jong, N. C. van der Vaart, C. J. P. M. Harmans, and C. T. Foxon, Phys. Rev. Lett. 69, 1592 (1992)

[Keller1996] M. W. Keller, J. M. Martinis, N. M. Zimmerman, and A. H. Steinbach, Appl. Phys. Lett. 69, 1804 (1996)

[Keller1999] M. W. Keller, A. L. Eichenberger, J. M. Martinis, and N. W. Zimmerman, Science 285, 1706 (1999)

[Klass1991] U. Klass, Z. Phys. B 82, 351 (1991)

[Klein1996] D. L. Klein, R. Roth, A. P. Alivisatos, A. K. Lim, and P. L. McEuen, Nature 389, 699 (1997)

[Klein1997] D. L. Klein, P. L. McEuen, J. E. Bowen Katari, R. Roth, and A. P. Alivisatos, Appl. Phys. Lett. 68, 2574 (1996)

[Klitzing1980] K. v. Klitzing, G. Dorda, and M. Pepper, Phys. Rev. Lett. 45, 494 (1980)

[Knott1994] R. Knott, Solid State Electronics 37, 689 (1994)

[Korotkov1996] A. N. Korotkov, in *Molecular Electronics*, ed. by J. Jortner and M. A. Ratner, Blackwell (1996)

[Korotkov1998] A. N. Korotkov, Appl. Phys. Lett. 72, 3226 (1998)

[Kouwenhoven1997] L. P. Kouwenhoven, C. M. Marcus, P. L. McEuen, S. Tarucha, R. M. Westervelt, and N. S. Wingreen, in *Mesoscopic Electron Transport*, pp. 105-214, Proceedings of the NATO Advanced Study Institute, ser. E, vol. 345, Eds. L. P. Kouwenhoven, G. Schön, and L. L. Sohn, Kluwer (1997)

[Kubo1957] R. Kubo, J. Phys. Soc. Jpn. 12, 570 (1957)

[Kuhl1998] U. Kuhl and H.-J. Stöckmann, Phys. Rev. Lett. 80, 3232 (1998)

[Kulik1975] I. O. Kulik and R. I. Shekter, Sov. Phys. JETP 41, 308 (1975)

[Kumar1990] A. Kumar, S. E. Laux, and F. Stern, Phys. Rev. B 42, 5166 (1990)

[Landauer1957] R. Landauer, IBM J. Res. Dev. 1, 223 (1957)

[Laughlin1981] R. B. Laughlin, Phys. Rev. B 23, 5632 (1981)

[Lee1986] P. Lee, Physica 140A, 169 (1986)

[Leonard1993] D. Leonard, M. Krishnamurthy, C. M. Reaves, S. P. Denbaars, and P. M. Petroff, Appl. Phys. Lett. 63, 3203 (1993)

[Likharev1988] K. K. Likharev, IBM J. Res. Dev. 32, 144 (1988)

[London1938] F. London, Nature 141, 643 (1938)

[Lotkhov2001] S. V. Lotkhov, S. A. Bogoslovsky, A. B. Zorin, and J. Niemeyer, Appl. Phys. Lett. 78, 946 (2001)

[Louie1977] S. G. Louie, and M. L. Cohen, Phys. Rev. B 15, 2154 (1977)

[Lounasmaa1974] O. V. Lounasmaa, *Experimental Principles and Methods Below 1 K*, Academic Press (1974)

[Ludoph2000] B. Ludoph and J. M. van Ruitenbeek, Phys. Rev. B 61, 2273 (2000)

[Luscher2001] S. Lüscher, T. Heinzel, K. Ensslin, W. Wegscheider, and M. Bichler, Phys. Rev. Lett. 86, 2118 (2001)

[Maan1984] J.C. Maan, in *Two Dimensional Systems, Heterostructures, and Superlattices*, p. 183, eds. G. Bauer, F. Kuchar, and H. Heinrich, Springer Verlag, Berlin (1984)

[Maao1994] Frank A. Maao, I. V. Zozulenko, and E. H. Hauge, Phys. Rev. B 50, 17320 (1994)

[Mailly1993] D. Mailly, C. Chapelier, and A. Benoit, Phys. Rev. Lett. 70, 2020 (1993)

[Mamin1990] H. J. Mamin, P. H. Guethner, and D. Rugar, Phys. Rev. Lett. 65, 2418 (1990)

[Marrian1993] For a review, see *Technology of Proximal Probe Lithography*, Ed. C.R.K. Marrian, SPIE Optical Engineering Press, Bellingham, WA (1993)

[Martinis1994] J. M. Martinis, M. Nahum, and H. D. Jensen, Phys. Rev. Lett. 72, 904 (1994)

[Marzin1989] J.-Y. Marzin and J.-M- Gerard, Phys. Rev. Lett. 62, 2172 (1989)

[Maue1935] A.W. Maue, Z. Physik 94, 717 (1935)

[McClintock1984] P.V.E. McClintock, D.J. Meredith, and J.K. Wigmore, *Matter at Low Temperatures*, Blackie and Son Ltd., (1984)

[McEuen1992] P. L. McEuen, E. B. Foxman, Jari Kinaret, U. Meirav, M. A. Kastner, N. S. Wingreen, and S. J. Wind, Phys. Rev. B 45, 11419 (1992)

[Meirav1989] U. U. Meirav, M. A. Kastner, M. Heiblum, and S. J. Wind, Phys. Rev. B 40, 5871 (1989)

[Meirav1990] U. Meirav, M. A. Kastner, and S. J. Wind, Phys. Rev. Lett. 65, 771 (1990)

[Mehta1991] M.L. Mehta, *Random Matrices*, Academic Press (1991)

[Mohanty1997] P. Mohanty, E. M. Q. Jariwala, and R. A. Webb, Phys. Rev. Lett. 78, 3366 (1997)

[Molenkamp1990] L. W. Molenkamp, A. A. M. Staring, C. W. J. Beenakker, R. Eppenga, C. E. Timmering, J. G. Williamson, C. J. P. M. Harmans, and C. T. Foxon, Phys. Rev. B 41, 1274 (1990)

[Monch2001] W.Mönch, *Semiconductor Surfaces and Interfaces*, Springer (2001)

[Montie1991] E. A. Montie, E. C. Cosman, G. W. 't Hooft, M. B. van der Mark, and C. W. J. Beenakker, Nature 350, 594 (1991)

[Moon1997] J. S. Moon, J. A. Simmons, and J. L. Reno, Appl. Phys. Lett. 71, 656-658 (1997)

[Nihey1995] F. Nihey, S. W. Hwang, and K. Nakamura, Phys. Rev. B 51, 4649 (1995)

[Odom1998] T. W. Odom, J. Huang, P. Kim, and C. M. Lieber, Nature 391, 62 (1998)

[Painter1970] G. S. Painter and D. E. Ellis, Phys. Rev. B1, 4747 (1970)

[Palmer1982] R. G. Palmer, Adv.Phys. 31, 669 (1982)

[Park1999] H. Park, A. K. Lim, A. P. Alivisatos, J. Park, and P. L. McEuen, Appl. Phys. Lett. 75, 301 (1999)

[Park2000] H. Park, J. Park, A. K. Lim, E. H. Anderson, A. P. Alivisatos, and P. L. McEuen, Nature 407, 57 (2000)

[Parker1968] G. H. Parker and C. A. Mead, Phys. Rev. Lett. 21, 605 (1968)

[Patel1998] S. R. Patel, S. M. Cronenwett, D. R. Stewart, A. G. Huibers, C. M. Marcus, C. I. Duruz, J. S. Harris, K. Campman, and A. C. Gossard, Phys. Rev. Lett. 80, 4522, (1998)

[Pauling1960] L. Pauling, *The Nature of the Chemical Bond*, Cornell University Press, (1960)

[Pauw1958] L. J. van der Pauw, Philips Research Reports 13, 1-9 (1958)

[Pedersen2000] S. Pedersen, A. E. Hansen, A. Kristensen, C. B. Srensen, and P. E. Lindelof , Phys. Rev. B 61, 5457 (2000)

[Petroff1984] P. M. Petroff, R. C. Miller, A. C. Gossard, and W. Wiegmann, Appl. Phys. Lett. 44, 217 (1984)

[Petta2001] J.R . Petta and D. C. Ralph, Phys. Rev. Lett. 87, 266801 (2001)

[Pfannkuche1992] D. Pfannkuche and R. R. Gerhardts, Phys. Rev. B 46, 12606 (1992)

[Pfeiffer1989] L. Pfeiffer, K. W. West, H. L. Stormer, and K. W. Baldwin, Appl. Phys. Lett. 55, 1888 (1989)

[Picciotto2001] R. de Picciotto, H. L. Stormer, L. N. Pfeiffer, K. W. Baldwin, and K.W . West, Nature 411, 52 (2001)

[Pinczolits1998] M. Pinczolits, G. Springholz, and G. Bauer, Appl. Phys. Lett. 73, 250 (1998)

[Porath2000] D. Porath, A. Bezryadin, S. de Vries, and C. Dekker, Nature 403, 635 (2000)

[Pothier1992] H. Pothier, P. Lafarge, C. Urbina, D. Esteve, and M. H. Devoret, Europhys. Lett. 17, 249 (1992)

[Prange1990] R.E. Prange and S. M. Girvin (Eds.), *The Quantum Hall Effect*, Springer (1990)

[Quinn1989] T. Quinn, Metrologia 26, 69 (1989)

[Ralph1995] D. C. Ralph, C. T. Black, and M. Tinkham, Phys. Rev. Lett. 74, 3241 (1995)

[Reed1997] M. A. Reed, C. Zhou, C. J. Muller, T. P. Burgin, and J. M. Tour, Science 278, 252 (1997)

[Reif1985] F. Reif, *Fundamentals of Statistical and Thermal Physics*, McGraw-Hill Book Co., (1985)

[Richardson1988] R.C. Richardson and E.N. Smith, *Experimental Techniques in Condensed Matter Physics at Low Temperatures*, Addison-Wesley (1988)

[Ridley1999] B.K. Ridley, *Quantum Processes in Semiconductors*, Oxford University Press (1999)

[Rose-Innes1973] A.C. Rose-Innes, *Low Temperature Laboratory Techniques*, English University Press (1973)

[Rychen1998] J. Rychen, T. Vancura, T. Heinzel., R. Schuster, and K. Ensslin, Phys. Rev. B 58, 3568 (1998)

[Saito1992] R. Saito, M. Fujita, G. Dresselhaus, and M. S Dresselhaus, Appl. Phys. Lett. 60, 2204 (1992)

[Saito1998] R. Saito, G. Dresselhaus, and M. S Dresselhaus, *Physical Properties of Carbon Nanotubes*, Imperial College Press (1998)

[Salis1997] G. Salis, B. Graf, K. Ensslin, K. Campman, K. Maranowski, and A. C. Gossard, Phys. Rev. Lett. 79, 5106 (1997)

[Salis1998] G. Salis, B. Ruhstaller, K. Ensslin, K. Campman, K. Maranowski, and A. C. Gossard, Physical Review B 58, 1436 (1998)

[Salis1999] G. Salis, P. Wirth, T. Heinzel, T. Ihn, K. Ensslin, K. Maranowski, and A. C. Gossard, Physical Review B 59, R5304 (1999)

[Saxler2000] A. Saxler, P. Debray, R. Perrin, S. Elhamri, W. C. Mitchel, C. R. Elsass, I. P. Smorchkova, B. Heying, E. Haus, P. Fini, J. P. Ibbetson, S. Keller, P. M. Petroff, S. P. DenBaars, U. K. Mishra, and J. S. Speck, J. Appl. Phys. 87, 369 (2000)

[Scheer1997] E. Scheer, P. Joyez, D. Esteve, C. Urbina, and M. H. Devoret, Phys. Rev. Lett. 78, 3535 (1997)

[Schlosser1996] T. Schlösser, K. Ensslin, J.P. Kotthaus, and M. Holland, Europhys. Lett. 33, 683 (1996)

[Shockley1939] W. Shockley, Phys. Rev. 56, 3175 (1939)

[Schonhammer1996] K. Schönhammer and V. Meden, Am. J. Phys. 64, 1168 (1996)

[Schottky1938] W. Shottky, Naturwissenschaften 26, 843 (1938)

[Schuh1985] B. Schuh, J. Phys. A 18, 803 (1985)

[Schuster1997] R. Schuster, K. Ensslin, J.P. Kotthaus, G. Böhm, and W. Klein, Phys. Rev. B 55, 2237 (1997)

[Scott-Thomas1989] J. H. F. Scott-Thomas, S. B. Field, M. A. Kastner, H. I. Smith, and D. A. Antoniadis, Phys. Rev. Lett. 62, 583 (1989)

[Seeger1997] K. Seeger, *Semiconductor Physics, An Introduction*, Springer Series in Solid State Sciences (1997)

[Senz2002] V. Senz, PhD thesis, ETH Zürich (2002)

[Sharvin1981] D. Y. Sharvin and Y. V. Sharvin, JETP Lett. 34, 272 (1981)

[Shepard1992] K. L. Shepard, M. L. Roukes, and B. P. van der Gaag, Phys. Rev. Lett. 68, 2660 (1992)

[Sherbakov1996] A. G. Scherbakov, E. N. Bogachek, and U. Landman, Phys. Rev. B 53, 4054 (1996)

[Shirakashi1998] J. Shirakashi, K. Matsumoto, N. Miura, and M. Konagai, Appl. Phys. Lett. 72, 1893 (1998)

[Simmel1997] F. Simmel, T. Heinzel, and D. A. Wharam, Europhys. Lett. 38, 123 (1997)

[Simon1994] S. Simon and B. Halperin, Phys. Rev. Lett. **73**, 3278 (1994).

[Simonyi1963] See, for example, K. Simonyi, *Foundationsof Electrical Engineering*, Pergamon Press, (1963)

[Sivan1996] U. Sivan, R. Berkovits, Y. Aloni, O. Prus, A. Auerbach, and G. Ben-Yoseph, Phys. Rev. Lett. 77, 1123, (1996)

[Smit2002] R. M. H. Smit, Y. Noat, C. Untiedt, N. D. Lang, M. C. van Hemert, and J. M. van Ruitenbeek, Nature 419, 906 (2002)

[Smith1986] T. P. Smith, W. I. Wang, and P. J. Stiles , Phys. Rev. B 34, 2995 (1986)

[Springholz1998] G. Springholz, V. Holy, M. Pinczolits, and G. Bauer, Science 282,734 (1998)

[Springsguth1997] D. Springsguth, R. Ketzmerick, and T. Geisel, Phys. Rev. B 56, 2036 (1997)

[Stauffer1995] D. Stauffer and A. Aharony, *Introduction to percolation theory*, Wiley-VCH (1995)

[Stillman1976] G. E. Stillman and C. M. Wolfe, Thin Solid Films 31, 69 (1976)

[Stormer1982] H. Stormer, A. C. Gossard, and W. Wiegmann, Solid State Comm. 41, 707 (1982)

[Stormer1986] H. L. Stormer, J. P. Eisenstein, A. C. Gossard, W. Wiegmann, and K. Baldwin, Phys. Rev. Lett. 56, 85 (1986)

[Stormer1992] H. Stormer, Solid State Commun. 84, 95 (1992)

[Stranski1939] I. N. Stranski and L. Krastanov, Akad. Wiss. Lit. Mainz Math.-Naturwiss. Kl. IIb 146, 797 (1939)

[Sze1985] S. M. Sze, *Semiconductor devices : physics and technology* Wiley, New York (1985)

[Tamm1932] I. E. Tamm, Phys. Z. Sowjetunion. 1, 733 (1932)

[Tans1997] S. J. Tans, M. H. Devoret, H. Dai, A. Thess, R. E. Smalley, L. J. Geerligs, and C. Dekker, Nature 386, 474 (1997)

[Tarucha1996] S. Tarucha, D. G. Austing, T. Honda, R. J. van der Hage, and L. P. Kouwenhoven, Phys. Rev. Lett. 77, 3613 (1996)

[Tersoff1984] J. Tersoff, Phys. Rev. B 30, 4874 (1984)

[Thomas1996] K. J. Thomas, J. T. Nicholls, M. Y. Simmons, M. Pepper, D. R. Mace, and D. A. Ritchie , Phys. Rev. Lett. 77, 135 (1996)

[Thornton1989] T. J. Thornton, M. L. Roukes, A. Scherer, and B. P. Van de Gaag, Phys. Rev. Lett. 63, 2128 (1989)

[Thouless1981] D. J. Thouless, J. Phys. C 14, 3475 (1981)

[Thouless1982] D. J. Thouless, M. Kohmoto, M. P. Nightingale, and M. de Nijs, Phys. Rev. Lett. 49, 405 (1982)

[Timp1988] G. Timp, Surf. Sci. 196, 68 (1988)

[Topinka2000] M. A. Topinka, B. J. LeRoy, S. E. J. Shaw, E. J. Heller, R. M. Westervelt, K. D. Maranowski, and A. C. Gossard, Science 289, 2323 (2000).

[Tsui1982] D. C. Tsui, H. L. Stormer, and A. C. Gossard , Phys. Rev. Lett. 48, 1559 (1982)

[Tuttle1989] G. Tuttle, H. Kroemer, and J. H. English, J. Appl. Phys. 65, 5239 (1989)

[Ullmo2001] D. Ullmo and H. Baranger, Phys. Rev. B 64, 245324 (2001), and references therein.

[Umansky1997] V. Umansky, R. de Picciotto, and M. Heiblum, Appl. Phys. Lett. 71, 683 (1997)

[Umbach1986] C. P. Umbach, C. Van Haesendonck, R. B. Laibowitz, S. Washburn, and R. A. Webb, Phys. Rev. Lett. 56, 386 (1986)

[Vallejos1998] R. O. Vallejos, C. H. Lewenkopf, and E. R. Mucciolo, Phys. Rev. Lett. 81, 677, (1998); see also G. Hackenbroich and H. Weidenmüller, Phys. Rev. Lett. 79, 127, (1997)

[Vries1995] D. K. de Vries and A. D. Wieck, Am. J. Phys. 63, 1074-1078 (1995)

[Volmer1926] M. Volmer and A. Weber, Z. Phys. Chem. 119, 277 (1926)

[Wallace1947] P. R. Wallace, Phys. Rev. 71, 622 (1947)

[Walukiewicz1984] W. Walukiewicz, Phys. Rev. B 30, 4571 (1984)

[Warren1985] A. C. Warren, IEEE Electron.Dev. Lett. 6, 294 (1985)

[Webb1985] R. A. Webb, S. Washburn, C. P. Umbach, and R. B. Laibowitz, Phys. Rev. Lett. 54, 2696 (1985)

[Wees1988] B. J. van Wees, H. van Houten, C. W. J. Beenakker, J. G. Williamson, L. P. Kouwenhoven, D. van der Marel, and C. T. Foxon, Phys. Rev. Lett. 60, 848 (1988).

[Wees1988a] B. J. van Wees, L. P. Kouwenhoven, H. van Houten, C. W. J. Beenakker, J. E. Mooij, C. T. Foxon, and J. J. Harris , Phys. Rev. B 38, 3625 (1988)

[Wegscheider1997] W. Wegscheider, G. Schedelbeck, G. Abstreiter, M. Rother, and M. Bichler, Phys. Rev. Lett. 79, 1917 (1997)

[Wegscheider1998] W. Wegscheider, in *Optics of Semiconductor Quantum Wires and Dots*, ed. by G. W. Bryant, Gordon and Breach Science Publishing (1998)

[Wei1997] Y. Y. Wei, J. Weis, K. v. Klitzing, and K. Eberl, Appl. Phys. Lett. 71, 2514 (1997)

[Weiss1989] D. Weiss, K. v. Klitzing, K. Ploog, and G. Weimann, Europhys. Lett. 8, 179 (1989)

[Weiss1991] D. Weiss, M. L. Roukes, A. Menschig, P. Grambow, K. von Klitzing, and G. Weimann, Phys. Rev. Lett. 66, 2790 (1991)

[Weiss1993] D. Weiss, K. Richter, A. Menschig, R. Bergmann, H. Schweizer, K. v. Klitzing, and G. Weimann, Phys. Rev. Lett. 70, 4118 (1993)

[Wharam1988] D. A. Wharam, T. J. Thornton, R. Newbury, M. Pepper, H. Ahmed, J. E. F. Frost, D. G. Hasko, D. C. Paacock, D. A. Ritchie, and G. A. C. Jones, J. Phys. C 21, L209 (1988)

[Wharam1988a] D. A. Wharam, M. Pepper, H. Ahmed, J. E. F. Frost, D. G. Hasko, D. C. Paacock, D. A. Ritchie, and G. A. C. Jones, J. Phys. C 21, L887 (1988)

[Wildoer1998] J. W. G. Wildoer, L. C. Venema, A. G. Rinzler, and R. E. Smalley, and C. Dekker, Nature 391, 59 (1998)

[Wilks1970] J. Wilks, *An Introduction to Liquid Helium*, Claredon Press (1970)

[Willett1987] R. Willett, J. P. Eisenstein, H. L. Strmer, D. C. Tsui, A. C. Gossard, and J. H. English, Phys. Rev. Lett. 59, 1776 (1987)

[Williams1990] R. E. Williams, *Modern GaAs Processing Methods*, Artech House (1990)

[Winkler1989] R. W. Winkler, J. P. Kotthaus, and K. Ploog, Phys. Rev. Lett. 62, 1177 (1989)

[Worschech1999] L. Worschech, F. Beuscher, and A. Forchel, Appl. Phys. Lett. 75, 578 (1999)

[Xie1995] Q. Xie, A. Madhukar, P. Chen, and N. P. Kobayashi, Phys. Rev. Lett. 75, 2542 (1995)

[Yacoby1990] A. Yacoby and Y. Imry, Phys. Rev. B 41, 5341 (1990)

[Yacoby1991] A. Yacoby, U. Sivan, C. P. Umbach, and J. M. Hong, Phys. Rev. Lett. 66, 1938 (1991)

[Yacoby1996] A. Yacoby, H. L. Stormer, N. S. Wingreen, L. N. Pfeiffer, K. W. Baldwin, and K. W. West, Phys. Rev. Lett. 77, 4612 (1996).

[Yoo1997] M. J. Yoo, T. A. Fulton, H.F. Hess, R. L. Willett, L. N. Dunkelberger, R. J. Chichester, L. N. Pfeiffer, and K. W. West, Science 276, 579 (1997)

[Yoshioka2002] D. Yoshioka, *The Quantum Hall Effect*, Springer (2002)

[Yu1999] P. Yu and M. Cardona, *Fundamental of Semiconductors*, Springer, Berlin (1999)

[Zak1985] J. Zak, Phys. Rev. B 32, 2218 (1985)

[Zaremba1992] E. Zaremba, Phys. Rev. B 45, 14143 (1992)

[Ziman1992] J.M. Ziman, *Elements of Advanced Quantum Theory*, Cambridge University Press (1992)

[Ziman1995] J.M. Ziman, *Principles of the Theory of Solids*, Cambridge University Press (1995)

[Zimmerli1992] G. Zimmerli, T. M. Eiles, R. L. Kautz, and John M. Martinis, Appl. Phys. Lett. 61, 237 (1992)

[Zimmerman2001] N. M. Zimmerman, W. H. Huber, A. Fujiwara, and Y. Takahashi, Appl. Phys. Lett. 79, 3188 (2001)

Index

0.7 structure, 181

acceptor, 41
accumulation, 78
addition energy, 255
addition spectrum, 239
Aharonov-Bohm effect, 17, 203
Al[Ga]As, 32
Altshuler-Aronov-Spivak oscillations, 205, 283
ambipolar devices, 78
angle evaporation technique, 108
antidots, 279
antilocalization, 222
artificial atom, 12

ballistic transport, 13
Ballistic wires, 171
band alignment, 70, 77
band bending, 68
band gaps, 26
bending of energy bands, 68
Berggren model, 169
Bloch theorem, 26
Boltzmann distribution, 36
Boltzmann equation, 45
Boltzmann model, 23, 43
bonding, 110
boundary scattering, 169
break junctions, 190
Bridgman growth technique, 96
Brillouin zones, 24
Brownian motion, 132

capacitance spectroscopy, 142
carbon nanotubes, 192, 272
charge neutrality level, 68
chemical potential, 34

cleaved edge overgrowth, 99
conductance quantization, 171
constant interaction model, 253
contact resistance, 176
correlation function, 132
cotunnelling, 245
Coulomb blockade, 225
Coulomb staircase, 232
cryostats, 117
crystal growth, 95
crystal structure:Si,GaAs, 24
current sources, 123
current-to-voltage converter, 126
cyclotron radius, 135
Czochralski growth, 95

density of states, 33
dephasing time, 134
diamagnetic shift, 158
differential conductance, 128
diffusion constant, 132
diffusion equation, 132
dilution refrigerator, 119
direct band gap, 32
donor, 41
doping, 40
double barrier, 217
Drude model, 45

edge channels, 189
edge states, 166, 185
effective density of states, 36
effective mass, 12
effective mass in heterostructures, 87
effective mass in MOSFETs, 80
Einstein relation, 133
elastic mean free path, 131
electrometer, 243

electron beam lithography, 11, 105
electron waveguide, 180
electron-phonon scattering, 51
energy bands, 26
envelope wave equation, 40
envelope wave functions, 36
ergodicity, 210
etching, 106

fcc lattice, 24
Fermi energy, 34
Fermi wavelength, 12, 131
Fermi-Dirac distribution, 34
field effect transistor, 78
filling factor, 140
flux cancellation, 222
Fock-Darwin model, 181
four-probe measurement, 128
Fourier transformation, 291
fractional quantum Hall effect, 144
freezeout regime, 43
Friedel oscillations, 55
fullerene, 272

Ga[Al]As, 81
GaAs, 24
GaAs: band structure, 27
GaAs: holes, 31
Gaussian ensembles, 266
graphite, 25, 27
guiding center, 147

Hall bar, 50
Hall effect, 50
heavy holes, 31
HeII, 112
helium, 110
helium cryostats, 117
HEMT, 81
heterointerface, 75, 81
Hofstadter butterfly, 284
Hund's rules, 261

impurity scattering, 51
InAs/AlSb quantum wells, 83
indirect band gap, 32
induced gap states, 70, 76
interaction parameter, 135
interfaces, 69

intrinsic carrier concentration, 35
intrinsic regime, 43
inversion, 78

kp model, 30
Kramers degeneracy, 31
Kubo formula, 133, 280

Landau level, 139
Landau quantization, 137
Landauer formula, 176
Landauer-Büttiker formalism, 184
Langevin equation, 132
LEC growth technique, 96
lift-off, 104
light holes, 31
lithography, 11, 102
localization length, 135
Luttinger parameters, 31

magnetic length, 135, 138
magnetic mass, 168
magneto-resistivity tensor, 49
mask aligner , 102
mass action law, 36
MBE, 97
mesoscopic transport, 12
metal organic chemical vapor deposition, 96
metal-insulator transition, 141
metallization, 108
microchip, 11
mobility in HEMTs, 83
MOCVD, 96
modulation doping, 20, 81
molecular beam epitaxy, 97
molecular electronics, 21
Moore's law, 11
MOS interface, 79
MOSEFT, 78
MOSFET, 20

nearly free electron model, 26

Ohmic contact, 74
Onsager-Casimir symmetry relation, 50
operational amplifier, 125
optical lithography, 11
optical lithography , 102
overcut profile, 104

Index

pentacene, 87
persistent current, 315
phase coherence, 17
phase coherence length, 134
photoresist, 104
PI controller, 130
pinning, 43
pinning of the Fermi level, 68
Poisson distribution, 215
polythiophene, 87
quantized chaos, 264, 282
quantum dot, 12, 17, 249
quantum film, 12
quantum Hall effect, 15, 143
quantum point contact, 14, 182
quantum point contacts, 166
quantum scattering length, 131
quantum scattering time, 131
quantum well, 78
quantum wire, 12, 165, 210
Quantum wires, 165
quasi-2DEG, 153
random matrix theory, 266
RHEED, 97

S - matrix, 216
sample holders, 122
saturation regime, 43
scanning probe lithography, 106
scattering in heterostructures, 88
scattering mechanisms, 50
Schottky barrier, 69
Schottky diode, 73
Screening, 53
screening in two dimensions, 88
self-assembled quantum dots, 99
semi-insulating, 43
semiconductor: definition, 27
SET transistor, 236
short period superlattice, 98
Shubnikov-de Haas oscillations, 15, 144

Si, 24
Si-MOSFET, 78
Si: band structure, 27
Si: holes, 31
Si[Ge] quantum wells, 83
silicon, 20
single electron box, 247
single electron pump, 243
single electron tunnelling, 17, 225
single electron turnstile, 247
size quantization, 16
skipping orbits, 148, 185
spin-orbit splitting, 30
split gates, 171
subband, 79
superlattices, 19, 277
superleak, 113
surface bands, 66
surface recombinations, 66
surface states, 60

thermal length, 135
Thomas-Fermi screening, 54
tight binding model, 26
transconductance, 125, 128
transistor amplifier, 124
triangular interface potential, 78
two-probe measurement, 128

undercut profile, 104
universal conductance fluctuations, 209

valley degeneracy, 29
velocity autocorrelation function, 132
virtual ground, 127
voltage amplifier, 125
voltage divider, 123
voltage sources, 123

warped spheres, 31
weak localization, 206
Weiss oscillations, 277

337